Recent Developments in Wood Polymer Composites

Recent Developments in Wood Polymer Composites

Guest Editor

Ming Zhang

MDPI

Basel • Beijing • Wuhan • Barcelona • Belgrade • Novi Sad • Cluj • Manchester

Guest Editor
Ming Zhang
College of Materials Science
and Engineering
Beihua University
Jilin
China

Editorial Office
MDPI AG
Grosspeteranlage 5
4052 Basel, Switzerland

This is a reprint of the Special Issue, published open access by the journal *Polymers* (ISSN 2073-4360), freely accessible at: www.mdpi.com/journal/polymers/special_issues/18Z2Q2B9RZ.

For citation purposes, cite each article independently as indicated on the article page online and using the guide below:

Lastname, A.A.; Lastname, B.B. Article Title. *Journal Name* **Year**, *Volume Number*, Page Range.

ISBN 978-3-7258-3742-7 (Hbk)
ISBN 978-3-7258-3741-0 (PDF)
https://doi.org/10.3390/books978-3-7258-3741-0

Contents

About the Editor

Ming Zhang

Dr. Ming Zhang is a full professor at Beihua University and an outstanding master's supervisor in Jilin Province. She received a joint PhD degree from Northeastern Forestry University and the University of Pennsylvania. Her research interests include biomimetic construction and the liquid wetting/guiding mechanism of wood-based nanocomposites and its application in water treatment, as well as the component disassembly, intelligent recombination, and structural regulation mechanism of plant biomass.

Dr. Zhang has successively led over 10 projects, including the National Natural Science Foundation of China, the "Outstanding Young Talent" Fund of Jilin Province, the "Outstanding Young Talent" Cultivation Program of Jilin City, the "Spring Seedling Talent" Cultivation Project, the Natural Science Foundation of Jilin Province, the Industrial Innovation Special Project of the Development and Reform Commission of Jilin Province, etc. She has published nearly 60 academic papers in journals such as *Chem Eng J.*, *Carbohyd Polym.*, etc. Over 40 of them are indexed by *SCI*, with nearly 30 in top journals, 9 of them with impact factors above 10, and 3 of them highly cited papers in *ESI*. Her H-index is 26, and her papers have been cited over 2,100 times. Additionally, she serves as the reviewer for nearly 30 international journals, including *Adv. Funct. Mater.*, *Chem. Eng. J.*, etc.

Dr. Zhang is the principal author of the publications "Key Technologies for the Preparation of Bionic Intelligent Biomass Composite Materials" and "Illustrated Bionic Technology". She serves as a board member of the Bio-based Materials Science Division of the China Forestry Society, a member of the Materials Expert Committee of the Viser Expert Database in Singapore and the International Association of Wood Anatomists, and is an expert in the CNKI Review Expert Database. She is also a young Editorial Board Member of journals such as *Bamboo Journal*, *Rare Metals*, *Materials*, and *Green Carbon*, as well as the Guest and Academic Editor for journals such as *Polymers* and *Journal of Polymer Materials*.

Preface

In the era of sustainable development, materials science is undergoing a rapid transformation at an unprecedented pace. Among the myriad of innovations, wood polymer composites have emerged as a beacon of progress that is continuously evolving and attracting intensive research endeavors. The academic community has responded with a proliferation of publications, each presenting exciting results and advancements that hold significant potential for practical applications. However, wood polymer composites are also confronted with some challenges such as synthetic procedures, scale-up productions, performance enhancement, multifunctionality, biomimetic structures and functions, intelligent properties, underlying mechanisms of action, and high-tech applications.

This reprint focuses on the theme of "Recent Developments in Wood Polymer Composites". It aims to provide a concise and penetrating exploration of the synthesis, structure, properties, and applications of wood polymer composites, thereby revealing their unique value in sustainable development and outlining the latest research advancements in this field. We anticipate that this reprint will serve as an invaluable reference for researchers, policymakers, and industry professionals, contributing to the ongoing evolution of wood polymer composites.

The reprint is designed for a multidisciplinary audience, including the professionals in forestry engineering, materials science, environmental science, and chemical engineering, as well as policymakers and business leaders with an interest in sustainable materials. For researchers, it offers state-of-the-art findings and innovative perspectives. For policymakers, it provides scientific evidence for formulating relevant industrial policies. For business professionals, it reveals the broad prospects and commercial value of wood polymer composites in practical applications.

As the Guest Editor of this Special Issue, I express my gratitude to the authors for their high-quality contributions, the reviewers for their valuable feedback that enhanced the submitted works, and the administrative staff of MDPI publications for their support in completing this project. Special thanks go to the Managing Editor of this Special Issue, Mr. Paul Bian, for his excellent collaboration and valuable assistance. Moreover, we are deeply appreciative of the insights and suggestions from our collaborators; the financial support from funding agencies (the National Natural Science Foundation of China (32401508), the Natural Science Foundation of Jilin Province (YDZJ202201ZYTS441), and the Jilin Province Development and Reform Commission Industrial Innovation Special Fund Project (2023C038-2)); the advanced facilities and technological platforms provided by laboratories; and the selfless help and encouragement from our colleagues. Without this collective support, this reprint would not have come to fruition.

Ming Zhang
Guest Editor

polymers

MDPI

Review

Research Advances in Wood Composites in Applications of Industrial Wastewater Purification and Solar-Driven Seawater Desalination

Dongsheng Song [1], Dingqiang Zheng [1], Zhenghui Li [1], Chengyu Wang [2], Jian Li [2] and Ming Zhang [1,*]

[1] Key Laboratory of Wooden Materials Science and Engineering of Jilin Province, School of Material Science and Engineering, Beihua University, Jilin 132013, China; sds230217@163.com (D.S.); 13982758633@163.com (D.Z.); lizhenghui202304@163.com (Z.L.)
[2] Key Laboratory of Bio-Based Material Science and Technology, Ministry of Education, School of Material Science and Engineering, Northeast Forestry University, Harbin 150040, China; wangcy@nefu.edu.cn (C.W.); nefujianli@163.com (J.L.)
* Correspondence: mattzhming@163.com

Abstract: In recent years, the ecosystem has been seriously affected by sewage discharge and oil spill accidents. A series of issues (such as the continuous pollution of the ecological environment and the imminent exhaustion of freshwater resources) are becoming more and more unmanageable, resulting in a crisis of water quality and quantity. Therefore, studies on industrial wastewater purification and solar-driven seawater desalination based on wood composites have been widely considered as an important development direction. This paper comprehensively analyzes and summarizes the applications of wood composites in the fields of solar-driven seawater desalination and polluted water purification. In particular, the present situation of industrial wastewater containing heavy metal ions, microorganisms, aromatic dyes and oil stains and related problems of solar-driven seawater desalination are comprehensively analyzed and summarized. Generally, functional nanomaterials are loaded into the wood cell wall, from which lignin and hemicellulose are selectively removed. Alternatively, functional groups are modified on the basis of the molecular structure of the wood microchannels. Due to its three-dimensional (3D) pore structure and low thermal conductivity, wood is an ideal substrate material for industrial wastewater purification and solar-driven seawater desalination. Based on the study of objective conditions such as the preparation process, modification method and selection of photothermal conversion materials, the performances of the wood composites in filtration, adsorption and seawater desalination are analyzed in detail. In addition, this work points out the problems and possible solutions in applying wood composites to industrial wastewater purification and solar-driven seawater desalination.

Keywords: wood composites; polymers; nano/micropore structure; industrial wastewater purification; solar-driven seawater desalination

Citation: Song, D.; Zheng, D.; Li, Z.; Wang, C.; Li, J.; Zhang, M. Research Advances in Wood Composites in Applications of Industrial Wastewater Purification and Solar-Driven Seawater Desalination. *Polymers* **2023**, *15*, 4712. https://doi.org/10.3390/polym15244712

Academic Editor: María del Carmen García Payo

Received: 24 November 2023
Revised: 10 December 2023
Accepted: 11 December 2023
Published: 14 December 2023

1. Introduction

Wood is composed of various tissue structures, cell morphologies, pore structures and chemical compositions. Therefore, it is a kind of polymer-based natural composite with a hierarchical and porous structure. Meanwhile, it also has obvious anisotropy, from the meter-level trunk to the decimeter- and centimeter-level wood fibers, millimeter-level annual rings and micron-level wood cells. Up to the cellulose nanofibrils, it has an extremely delicate and orderly multi-scale hierarchical structure [1,2].

Wood can be divided into coniferous wood and broadleaf wood. Coniferous wood mainly includes axial tracheids, wood rays, axial parenchyma and resin canals. Broadleaf wood mainly includes conduits, wood fibers, axial parenchyma and wood rays (a few types contain certain tracheids). It is these wood cells with different shapes, sizes and arrangements that form wood through an orderly and close combination. Then, they can create wood's unique pore structure [3]. According to their size, voids in wood can be divided into macropores, micropores and mesopores: (1) Macropores refer to pores that can be seen by the naked eye. Examples include wood cells (width: 50~1500 μm; length: 0.1~10 mm), vessels (20~400 μm), tracheids (15~40 μm) and intercellular canals (50~300 μm). (2) A micropore is a void with the order of magnitude of a molecular chain cross-section as the maximum starting point. For instance, the cross-section of the cellulose molecular chain is of the order of magnitude of a micropore. (3) Mesopores refer to voids with one, two or three dimensions in the nanometer scale (1~100 nm). For instance, there are marginal pores (10 nm~8 μm), simple pit pores (50~300 nm) and wood cell wall gaps (2~10 nm) in a dry or wet state and microfibril gaps (1~10 nm) in a swollen state in coniferous wood [4].

The wood cell wall is made up of approximately 45% cellulose (linear polymer composed of β-D-glucose) as the skeleton, approximately 30% hemicellulose (heterogeneous polymer composed of different types of monosaccharides) in a bonding role [5] and approximately 25% lignin (a complex, amorphous, 3D reticulated phenolic polymer composed of phenylpropane units) in a penetrating role. Wood is gradually assembled from monomolecular cellulose (~0.52 nm), elementary fibrils (2~3 nm), microfibrils (10~30 nm), macrofibrils (~10 μm) and cell wall lamellae (S1, S2 and S3 layers in primary wall and secondary wall) by inherent physical and chemical interactions [6]. After physical modification, chemical modification or physical/chemical combination modification, it will provide an important substrate and template for the bionic preparation of high-performance, high-value and multifunctional novel materials. The further development of functional and intelligent wood is bound to create an unlimited potential for novel material fields such as selective adsorption and separation, catalyst loading, water purification, seawater desalination, photoelectric devices and sensing devices [7–9].

2. Application of Wood Composites in Water Purification

Water is the source of life and the link that binds all living beings on this planet. The rapid development of industry and the rapid growth of population have caused serious water pollution issues. More attention should be drawn to the deterioration of the global ecological environment and the scarcity of freshwater resources. Determining how to treat wastewater from different fields (metallurgy, mining, chemical industry, leather industry, batteries, etc.), nuclear energy, agriculture, shipping and so on is an important and urgent research topic [10–14]. Nowadays, the common materials for polluted water treatment are activated carbon, bentonite, diatomite, geopolymer, fly ash, resin and so on. However, they generally have some problems, such as high price, low treatment speed, poor recycling, poor hydrophilic pollutant removal efficiency, single pollutant type, small adsorption capacity, easily causing secondary pollution, and easy oxidation. The unique pore structure of wood is very beneficial for fluid to flow through. Meanwhile, it absorbs the tiny particles in the intercepted fluid. More importantly, it is green, light in weight, good in toughness, impact-resistant and renewable. It has great potential in the field of high-flux wastewater treatment [15,16]. Under the background of the global resource crisis, wood is increasingly being used as a raw material. After functionalization, novel wood composite filter membrane sand adsorption materials can be obtained to remove heavy metal ions, microorganisms, aromatic dyes, oil stains and other pollutants from wastewater. Apparently, this is of great and positive significance to ecological protection and resource recovery and reuse.

2.1. *Adsorption of Heavy Metal Ions*

Because of their good solubility and stability, heavy metal ions in water exhibit the characteristics of high toxicity, non-degradation and biological enrichment in the ecosystem. If water containing heavy metal ions is discharged into the environment without treatment, it will cause serious harm to human health and the safety of other organisms [17].

Nowadays, the common methods to remove heavy metal ions from industrial wastewater include chemical precipitation, lime condensation, ion exchange, reverse osmosis and solvent extraction. However, they generally have problems such as complicated operations and high costs [18]. Therefore, an ideal choice is to treat heavy metal ions with adsorbents for the deep purification of water. Moreover, an adsorbent should meet the following standards: (1) low-cost and reusable; (2) effective and rapid; (3) selective and economically feasible [19]. The microstructure of wood contains a large number of hollow cells, which are interconnected and form interconnected channels, displaying a certain water flux. Moreover, wood is a typical multi-group ligand that can purify wastewater by adsorbing various heavy metal ions: (1) O– and COO– on the wood will react with heavy metal ions (Mn^+); (2) the negative polar bond in –OH, –NH, –OCH_3 and –C=O in the wood will generate electrostatic attraction with heavy metal ions; (3) –OH and –COOH in the wood will exchange ions with heavy metal ions, and H^+ will be released into the water.

Sawdust is cheap and contains cellulose and lignin, which can absorb a variety of heavy metal ions. Therefore, it has a broad application prospect in the field of wastewater treatment. Ahmad et al. [20] ground sawdust into wood powder. Then, formaldehyde was used for the methylation reaction of wood powder to produce an adsorbent. The results show that the maximum removal rates for Cu^{2+} and Pb^{2+} are 99.39% and 94.61% when the adsorption material is in a solution with successive pH values of 7.0 and 6.6. Too high or too low a pH value will reduce the adsorption capacity of materials. This is because ion exchange and hydrogen bonding are the key to the efficiency of the removal of heavy metal ions by the adsorption material. In a water environment with a lower pH value, H^+ competes with heavy metal cations for adsorption sites on adsorption materials. In a water environment with a higher pH value, OH^- will form soluble hydroxyl complexes with heavy metal cations, and the electrostatic interaction between heavy metal cations and adsorption materials will be weakened. Chen et al. [6] modified fast-growing eucalyptus bark (MEUB) with a formaldehyde and sulfuric acid solution to prepare a wood adsorbent. The result shows that MEUB has different action mechanisms on Pb^{2+}, Ni^{2+} and Cr^{6+} in wastewater. The adsorption of Pb^{2+} and Ni^{2+} is physical adsorption (no variation after heating). The adsorption of Cr^{6+} needs to overcome a certain activation energy, thus being classified as chemical adsorption (promoted variation after heating).

In order to improve the adsorption capacity, adsorption efficiency and selectivity of wood as adsorbent, other functional groups or inorganic nanomaterials can be further grafted or loaded in the wood channels. He et al. [21] prepared a 3D wood microfilter by modifying wood for the removal of heavy metal pollutants in wastewater. Specifically, a green deep eutectic solvent was used to remove lignin from beech wood. Then, carboxyl and sulfhydryl groups (-SH) were grafted on the surface of cellulose by sequentially using citric acid and l-cysteine. Finally, a 3D wood microfilter with an abundance of pores and adsorption sites was formed (Figure 1a). The adsorption kinetics and adsorption isotherms of heavy metal ions (Cu^{2+} and Cd^{2+}) on the 3D wood microfilter were systematically investigated. The results showed that the 3D wood microfilter had a fast adsorption rate and high saturation capacity for both Cu^{2+} and Cd^{2+}. Based on the advantages of easy multilayer assembly, a three-layer wood microfilter was designed to achieve the high flux rate (1.53×10^3 $L \cdot m^{-2} \cdot h^{-1}$) and high removal efficiency (>98%) for heavy metal ions in wastewater.

Yang et al. [22] prepared a sulfhydryl functionalized wood (SH-wood) membrane with a three-dimensional mesoporous structure and low-tortuosity lumens. This SH-wood membrane serves as a multisite metal trap that achieves a high removal efficiency towards heavy metal ions from wastewater (Figure 1b,c). Benefiting from the unique microstructure of wood, the as-prepared membrane exhibits a high saturation absorption capacity of 169.5 $mg \cdot g^{-1}$, 384.1 $mg \cdot g^{-1}$, 593.9 $mg \cdot g^{-1}$ and 710.0 $mg \cdot g^{-1}$ for Cu^{2+}, Pb^{2+}, Cd^{2+}, and Hg^{2+}, respectively. Meanwhile, the SH-wood membrane can be easily regenerated at least eight times without apparent performance loss. Furthermore, an SH-wood filter with stacking multilayers was designed. Because of its high heavy metal ion absorption capability, the multilayer SH-wood filter can effectively remove diverse heavy metal ions from real wastewater, meeting the WHO standards and displaying a high flux rate of 1.3×10^3 L·m^{-2}·h^{-1} (Figure 1d). A further research result shows that the SH-wood membrane can be reused at least eight times. The cost of wastewater treatment is approximately USD 1 per ton, and thus, this treatment option has better economic and practical potential than traditional heavy metal ion adsorbents (activated carbon, clay, etc., displaying drawbacks such as easily being influenced by the environment, unstable treatment effect and fewer recycling times). Apparently, the SH-wood device has more development potential [23].

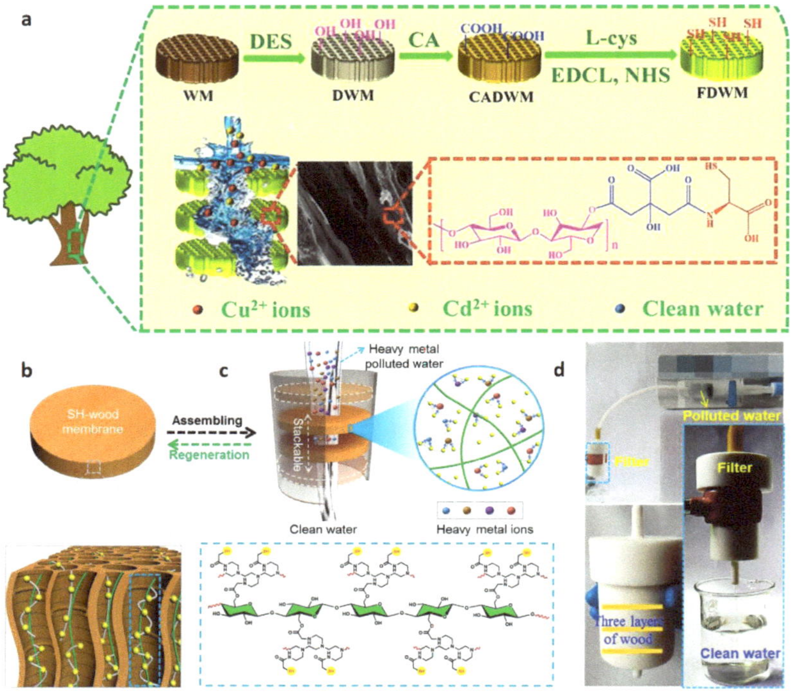

Figure 1. (**a**) A 3D wood microfilter for fast and efficient removal of heavy metal ions from wastewater. Schematic of SH-wood stacks for heavy metal ion removal from aqueous solution [21]. Reprinted with permission from Ref. [21]. Copyright 2020 ACS Publications Ltd. (**b**) SH-wood membrane, including a magnified drawing of its microstructure and chemical composition [22]. (**c**) The multilayer device for large-scale heavy metal ion removal. The magnified schematic shows that the heavy metal ions can combine with −SH groups when the polluted water flows through the channels of modified wood [22]. (**d**) Photos of the experimental setup for the filtration of heavy-metal-polluted water and the clean water flowing out through the three-layer SH-wood membrane [22]. Reprinted with permission from Ref. [22]. Copyright 2023 ACS Publications Ltd.

Cai et al. [24] synthesized UiO-66-NH_2 (a metal–organic framework (MOF) material with photocatalytic activity) in situ in the wood channels to produce a UiO-66-NH_2/wood composite membrane. Then, three layers of composite membranes were assembled together as a composite filter. Compared with the results of Yang et al. [22], the treatment rate of the device towards simulated wastewater containing Cu^{2+} and Hg^{2+} is 1.3×10^2 L·H^{-1}·m^{-2}. The Cu^{2+} and Hg^{2+} removal rates of the sample are over 90%, and the treated water still meets the drinking water standard. Wang et al. [25] used pine as a template and prepared metal oxides (NiO and NiO/Al_2O_3) in it using the impregnation and calcination method. The result shows that NiO and NiO/Al_2O_3 have good adsorption towards Pb^{2+} in simulated wastewater, and their removal rates can be over 99%. Vitas et al. [26] modified beech wood with 3 mmol·g^{-1} –COOH groups by optimizing the reaction conditions of esterification using anhydride. The result shows that the modified beech wood can be used as a biological adsorbent to remove 95% of Cu^{2+} from low-concentration solutions (100~500 mg·L^{-1}). Raman spectroscopy and energy spectrum images confirmed –COOH was mostly located on the wood cell walls. Liu et al. [27] compared a wood membrane with a pore size of several tens of microns (μm) with a polymer membrane with a pore size of 0.45 μm. Compared with the polymer membrane with a cobweb structure, the removal efficiency of the wood membrane towards Fe^{3+}, Cu^{2+} and Mn^{2+} by gravity-driven filtration was improved. This is because the sponge-like structure of the polluted layer of the wood membrane prolonged the retention time of heavy metal ions. The carboxylic group (–COOH) content of the fouling layer for the wood membrane was greater than that for the polymer membrane. Additionally, the heavy metal captured more microbes on the surface of the wood membrane compared with that of the polymer membrane. Wood membranes provide a promising route to producing facile, biodegradable and sustainable membranes as a green alternative to polymer membranes in heavy metal removal from drinking water. After the pyrolysis of a urea-impregnated wood sponge in an argon atmosphere, Gu et al. [28] successfully prepared a novel adsorbent, namely a lignosulfonate (LS) functionalized g-C_3N_4/carbonized wood sponge (denoted as LS-C_3N_4/CWS). As expected, the as-prepared LS-C_3N_4/CWS shows excellent decontamination capability toward Pb^{2+}, Cd^{2+} and Cu^{2+} with high adsorption capacities of 659.6 mg·g^{-1}, 329.1 mg·g^{-1} and 173.5 mg·g^{-1}, respectively, which is superior to that of most of the reported wood-based adsorbents and nanomaterials. Moreover, the LS-C_3N_4/CWS can be readily recovered, and it maintains a high removal efficiency after ten cycles of adsorption–regeneration, displaying excellent recyclability. Significantly, the LS-C_3N_4/CWS can be directly utilized as an ultrafiltration membrane to continuously treat a large volume of simulated wastewater (9550 mL·g^{-1} for Pb^{2+}, 1500 mL·g^{-1} for Cd^{2+}, and 8700 mL·g^{-1} for Cu^{2+}). After filtration, it can result in a lower concentration than the permitted concentration in drinking water.

Compared with traditional methods, modified wood composites show better results in terms of lifetime and throughput. Thanks to the unique microstructure of wood, its efficiency of removal of heavy metal ions in wastewater can reach at least 90%. The mechanical properties of wood composites are particularly prominent as well, with excellent repeatability, strong anti-fouling ability and stable performance. Moreover, a high flux of modified wood composites, 1300~5000 L·m^{-2}·h^{-1}, can be maintained. Most importantly, modified wood composites are green, environmentally friendly and biodegradable.

2.2. *Disinfection and Sterilization*

According to the statistics of the World Health Organization (WHO), approximately 1.6 million people die of diarrhea due to the lack of safe drinking water and basic sanitation facilities every year. Sterilization and disinfection of drinking water can effectively prevent diseases from being spread in water. The pore structure of wood has a natural barrier to larger colonies. After combination with antibacterial nanoparticles (NPs) (e.g., Ag NPs), a wood water filter with outstanding bacteria removal ability can be made. This is because Ag enters bacterial cells in the form of particles through endocytosis and is continuously released in the form of Ag^+. Specifically, Ag^+ will cross-link or catalyze DNA molecules to form free radicals. Then, the proteins are denatured, the electron donors on DNA molecules are inhibited, and the DNA molecular chains break. In addition, Ag^+ can combine with sulfhydryl and amino groups in cells, which will destroy the activity of cell synthetases. All the above procedures make bacteria and other microorganisms lose their ability to reproduce. When the bacteria die, Ag^+ will be released and repeatedly perform the function of sterilization [29,30].

Macior et al. [31] used poly(methyl methacrylate) (PMMA) and poly (2-(dimethylamino) ethyl methacrylate) (PDMAEMA) to functionalize and endow wood with antibacterial properties. The antibacterial result showed that no bacterial growth was observed where the wood block was in direct contact with *S. aureus* and *E. coli* inocula. Therefore, it showed that the modified wood block had good antibacterial performance (Figure 2a). Dai et al. [32] used nano-silver, which is characterized by antimicrobial properties, to modify wood (Figure 2b). The results show that when the retention rate of silver reaches 0.324 $g \cdot m^{-2}$, the bacteriostatic rates of wood for Aspergillus Niger, Penicillium Citrinum and Trichoderma Viride were significantly improved, and the EX-Fr values reached 80%, 75% and 80%, respectively. Boutilier et al. [2] removed bark from pine branches, selected the part (xylem) rich in transport tissue, and inserted it into a catheter. Then, a wastewater filter was prepared based on the physical barrier of the wood structure. The result shows that the sample can filter out the bacteria in polluted water, and its removal rate exceeds 99.9%. The filtration and sterilization mainly occur in the first 2~3 mm part of the wood xylem (Figure 2c). Approximately 4 L of purified water can be obtained through 1 cm^2 of filtration area every day, which is enough to meet a person's normal drinking water demand. Che et al. [33] prepared an antibacterial wood filter by in situ synthesis of Ag NPs in mesoporous wood. When the mass fraction of Ag NPs in the filter is 1.25%, it can not only remove *E. coli* (6.0 orders of magnitude) and *S. aureus* (5.2 orders of magnitude) in polluted water (Figure 2d), but also remove cationic water-soluble aromatic dyes such as methylene blue (MB, 98.5%).

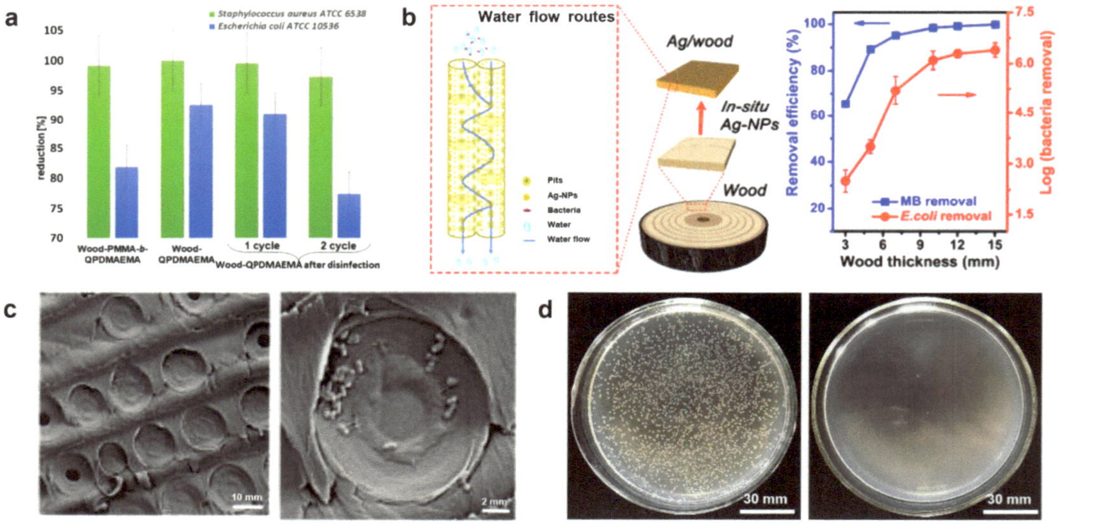

Figure 2. (**a**) Percent of reduction (PR) in *S. aureus* and *E. coli* culture density after incubation on modified wood blocks in reference to control wood sample [31]. Reprinted with permission from Ref. [31]. Copyright 2022 MDPI Ltd. (**b**) Filtration of model bacteria by the xylem filter [32]. Reprinted with permission from Ref. [32]. Copyright 2022 MDPI Ltd. (**c**) SEM images showing bacteria accumulated on the margo pit membranes after filtration. Scale bars are 10 mm and 2 mm [2]. Reprinted with permission from Ref. [2]. Copyright 2014 Public Library of Science. (**d**) Natural-wood- and Ag/wood-filtrated *E. coli* suspension [33]. Reprinted with permission from Ref. [33]. Copyright 2019 ACS Publications Ltd.

Electroporation sterilization technology applies a pulsed strong electric field to microorganisms and bacteria. It will destroy the cell membrane of bacteria and cause an osmotic imbalance inside and outside the cell membrane, eventually leading to the death of bacteria. There is no toxic by-product in the sterilization process. However, the high energy consumption and high risk of this technology limit its use in wastewater treatment [34].

Notably, researchers found that introducing 1D nanomaterials into conductive materials can solve the problems of energy consumption and safety [35]. Yang et al. [36] uniformly loaded Ag NPs into wood pores using the impregnation method, and the wood was further carbonized in a high-temperature tubular furnace. Then, Ag NP/carbonized wood membrane (3D Ag NP/WCM) composites with a three-dimensional mesoporous structure were obtained. The results show that the structure of nanofibers in carbonized wood is clearer. When a voltage is applied, the nanofibers will produce a peaking effect and greatly enhance the surrounding electric field, which can destroy the cell membrane of bacteria and lead to their inactivation. After electroporation, the damaged bacterial cells are more conducive to the invasion of Ag NPs in the carbonized wood and promote the sterilization process. The 3D Ag NP/WCM composite can be used in the condition of low voltage (4 V), low energy consumption (2 $J \cdot L^{-1}$) and high flux (3.8×10^3 $L \cdot h^{-1} \cdot m^{-2}$). It has a high bacteria removal rate (over 99.999%) and good stability (over 12 h). Compared with traditional electroporation sterilization technology, the wood composite not only avoids high energy consumption, but also reduces the safety risk of operation. It is a green, economical, fast, renewable and high-flux sterilization material for water treatment. Du et al. [37] used the plate counting method to investigate the antibacterial effect of Ag@Wood. The result shows that the original wood has no antibacterial activity. In contrast, Ag@Wood exhibited the antibacterial efficiencies of 99.97%, 99.98% and 99.98% towards *E. coli*, *S. aureus* and *B. subtilis*, respectively, displaying a remarkable antibacterial activity. This is because Ag NPs in wood can bind closely to hydrosulfonyl in zymoprotein from bacteria, which could coagulate the protein, destroy the activity of bacterial synthase, and finally limit the proliferation and development of the bacteria. In addition, the dissolved Ag^+ in Ag@Wood will further combine with bacterial membrane proteins, interfere with the synthesis of peptidoglycan and hinder the synthesis of the cell wall, leading to the leakage of substances in the bacterial membrane. Last but not least, Ag^+ is released from the inactivated bacteria and continues its bactericidal activities.

Nowadays, traditional methods for sterilization are chlorination disinfection, ozone disinfection, heavy metal ion disinfection, etc. They work by decomposing the organic matter, bacteria and microorganisms in water through a hydroxide reaction or peroxidation. However, they usually have some issues such as cancer-causing by-products, high cost and difficult maintenance. Compared with these traditional sterilization and disinfection methods, wood composites have the characteristics of high efficiency, simplicity, stability, low cost, environmental protection and so on, and they are basically not affected by the surrounding temperature and pH. In addition, wood is green and rich in cellulose, hemicellulose and lignin, and it does not produce secondary pollution or toxic carcinogens. Moreover, the wood microstructure can be resistant to many species (e.g., *Aspergillus Niger*, *E. coli*, *S. aureus* and *B. subtilis*) and achieve a water flux of up to 3.8×10^3 $L \cdot h^{-1} \cdot m^{-2}$. Wood composites are capable of removing at least 75% or even up to 100% of bacteria from wastewater. Therefore, wood composites have considerable development prospects in the field of sterilization and disinfection.

2.3. Removal of Aromatic Dyes

Printing and dyeing wastewater contains a lot of aromatic dyes, which are very difficult to remove. In addition, it also has the characteristics of dark color, high chemical oxygen demand (COD), high biological oxygen demand (BOD), complex and changeable composition, large discharge, wide distribution and difficult degradation. If the industrial wastewater is discharged without treatment, it will inevitably bring serious harm to the ecological environment due to its toxicity [38]. Therefore, removing these dye pollutants from water resources and wastewater is vital and important [39]. The natural pore structure of wood has a strong physical adsorption effect on aromatic dyes in wastewater. In addition, when the printing and dyeing wastewater flows through the pores of wood, its hydrodynamic effect is enhanced. In order to increase the time and opportunity of aromatic dye contact with active sites, functional nanomaterials or groups are loaded or grafted in the pore channels of wood.

Chen et al. [16] synthesized Pd NPs in situ in basswood microchannels by using a hydrothermal method to prepare a Pd NP/wood membrane (Figure 3a). Specifically, cellulose, with rich hydroxyl groups, can immobilize Pd NPs; thus, the wood changed from yellow to black at first. This is because the plasma effect produced by Pd NPs fixed on the surface of a wood microchannel absorbs a lot of light. When wastewater containing MB flowed through the wood microchannels, MB was degraded by Pd NPs. The color changed from blue to colorless, and the MB degradation efficiency was over 99.8%. The interaction between MOFs and aromatic dyes can be used to treat different aromatic dyes in wastewater [40]. Guo et al. [41] used $ZrCl_4$, terephthalic acid and acetic acid as precursors for the in situ synthesis of UiO-66 MOF nanoparticles in three-dimensional mesoporous wood using the hydrothermal reaction method to obtain a UiO-66/wood membrane (Figure 3b–d). Wood membrane filters for wastewater treatment can be obtained by changing the size and layers of this UiO-66/wood membrane according to actual needs. The results show that the flux of the filter assembled with three pieces of wood membrane is 1.0×10^3 $L \cdot m^{-2} \cdot h^{-1}$. The removal rates for cationic water-soluble aromatic dyes such as rhodamine 6G (Rh6G), propranolol and bisphenol A exceeded 96%, offering a rapid, multi-effect and recyclable method for removing aromatic dyes in this field. Wood has an abundance of nutrient transportation channels. Meanwhile, it is also a natural water purifier. Nevertheless, the main porous structure of initial wood is not enough to effectively separate small molecules such as aromatic dye pollutants. Meanwhile, fouling in a filter will block the channels and cause poor water flux, which will restrict its large-scale application. Liu et al. [42] combined Fenton-like catalysis based on Mn_3O_4 loading with the microchannels of fir wood for water transfer and wastewater purification (the interfacial area was estimated to be up to 6×10^4 m^2/m^3). The results showed that Mn_3O_4/TiO_2/wood exhibited remarkable catalytic efficiency in the degradation of MB. The pollution problem in the Fenton-like catalysis process can be significantly alleviated. Goodman et al. [43] fixed graphene nanosheets (GnPs), which were treated with lignin, in basswood by vacuum impregnation to produce a GnP wood filter. When the water flux is 364 $L \cdot m^{-1} \cdot h^{-1}$, its MB adsorption capacity in a 10 $mg \cdot L^{-1}$ MB solution reaches up to 46 $mg \cdot g^{-1}$. Further exploration found that the aromatic dyes in the GnP wood filter and generated waste after adsorption can be effectively removed by the solvent exchange method. After five cycles of adsorption, its regeneration efficiency is still more than 80%.

However, the efficiency of the removal of high-concentration aromatic dyes by conduits in wastewater is generally low, which is the same for wood-based devices with tracheid channels [44]. In solvothermal conditions, Cui et al. [45] introduced polyoxometalate-based metal–organic frameworks (POMOFs) into natural wood (POMOF/wood) for effectively removing aromatic dyes and capturing iodine. Keggin-type POM anions with a highly negative charge were encapsulated to adjust the charge of the UiO-66 MOF, and the charge overcompensation in the POMOFs allowed them to efficiently adsorb cationic dyes. Benefiting from wood's unique microstructure, the removal efficiencies of POMOF/wood towards MB and gentian violet (GV) (with a permeance of 1.0×10^4 $L \cdot m^{-2} \cdot h^{-1} \cdot bar^{-1}$) reach up to 94.07% and 95.23%, respectively. Furthermore, POMOF/wood has a high capacity for capturing iodine, with a maximum adsorption of 1232.8 $mg \cdot g^{-1}$ in vapor. Cheng et al. [38] synthesized Ag NPs in balsa wood to prepare a dual-function Ag/wood filter that can simultaneously remove aromatic dyes and separate oil and water (Figure 3d). The results show that Ag NPs anchored on wood channels act as the catalytic sites for MB degradation in wastewater. The superhydrophilicity and underwater superoleophobicity of Ag/wood can effectively separate oil from water when the flux is 2600 $L \cdot h^{-1} \cdot m^{-2}$ driven by gravity. The MB removal efficiency of the Ag/wood filter (thickness: 6 mm) can reach 94.0%, and its oil–water separation efficiency is over 99%.

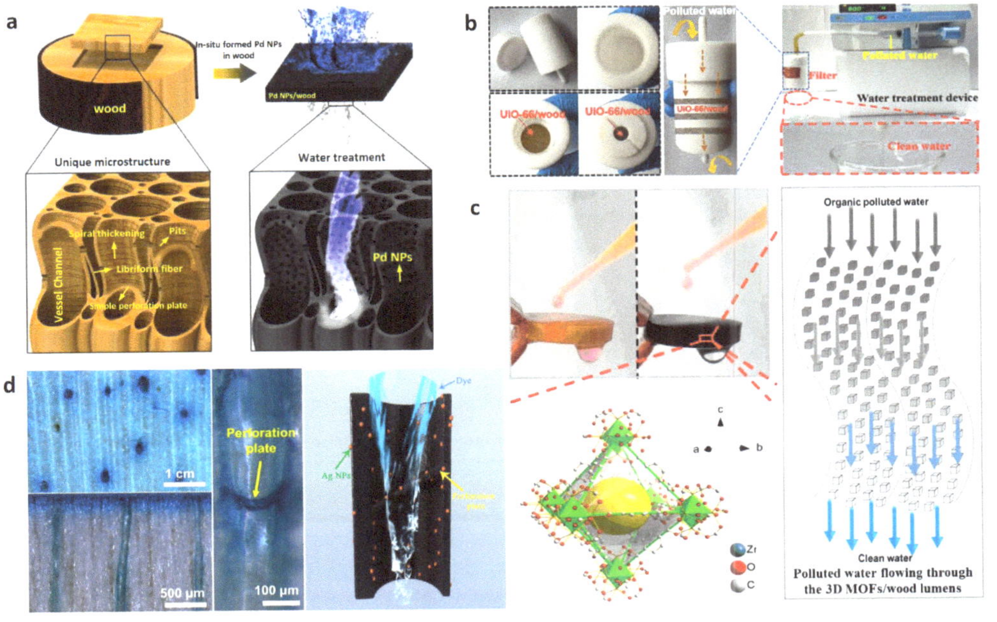

Figure 3. (**a**) The magnified image shows the Pd NPs in the wood channels and the color change (blue to colorless) when the MB solution flows through the Pd NP/wood membrane [16]. Reprinted with permission from Ref. [16]. Copyright 2017 ACS Publications Ltd. (**b**) Propranolol removal performance of the all-in-one three-layer filter based on the UiO-66/wood 258 membrane [41]. (**c**) Rh6G removal performance of the UiO-66/wood membrane [41]. Reprinted with permission from Ref. [41]. Copyright 2019 ACS Publications Ltd. (**d**) Water transport pathways within the wood filter [38]. Reprinted with permission from Ref. [38]. Copyright 2020 ACS Publications Ltd.

Liu et al. [46] used natural wood and MoS_2 as an effective photocatalyst to remove organic pollutants from water under solar illumination. The results show that the photocatalytic degradation activity of wood/MoS_2 composites is obviously better than that of natural wood under one solar irradiation. The improvement of photocatalytic degradation performance is attributed to (1) the increased specific surface area and (2) the catalytically active sites introduced from the edge. These factors accelerate the adsorption and degradation of organic pollutants. The synergistic effect of adsorption and photocatalysis also ensures the high versatility of the composite catalyst in the degradation of aromatic dyes. The kinetic constants of photocatalytic degradation for rhodamine B (RhB), methyl orange (MO) and MB are 0.040, 0.035 and 0.032 min^{-1}, respectively. The degradation performance can be well maintained after three cycles of testing. Considering the three-dimensional mesoporous structure of wood and the catalysis of anchored Ag/AgCl NPs, Zhang et al. [47] prepared a Janus PPy@Ag/AgCl@Wood membrane that can be applied to the purification of dye-contaminated wastewater. First, the degradation capacity of Janus mesoporous wood-based membrane was evaluated utilizing MB as the model dye. A blue MB/$NaBH_4$ solution (10 $mg \cdot L^{-1}$/100 $mg \cdot L^{-1}$, pH = 10) faded to colorless after permeating the membrane, showing a degradation efficiency of 86.4%. The main reason is that MB is a classic cationic dye, and cellulose in wood cell walls with hydroxyl groups usually carries many negative charges. Due to physical adsorption (e.g., Van der Waals' force, hydrogen-bond interaction, electrostatic attraction), some MB is directly removed. The MB/$NaBH_4$ solution and its filtrate showed a blue shift. It was proved that when the solution passes through the microchannels of the Janus wood, it is catalyzed by the anchored Ag/AgCl NPs. Therefore, the N-demethylation reaction occurred.

The adsorbents used for aromatic dye removal in industrial wastewater mainly include activated carbon, silicon polymers, macroporous resin and other materials with large specific surface areas. However, they have not been widely used due to their relatively high cost and poor degradation properties. In contrast, modified wood composites are green and low-cost, and their dye removal rate can reach more than 94%. At the same time, the water flux of wood composites can reach 3.0×10^3 L·m^{-2}·h^{-1}, greatly reducing the cost during the treatment, offering a good application prospect in the field of dye removal.

2.4. Oil–Water Separation

Oil pollution in wastewater mainly comes from petroleum exploitation, the chemical industry, steel factories, coking workshops, gas generating stations and other industrial departments. Its mass concentration is generally 5000~10,000 mg/L. Most of these oils float on the surface of rivers and oceans and form oil films, resulting in a lack of oxygen in water bodies. Eventually, this results in the death of a large number of aquatic organisms. When oil washes up on a beach, it will cause serious harm to the waterfowls, shrimps, crabs and other creatures on the beach [48]. Wood also has a good effect on oily wastewater separation; its pit structure is very beneficial to the demulsification of oil–water emulsions. In the filtration and separation of oil–water mixtures, the wettability of the solid material plays an important role and has become an accelerator in this field. The key points in preparing special wetting materials are (1) the biomimetic construction of a hierarchical micro–nanostructure on the surface of a substrate with low/high surface energy or (2) directly using low/high-surface-energy materials to biomimetically construct a hierarchical micro–nanostructure on the surface of a substrate.

Superhydrophobic wood nanocomposites can be obtained by removing lignin from wood, further loading nanomaterials and then carrying out polymer backfilling and silanization treatment. Then, the expected goal for oil–water separation with high efficiency, high precision and high controllability can be achieved [16,44,49]. Zhao et al. [50] modified wood with polymethylsiloxane (POMS) to obtain superhydrophobic wood materials with a water contact angle (WCA) of 153° (Figure 4a). POMS-modified wood has good oil absorption and oil–water separation performances. However, the accurate filtration efficiency and reusability of POMS-modified wood still need further discussion. Aside from cellulose, which is rich in hydroxyl groups, lignin and hemicellulose also exist in wood. They also contain -NH_2 and -OH groups, resulting in the good hydrophilicity of wood. When wood is soaked in water, a hydrophilic and oil-repellent water film is formed on its surface. When an oil–water mixture is dripped on a wood surface, the water permeates into the wood, while the oil is excluded, showing a good superoleophobicity underwater. Wang et al. [51] prepared a Ag NP/wood membrane using a simple one-step hydrothermal method. The results show that the loading of Ag NPs increases the micro/nanoroughness of wood, which is beneficial to the demulsification of oil-in-water emulsions. Even after 10 cycles of testing, the oil-in-water emulsion separation efficiency was still over 90%. Within 5 min, its photocatalytic degradation rate for MB was 97.21%. Guan et al. [52] selectively removed the lignin and hemicellulose of wood without further filling treatment. After modification by methylsilylation (Figure 4b–d), a highly porous hydrophobic wood sponge (SWS) with enhanced mechanical elasticity, low density and an oleophilic property was directly prepared. The "wood sponge" has a high oil absorption capacity of 41 g/g and excellent recovery capacity. Moreover, the assembled filter can continuously separate oily wastewater with the flux of 84.7 L·h^{-1}·g^{-1}.

After selective removal of hemicellulose and lignin, the basic framework of cellulose with hierarchical high porosity and low density can be prepared easily [53,54]. Fu et al. [55] used a $NaClO_2$ solution to remove the lignin of balsa wood. After freeze-drying, a porous delignified wood template with high hydrophilic and oleophobic properties was obtained. Then, it was impregnated with epoxy resin/amine/acetone solution. After curing, a hydrophobic and oleophilic wood composite with a unique pore structure was prepared. The product shows an outstanding compression strength (263 MPa) and oil absorption effect

(15 g/g), and it can absorb oil pollution on and under the surface of water at the same time. Wang et al. [56] coated one side of delignified wood with a dodecyl mercaptan solution. After ultraviolet radiation induction, a Janus wood membrane with asymmetric wetting properties and unidirectional water transmission was prepared, and it was suitable for selectively separating the mixtures of light oil/water and heavy oil/water (the separation efficiency was higher than 99.3%). Blanco et al. [57] directly used spruce (thickness: 1 mm) for oil–water separation. Under gravity, its flux (3500 $L \cdot m^{-2} \cdot h^{-1}$) and efficiency (>99%) for separating oil–water mixtures were investigated (Figure 4e). After the loading of Ag NPs in wood, superhydrophilic and underwater superoleophobic wood nanocomposites were prepared, the surface hydrophilicity of which could be further improved. After in situ auxiliary modification of photothermal materials (graphene) and transparent hydrophobic materials on delignified and hemicellulose wood, Chao et al. [58] prepared a compressible and resilient photothermal wood aerogel (Figure 4f). Based on the characteristics of the decrease in viscosity and increase in fluidity of crude oil with an increase in temperature, as well as the capillary force of the wood aerogel, its adsorption capacity for crude oil can reach 0.801 $g \cdot cm^{-3}$. The transparent hydrophobic coating of the aerogel endows it with selective adsorption of the oil phase. An intelligent infiltration effect of the crude oil phase at different temperatures can be realized at the same time. Moreover, the adsorbed crude oil can be released and collected by simple mechanical extrusion. The material can be compressed and recycled more than 10 times. It solves the problems of the high energy consumption, complicated separation process, unsatisfactory treatment effect and secondary pollution caused by traditional oil–water separation methods (gravity separation, centrifugation, air flotation, in situ combustion, bioremediation and flocculation, etc.).

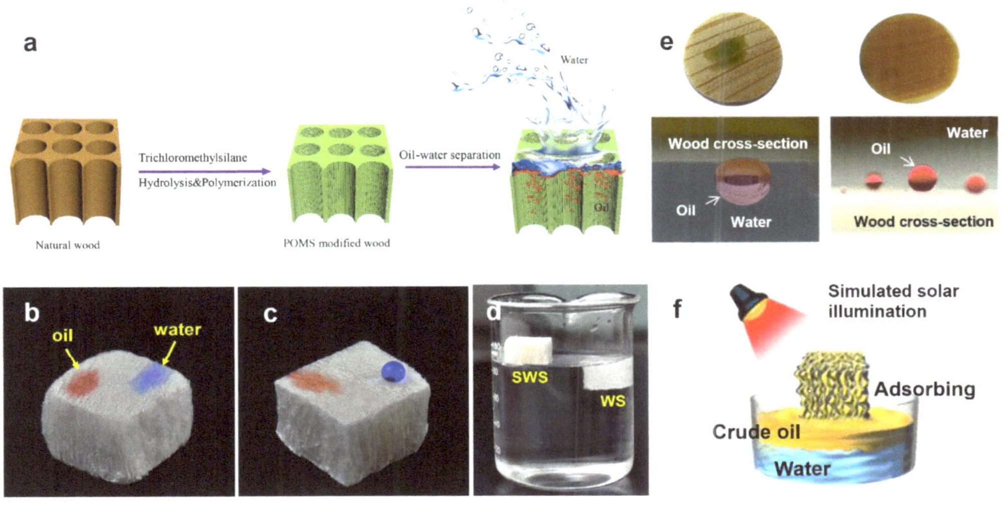

Figure 4. (**a**) Schematic illustration of the preparation procedure of POMS-modified wood for oil–water separation [50]. Reprinted with permission from Ref. [50]. Copyright 2020 Elsevier Ltd. Photographs of water and oil droplets on (**b**) natural wood and (**c**) silylated wood sponge (SWS) [52]. (**d**) Photographs of the SWS floating on the water surface in contrast with the wood sponge (WS) sinking in the water [52]. Reprinted with permission from Ref. [52]. Copyright 2018 ACS Publications Ltd. (**e**) Wetting properties of the wood cross-sections. Photographs of dyed light and heavy oils dripping on the wood surface underwater with high contact angles [57]. Reprinted with permission from Ref. [57]. Copyright 2017 Wiley-Blackwell Ltd. (**f**) Graphical illustration of the simulated solar illumination assisted crude oil adsorption [58]. Reprinted with permission from Ref. [58]. Copyright 2020 Elsevier Ltd.

Zhu et al. [59] fabricated a superhydrophobic methyltrimethoxysilane (MTMS)-modified wood aerogel (M-WA) using the vapor deposition method for oil–water separation. In order to build micro–nanoroughness and sufficient pores, balsa wood was chemically treated to remove lignin and hemicellulose and modified by MTMS vapor deposition. The superhydrophobic M-WA shows a high WCA of 151.8°. M-WA can selectively adsorb oil from an oil/water mixture, and its adsorption capacity for dichloromethane reaches 25.1 times its own weight. In addition, M-WA can be used as a filter to continuously separate an oil–water mixture, and its separation efficiency reaches 99.1%. Even after 20 cycles, the separation efficiency still remained at 98.5%, showing excellent recyclability. Chen et al. [60] proposed a superhydrophobic wood with excellent Joule heat and demulsification. Specifically, a carboxyl-modified carbon nanotube (cCNT) coating provides the conductive pathways, polyethylenimide (PEI) is used to support the demulsification procedure and polydimethylsiloxane (PDMS) is subsequently used for hydrophobic modification. A superhydrophobic wood with excellent Joule heat and demulsification performance was proposed. The as-prepared superhydrophobic PDMS/PEI-cCNT wood (WCA-155°) can withstand 50 cycles of compression tests under 40% strain. Oil can be separated and recovered from an oil–water mixture by an absorption–extrusion process. Due to the electrostatic interaction of PDMS/PEI-cCNT wood, a water-in-oil emulsion can be quickly demulsified with a separation efficiency greater than 99.7%. After the simple removal of lignin and hemicellulose from natural wood, cellulose nanofibers with hydrophilicity were prepared, and they were further used to produce a flexible cellulose aerogel with lamellar architecture. Wu et al. [61] used the superhydrophilicity of balsa-based cellulose aerogel to separate various aqueous oils. The separation efficiency of cellulose aerogel for immiscible water-based oil and water-in-oil emulsions can reach 99.97% and 98.45%, respectively. Shi et al. [62] prepared elastic wood (E-Wood) and then endowed it with excellent oil resistance by in situ polymerization of pyrrole for it to serve as an O/W emulsion filter. The mechanism by which PPy-E-Wood separates O/W emulsions is that water can easily wet and penetrate a PPy-E-Wood membrane and form a water film on its surface. In the separation, water quickly passes through PPy-E-Wood. However, non-polar micron oil droplets are repelled by the water film formed on the PPy-E-Wood surface. Therefore, the oil droplets are trapped at the top of PPy-E-Wood. Then, these larger oil droplets can be easily separated from the PPy-E-Wood surface by a simple rinse treatment. Its separation efficiency towards O/W (H/W and D/W) emulsions can be over 98%. The filtration is driven by gravity without external pressure. Even after 10 cycles of separation, its O/W emulsion separation efficiency still remained at 97.8%. Specifically, when PPy-E-Wood is applied to oily wastewater and oil-spilled seawater, it can effectively avoid the blockage of its exchange channels caused by oil sticking to its surface or bottom, resulting in the loss of solar evaporation performance.

Traditional oil–water separation technologies include the adsorption method, skimming method, gravity separation method, biological treatment method, air flotation method and centrifugal separation method. Most of these methods have low separation efficiency and poor selectivity. In particular, their separation effects are unsatisfactory for oil–water emulsions with a high dispersion and low concentration of oil droplets (droplet size of dispersed phase less than 20 μm). In addition, the excessive input cost leads to serious resource consumption. Wood composites can be made by modifying wood to make them have excellent hydrophilicity or lipophilicity. The separation efficiency and water flux of some wood composites can reach 99% and 3500 $L \cdot m^{-2} \cdot h^{-1}$, respectively. After 20 cycles, some wood composites still show a water/oil separation efficiency of 98.5%. In addition, some wood composites can withstand 50 cycles of compression tests under 40% strain.

3. Application of Wood Composites in Solar Desalination

The wood composites discussed in the previous section can be applied to wastewater purification without the interference of external factors (e.g., solar energy). Thanks to their structural advantages, wood composites enable simple adsorption and filtration.

However, when the salt content in brine is too high, it is difficult to desalinate seawater and obtain clean freshwater by simple adsorption or filtration of wood composites. The desalination of wood composites requires the input of external energy, for which clean and abundant solar energy is generally used. The specific process includes four steps: light absorption, photothermal conversion, interface evaporation and condensation collection. While wood itself has poor photothermal conversion performance, after modification with photothermal materials, wood composites can obtain good photothermal conversion and evaporation efficiency. Notably, water molecules are present in the liquid state in the process of wastewater purification using wood composites discussed in the previous section. This section discusses the application of wood composites in the solar interfacial evaporation of seawater with high salinity. Specifically, water molecules in seawater absorb heat from sunlight and then transform from the liquid state to the vapor state. Afterward, the desalting steam is condensed and collected as liquid freshwater. This process is the transformation of water molecules from the liquid state to the vapor state and then to the liquid state. This section focuses on the structural design of wood composites in solar desalination.

As a clean and renewable energy source, solar energy has a very broad application prospect in alleviating the shortages of energy and water resources. In solar desalination, sunlight is first captured by photothermal materials and then converted to heat. Further, the latent heat of the phase change (seawater from liquid to gas) can be overcome by the converted heat. Generally, the photothermal material is separated from the water body by a material with low thermal conductivity because the more converted heat is used to heat the fluid, the less heat is lost in this process. Meanwhile, the bottom water passes through the conduits in the substrate with low thermal conductivity and is further pumped to the top of the photothermal layer by its capillary force. Trees are the most extensive resources in the world [37,62,63], the conduits of which are beneficial to water transportation. Because of its outstanding transpiration capability, water can be pumped to a height of more than 100 m. Therefore, wood cut perpendicular to the growth direction is an ideal seawater desalination material. Moreover, the transpiration and sunlight absorption inside trees occur through xylem conduits and cavities [64]. Ninety percent of water absorbed by trees is emitted into the air through transpiration. The thermal conductivity of wood is very low and highly anisotropic along ($0.35\ W{\cdot}m^{-1}{\cdot}K^{-1}$) and perpendicular ($0.11\ W{\cdot}m^{-1}{\cdot}K^{-1}$) to its conduits, which is very beneficial for isolating unnecessary heat exchange between a photothermal layer and seawater, thus improving the heat management and evaporation performance.

The traditional methods for seawater desalination include multi-stage flash evaporation (MSF) and multi-effect distillation (MED), as well as reverse osmosis (RO) technology. However, most of them have serious shortcomings including a high dependence on fossil resources; energy consumption; greenhouse gas emissions; high technical requirements; high costs for investment, equipment operation and maintenance; less actual return; and even secondary pollution caused by the large consumption of chemicals and detergents. Wood composites with good photothermal conversion properties with different structural designs such as directly carbonized wood, wood/carbon nanomaterial composites, wood/semiconductor composites, wood/polymer composites and wood/precious metal composites have been prepared. In addition, sunlight provides an almost infinite source of clean heat energy. With different structural designs, a high evaporation rate ($1.351{\sim}4.31\ kg{\cdot}m^{-2}{\cdot}h^{-1}$) and photothermal conversion efficiency (87.4~122.2%) of wood composites have been achieved in solar desalination. Experimental data show that the photothermal layers on these wood composites after acid/alkaline and seawater immersion for 100 h and ultrasonic treatment for 2 h have no obvious changes. Apparently, these wood composites are extremely durable. Even after 100 cycles of freeze–thaw testing, the water evaporation rate of wood composites did not decrease significantly.

3.1. Directly Carbonized Wood

Wood emerges as a sustainable and promising precursor for carbon materials owing to its natural abundance and superb properties. The preparation of carbon materials with different morphologies (dots, spheres, nanowires, films and other 3D nanomaterials) and porous structures (disordered and ordered) is of great significance for various fields (e.g., adsorption, capacitors, catalysts, solar cells, sensors) [65]. Xue et al. [66] cut wood into cylinders and put them on an alcohol flame for carbonization. Then, they were immersed in cold water for rapid quenching. Finally, wooden cylinders with carbonized surfaces were obtained. Under 1 $kW{\cdot}m^{-2}$, their thermal conversion efficiency was as high as 72%. Also inspired by tree transpiration, Zhu et al. [19] designed a solar evaporation device with a double-layer structure. Firstly, wood was cut perpendicular to the wood growth direction and simply carbonized at 500 °C for 0.5 min. The thickness of the carbonized layer was only approximately 3 mm on the surface of the wood, which could be directly used for solar-driven seawater desalination. The results show the following: (1) The upper surface of carbonized wood can absorb 99% of light. (2) Under 10 $kW{\cdot}m^{-2}$, its photothermal conversion efficiency is 87%. (3) Under 10 $kW{\cdot}m^{-2}$, the amount of water evaporation increases linearly. (4) After 100 h of irradiation under 5 $kW{\cdot}m^{-2}$, the sample can still be used stably. (5) The sample can remain stable in seawater for a long time, and there is no salt accumulation on its surface. (6) Water can be directly extracted from the ground (sand and soil).

Kuang et al. [67] also used the above method to carbonize wood. The difference is that they used an electric drill to drill holes in the wood before carbonization (Figure 5a–c). Then, the carbonized layer on the wood surface was polished with sandpaper and then placed in 20 wt% NaCl solution for 6 h under one solar irradiation. The surface of the sample without drilled holes was completely covered with precipitated salt. However, there was no obvious salt deposition on the surface of the sample with drilled holes, displaying an excellent self-desalting ability. This is due to the rapid salt exchange among the pits on wood cell walls, wood microchannels and millimeter-sized drilling channels on wood. The increased salt concentration in the natural wood channel can be diluted at any time. Therefore, it will not block the steam discharge channel during the evaporating procedure of concentrated brine. It is always stable in long-term evaporation and has excellent performance under 1~5 solar irradiations. Its evaporation rate was 6.4 $kg{\cdot}m^{-2}{\cdot}h^{-1}$ when it was placed in high-salinity water (15 wt%) under six solar irradiations, showing excellent stability and durability. Chen et al. [68] developed a multifunctional solar evaporator composed of carbonized wood and biomass hydrogel modified by MXene. The evaporation rate and efficiency of the product were 3.71 $kg{\cdot}m^{-2}{\cdot}h^{-1}$ and 129.64%, respectively. Under two kinds of solar irradiation, the open circuit voltage is 1.8 mV. The high performance of the evaporator stems from (1) the high water transmission of natural wood structure, (2) the high solar absorption and heat conversion efficiency enhanced by carbonized wood and MXene, (3) the regulation of evaporation enthalpy and surface energy by MXene and (4) the temperature adjustment by the low-thermal-conductivity structural system.

Ghafurian et al. [69] utilized the natural water transport tissue of high-porosity wood (poplar) to maximize solar utilization for water vaporization. The inherent high porosity, low thermal conductivity and rapid capillary action of wood make it attractive in solar desalination. Natural wood cannot absorb broadband sunlight well. Therefore, these authors compared several innovative and potentially scalable technologies for wood surface modification, including laser carbonization, gold nanolayer deposition and their combination (Figure 5d). The use of a high-power laser is a fast and accurate method for engraving, cutting and carbonizing wood, enabling groove patterns to be generated rapidly and controllably without destroying the wood samples (Figure 5e). Under 3 kW/m^2, the best performance of gold coating carbonized by a hot plate is 4.02 $kg/m^2{\cdot}h$, which is twice that of untreated sample (~2 $kg/m^2{\cdot}h$), and the best performance of gold nanocoating is 3.54 $kg/m^2{\cdot}h$. In addition, the performances of samples after 10 cycles of thermal treatment were investigated. The result showed that these samples could evaporate stably at the speed of 3.3 kg/m^2·h. It can be concluded that surface modification provides wood with a scalable high performance. Therefore, fast-growing wood can be widely used in solar seawater desalination and/or low-temperature steam generation.

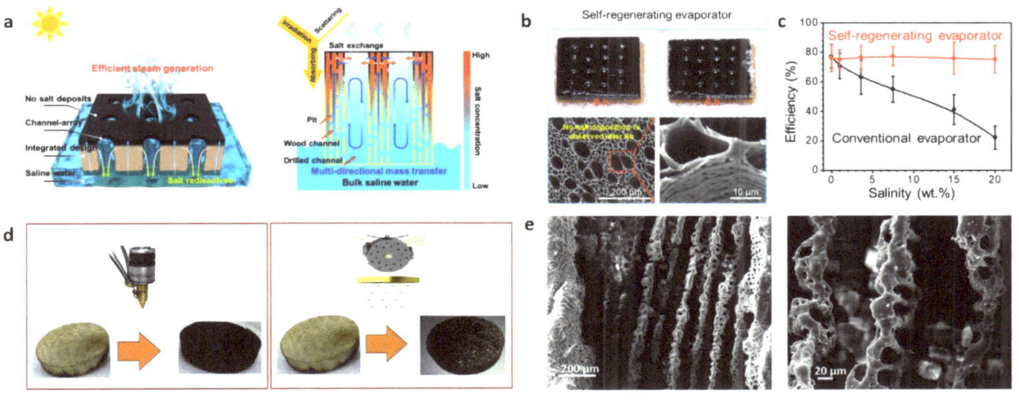

Figure 5. (**a**) Schematics of solar desalination by interfacial evaporation [67]. (**b**) Salt-free surface of the self-regenerating evaporator after 6 h of continuous testing in 20 wt% NaCl solution under one solar irradiation. SEM images showing the salt-free surface of self-regenerating evaporator [67]. (**c**) The influence of salt concentration on the steam generation efficiency of conventional and self-regenerating evaporators [67]. Reprinted with permission from Ref. [67]. Copyright 2019 Wiley-Blackwell Ltd. (**d**) Laser-treated wood (LT-Wood) and gold deposition wood (Au-Wood) [69]. (**e**) Laser treatment creates grooved patterns on the wood [69]. Reprinted with permission from Ref. [69]. Copyright 2020 Elsevier Ltd.

3.2. Wood/Carbon Nanomaterial Composites

Nanomaterials usually exhibit some unique electronic and optical properties. For instance, in the allotrope of a carbon nanotube, almost every band in the solar spectrum can excite electrons due to its large number of conjugated π bonds. This leads to various π-π^* transitions inside and a dark appearance. Meanwhile, when the incident light energy matches the electron transition in the molecule, the electron will absorb the light and rise from the highest occupied molecular orbital (HOMO) to the lowest unoccupied molecular orbital (LUMO). After absorbing these energies, electrons relax through electron–phonon coupling. The carbon allotrope not only absorbs the light, but also transfers the energy of excited electrons to the atomic lattice in the vibration mode. Macroscopically, it will increase the temperature of the material [70]. In addition, other carbon photothermal nanomaterials (such as carbon nanotubes (CNTs) [71], carbon dots (CDs) [72], graphene [73], graphene oxide (GO) [74] and reduced graphene oxide (rGO) [75]) are also promising photothermal materials for interfacial water evaporation.

Taking advantage of the low curvature of the pore structure and the anisotropic thermal conductivity of wood, Chao et al. [54] used delignified wood as a substrate and carbon dots with a photothermal effect prepared using the removed lignin to realize a solar evaporation system with an all-wood composition (Figure 6a–c). The results show that the evaporation rate and photothermal conversion efficiency of the solar evaporation system are 1.09 kg·m^{-2}·h^{-1} and 79.5% under 1 kW·m^{-2}. Liu et al. [74] obtained a wood-GO double-layer composite by dropping graphene oxide on the cross-section of wood. Under 5 kW·m^{-2}, the wood-GO composite and original wood were irradiated in dry and wet states. The results showed that wood-GO had a great temperature increase (Δ_{dry} = 43 °C, Δ_{wet} = 33 °C), while the original wood changed little (Figure 6d,e). A wood-GO composite was placed in simulated seawater with a salinity of 3 wt%. Under 12 kW·m^{-2}, its temperature reaches 67 °C in the first few seconds and remains constant. The generation of water vapor can be clearly seen on its surface. Compared with the original wood (10.08 kg·m^{-2}·h^{-1} and 59.5%), its evaporation efficiency and photothermal conversion efficiency can reach 14.02 kg·m^{-2}·h^{-1} and 82.8%, respectively. It not only realizes the effective utilization of renewable solar energy, but also realizes the full utilization of the internal circulation of wood-based materials.

To make use of the directional arrangement structure of natural wood fibers, Chao et al. [76] selectively removed lignin and hemicellulose and further modified the retained cellulose with photothermal coating (rGO) to prepare a wood-derived aerogel. Under 1 kW·m^{-2}, the material was hung between seawater tanks to carry out a "connecting bridge" seawater desalination. The results show that the evaporation rate and photothermal conversion efficiency are 1.351 kg·m^{-2}·h^{-1} and 90.89%, respectively. Compared with the traditional "close contact" solar evaporation, it avoids the heat loss and the decrease in light energy utilization rate caused by the close contact between photothermal materials and the water phase. It greatly improves its light energy utilization rate. Through the incomplete combustion of a paraffin candle flame, Hu et al. [77] constructed carbon nanoparticles on the surface of wood to prepare a novel double-layer solar evaporator. The results show that the carbon nanoparticle/wood composite with an interconnected porous structure has broadband and high light absorption (93~97%) and can convert solar energy into heat energy to heat the interface between water and air. Its principle is that water is effectively transported from the wood bottom to the surface of carbon nanoparticles by capillary force for sufficient evaporation water supply. Meanwhile, the abundant hydrophilic groups of wood form hydrogen bonds with water molecules, which weakens the hydrogen bonds between intermediate water molecules. Under 1 kW·m^{-2}, the water evaporation enthalpy is reduced by approximately 26%, while the evaporation rate and conversion efficiency are 2.06 kg·m^{-2}·h^{-1} and 90%, respectively, displaying good long-term stability, strong self-cleaning ability, and strong acid and alkali resistance.

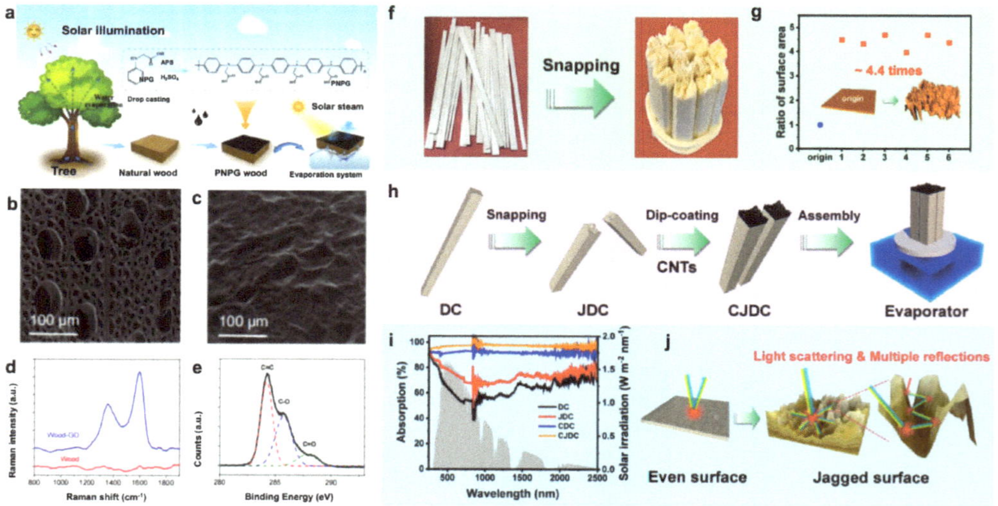

Figure 6. (**a**) Fabrication of PNPG wood solar evaporator [54]. SEM images of wood cross-section without (**b**) and with (**c**) GO on the surface of microporous structure [54]. Reprinted with permission from Ref. [54]. Copyright 2020 ACS Publications Ltd. (**d**) Raman spectrum of wood with (blue) and without (red) GO flake coating on wood surface [74]. (**e**) XPS spectra of wood and wood-GO composite [74]. Reprinted with permission from Ref. [74]. Copyright 2017 ACS Publications Ltd. Digital images of (**f**) pristine DC and JDC. (**g**) Top surface area ratios of JDC to DC. (**h**) Fabrication of CJDC evaporator. (**i**) Solar absorption spectra of DC, JDC, CDC and CJDC. (**j**) Schematic illustration

of jagged surface for enhancing solar absorption [78]. Reprinted with permission from Ref. [78]. Copyright 2023 Elsevier Ltd.

Zhang et al. [78] used CNTs to decorate disposable hydrophilic wood chopsticks, which were assembled into a 3D array for efficient water evaporation and purification. Figure 6f shows the fabrication and solar absorption performance of CNT-functionalized disposable wood chopstick (CDC) and CNT-decorated jagged disposable wooden chopstick (CJDC) evaporators. Firstly, the chopsticks were broken into fragments with a 3D serrated top surface. Compared with a traditional flat surface, its surface area is significantly increased (Figure 6h). In order to provide jagged disposable wood chopsticks (JDCs) with photothermal conversion capability, solar thermal CNTs were coated on the surface of JDCs using the impregnation method. Then, JDCs with high photothermal performance were obtained. To increase the top and side surface areas, a single CJDC unit was assembled into an integrated 3D evaporator for seawater desalination and water purification (Figure 6i). As shown in Figure 6j, a CJDC has a higher absorption of sunlight than a CDC. The CJDC can capture the incident sunlight more effectively by multiple reflections and scattering on its surface. The exposure height of the disposable chopstick evaporator decorated with CNTs was 4 cm. The evaporation rate and energy efficiency of the wood composites were 3.70 kg $m^{-2} \cdot h^{-1}$ and 122.2%, respectively. They have wide application prospects in solar-driven seawater desalination and industrial wastewater purification.

3.3. Wood/Semiconductor Material Composites

The band gap of a semiconductor material can determine its light absorption ability. When the light strikes the semiconductor surface, the concentration of carriers (electrons or holes) in the energy band will increase continuously. When the excited electrons jump back to the low-energy state, the energy will undergo a non-radiative relaxation process. Specifically, the photon energy will be converted into heat energy, thus affecting the photothermal conversion capability. Song et al. [79] coated Fe_3O_4/PVA on the surface of delignified basswood to obtain an Fe_3O_4/PVA/wood evaporator (Figure 7a). The wood was treated with $NaClO_2$ to remove hemicellulose and lignin for at least 10 h, while the cellulose content was basically preserved intact (Figure 7b). Delignified wood has better ink transferability than original wood (Figure 7c). In addition, delignified wood has better hydrophilicity than original wood. Polyvinyl alcohol (PVA) enhances the bonding force between a wood matrix and a semiconductor. Under 1 $kW \cdot m^{-2}$, the surface temperature of natural wood increased from 26 °C to 34 °C within 10 min, while the equilibrium temperature of Fe_3O_4/PVA/wood reached 63 °C. Compared with the wood composite without delignification, Fe_3O_4/PVA/wood shows a higher temperature. Fan et al. [80] embedded a hydrogen evolution semiconductor material (CdS) and photothermal material ($MoSe_2$) into porous delignified wood simultaneously to prepare a wood-mixed hydrogel. It can be used for water purification, hydrogen production and seawater desalination. By optimizing the structure and process design, the efficient pollutant removal, hydrogen production and steam generation of the product can be realized. Under one solar illumination, its hydrogen evolution rate, solar evaporation rate and energy conversion efficiency are 9.7 mmol $g^{-1} \cdot h^{-1}$, 1.92 $kg \cdot m^{-2} \cdot h^{-1}$ and 90.7%, respectively. The hydrogel packages the photocatalytic system, which can effectively prevent the evaporation of toxic volatile organic compounds (VOCs) and retain the ability of continuous and efficient steam production.

He et al. [81] soaked various woods (such as beech, cedar, pine, ash, oak, poplar and cudgel) in a tannic acid (TA) solution to obtain wood-TA. Afterward, wood-TA was immersed in $Fe_2(SO_4)_3$ solution to obtain wood-TA-Fe^{3+}. However, a polypropylene (PP) porous membrane, polyester fabric and polyurethane (PU) sponge in the same condition are blue-gray rather than black. After comparing the SEM images of original poplar, wood-TA and wood-TA-Fe^{3+}, it was found that many nanonodes appear in wood-TA-Fe^{3+} due to the coordination between the doped TA and Fe^{3+}. The rough wood surface and abundant pore structure further reduce the energy loss caused by light reflection. The results show that the photothermal layer of the sample after acid/alkaline and seawater immersion for 100 h, ultrasonic treatment for 2 h and 100 cycles of freeze–thaw testing has no obvious change. For complex water quality, it can effectively prevent oil droplets from adhering to the material surface, which will avoid the blocking of its waterway and a decrease in evaporation performance. In addition, the researchers also cut many grooves on the wood surface to evaluate wood-TA-Fe^{3+} with an uneven surface. The results show that the water evaporation rate of wood-TA-Fe^{3+} with surface modification reaches 1.85 $kg \cdot m^{-2} \cdot h^{-1}$, which is 4 times that of wood-TA-Fe^{3+}. Solar interfacial evaporation is considered as a promising strategy for solar-driven seawater desalination and industrial wastewater purification. Yan et al. [82] used delignified wood (DW) as a water transport substrate and lignosulfonate (LS)-modified nickel disulfide (NiS_2) as a light absorber (LS-NiS_2) to prepare an efficient LS-NiS_2/DW evaporator (Figure 7d). The results show that LS-NiS_2 has a high absorption rate (>95%) and photothermal conversion efficiency in a wide wavelength range. Therefore, the evaporator has a higher solar energy utilization rate. On the other hand, the hydrophilicity of DW facilitates the activation of water. The evaporation enthalpy of LS-NiS_2/DW (1274.4 $kJ \cdot kg^{-1}$) is lower than that of pure water. Under one solar irradiation, the evaporation rate of LS-NiS_2/DW reaches up to 2.80 $kg \cdot m^{-2} \cdot h^{-1}$, and its evaporation efficiency reaches 87.4%. It is worth noting that LS-NiS_2/DW shows a high evaporation rate (2.42~2.69 $kg \cdot m^{-2} \cdot h^{-1}$) in simulated seawater. Even after 24 h, no salt crystals are formed on its surface.

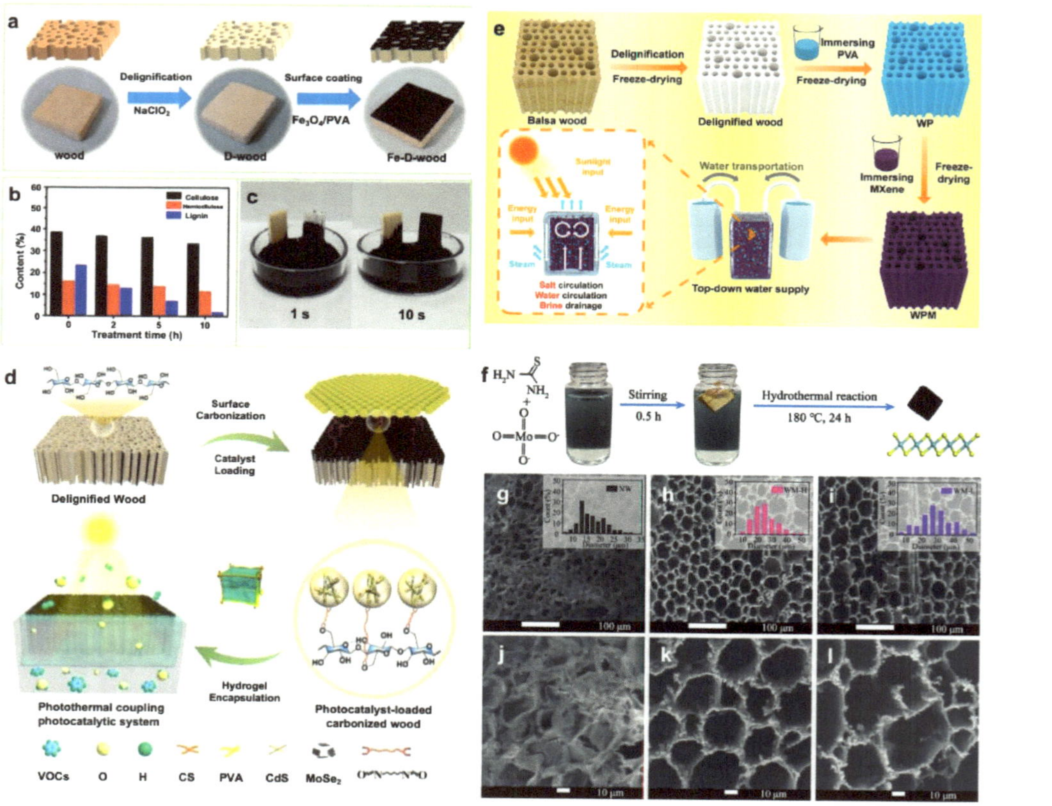

Figure 7. (**a**) Design and fabrication of Fe_3O_4/PVA-coated wood, (**b**) lignin content of wood over 10 h treatment with $NaClO_2$, and (**c**) photos of pen ink transportation in natural wood (left) and

delignified wood (right) [79]. Reprinted with permission from Ref. [79]. Copyright 2021 Elsevier Ltd. (**d**) Synthetic schematic illustration of the wood-based hydrogel coating with CdS-$MoSe_2$ for solar steam generation and hydrogen energy conversion [82]. Reprinted with permission from Ref. [82]. Copyright 2022 ACS Publications Ltd. (**e**) Preparation of WPM evaporator and its assembly with a top-down water supply device for solar steam generation and desalination [83]. Reprinted with permission from Ref. [83]. Copyright 2023 Elsevier Ltd. (**f**) Fabrication of MoS_2-coated wood. (**g–l**) SEM image of wood samples [84]. Reprinted with permission from Ref. [84]. Copyright 2021 Environmental Science Ltd.

In order to realize high-yield, high-salinity and long-term seawater desalination, Hu et al. [83] enhanced the hydrophilicity of balsa wood by delignification. Vacuum-assisted impregnation of polyvinyl alcohol (PVA) and a special top-down water supply device were adopted. Figure 7e illustrates the fabrication procedure for 3D PVA/MXene-decorated wood (WPM), which serves as a photothermal water evaporator with a top-down water supply design for solar-driven evaporation of seawater and high-salinity water. The presence of MXene nanosheets facilitates solar light absorption and solar–thermal conversion. Therefore, the evaporation rate of the PVA/MXene-decorated wood (WPM) solar evaporator under 1 $kW{\cdot}m^{-2}$ is as high as 4.31 $kg{\cdot}m^{-2}{\cdot}h^{-1}$. During the desalination of a 25 wt% NaCl solution for 8 h, its evaporation rate reached 3.83 $kg{\cdot}m^{-2}{\cdot}h^{-1}$. A sufficient down-top water supply dilutes the saline water with a high concentration on the surface of the WPM evaporator, thus avoiding salt deposition in the long-term solar-driven desalination process. The result shows that its average evaporation rate in 15 wt% NaCl solution under solar irradiation for more than 100 h is 4.24 $kg{\cdot}m^{-2}{\cdot}h^{-1}$. He et al. [84] reported a 3D wood membrane coated with MoS_2 (WM-H, with S trap) for effective seawater desalination. The MoS_2 was synthesized using thiourea and ammonium molybdate tetrahydrate (Figure 7f). High-resolution SEM images show the structural changes (Figure 7g–l). The pore size of NW is mainly distributed at approximately 10~25 μm. The specific surface area (SSA) increases from 0.22 $m^2{\cdot}g^{-1}$ (NW) to 9.92 $m^2{\cdot}g^{-1}$ (WM-H) and 10.45 $m^2{\cdot}g^{-1}$ (WM-L). The adsorption/desorption capacity of WM-H and WM-L is much higher than that of NW. After treatment, the pore structure of the wood samples becomes regular. The pore diameters of WM-H and WM-L (Figure 7g–i) increased to approximately 10~40 μm and 10~50 μm, respectively. Therefore, the vertically arranged channels of wood can rapidly provide water for its heating surface, which could support the high rate of steam generation. This is due to rapid water diffusion and powerful capillary pumping. The excellent photothermal MoS_2 provides enough heat for water evaporation. Under 1 $kW{\cdot}m^{-2}$, MoS_2-coated wood with S defects has an excellent evaporation rate and heat conversion efficiency of 1.46 $kg{\cdot}m^{-2}{\cdot}h^{-1}$ and 82.5%, respectively. The adsorption peaks of Na^+ on both sides of defective MoS_2 are respectively 2.17 and 1.49 times higher than those for MoS_2 without S defects.

3.4. Wood/Polymer Composites

Polydopamine (PDA) can be prepared by the self-polymerization of dopamine monomers under alkaline conditions. It has good adhesion and exhibits an absorption spectrum from ultraviolet light (UV) to near-infrared light (NIR). Therefore, it is an ideal photothermal coating material [85–87] that can evenly and stably adhere to the surface of a wood cavity on the basis of not blocking the pore channels. Polypyrrole (PPy) has good biocompatibility and environmental stability [14] and exhibits a high absorption rate of 90.8% in the whole solar spectrum [88,89]; it can be combined with the hydroxyl groups of cellulose through hydrogen bonds. The results show that PPy-wood shows almost full-spectrum light absorption and low incident angle sensitivity in the spectral range of 250~2500 nm (Figure 8a). Zou et al. [90] mixed arginine and a dopamine solution to obtain a black precipitate (APDA), which was coated on the surface of camphor wood to prepare APDA-wood (Figure 8b–e). Compared with traditional PDA, APDA has a narrower band gap and stronger light absorption capacity, in accordance with density functional theory (DFT). In addition, APDA has no obvious luminescence under the excitation of 365 nm, 500 nm and 808 nm, which indicates that the non-radiative transition is dominant. That is, the light absorbed by APDA will be converted into heat more quickly and effectively. Under one solar irradiation, the surface temperature of APDA-wood rises faster than that of pure wood and water. The surface temperature of APDA-wood can reach 38 °C within 5 min and stabilize at 40 °C, and the evaporation rate can reach 0.91 $kg \cdot m^{-2} \cdot h^{-1}$. In order to verify the actual seawater desalination, 3.5 wt% NaCl solution was used to simulate seawater. The results show that Na^+ in simulated seawater decreased by approximately 4 orders of magnitude after APDA-wood desalination. The concentrations of Ca^{2+}, Mg^{2+} and K^+ also decreased by at least 2~3 orders of magnitude, meeting the standards of the US Environmental Protection Agency (EPA) and the WHO. Even after 100 cycles, the water evaporation rate of APDA-wood did not decrease obviously.

In addition, the physical characteristics of wood endow PPy-wood with excellent heat insulation and water transmission performance. After the absorption of a pyrrole solution and a mixed solution of APS (ammonium sulfate) and HCl, Huang et al. [91] prepared black PPy-wood. In the comparison of the optical characteristics of PPy-wood and original wood in the spectral range of 250~2500 nm, the light absorption rate of the original wood (44.9%) was much lower than that of PPy (90.8%). More importantly, under the synergistic effect of PPy and wood, the light absorption rate of PPy-wood in the whole spectral range is as high as 97.5%. Under 1, 3, 5, 7 and 10 solar irradiations, the evaporation rates of PPy-wood were 1.33, 3.47, 5.85, 8.38 and 11.77 $kg \cdot m^{-2} \cdot h^{-1}$, respectively. They are much higher than the evaporation rates of pure water (0.50, 0.78, 1.19, 1.66 and 2.31 $kg \cdot m^{-2} \cdot h^{-1}$) under the same conditions. After treatment with strong acid (pH = 2), strong alkali (pH = 10), high temperature (100 °C) and ultrasound washing, the coating of PPy-wood did not show obvious peeling, which verified its good structural stability. In addition, the multiple scattering of light can be reduced by wood's rough surface. The light absorption efficiency of PPy-wood is over 93% at various angles (0~60°). Wang et al. [92] prepared PPy-wood by loading the photothermal PPy on balsa wood by in situ polymerization. After one solar irradiation for 1 h, the surface temperatures of water, wood and PPy-wood were 28.2 °C, 32.8 °C and 41.0 °C, respectively. It should be noted that the surface temperature of PPy-wood can reach 39.6 °C within 5 min of illumination. However, the surface temperatures of pure water and wood only increased slightly ($\Delta_{\text{pure water}} \approx 1.9$ °C, $\Delta_{\text{virgin wood}} \approx 6.5$ °C), further confirming that the PPy coating plays an important role in photothermal conversion. The evaporation rate (approximately 1.0 $kg \cdot m^{-2} \cdot h^{-1}$) and efficiency (over 70%) hardly changed in seven service cycles. After 45 days of long-term storage, the evaporation rate and efficiency of PPy-wood have no obvious variation.

Qu et al. [93] loaded polyaniline on the surface of natural wood using a spraying method. A low-cost, easily manufactured and high-performance wood evaporator was successfully assembled (Figure 8f). Because of the hydrogen bond between polyaniline and the wood surface, the solar absorption and photothermal conversion efficiency of the evaporator are greatly enhanced. The capillary effect of wood and the hydrophilicity of cellulose facilitate the continuous upward flow of water. Meanwhile, wood's low thermal conductivity causes the conversion of solar energy to heat at the interface of the wood block. In addition, the porosity and layering of wood blocks decrease the hydrogen bond density of water passing through. Under one solar irradiation, the high temperature of 68.3 °C can be reached within 30 min, verifying the good photothermal conversion effect of a PANI-wood block. In the same condition, the surface temperature of natural wood is relatively stable at 46.2 °C (Figure 8g). The temperature of the evaporator and natural wood tends to be stable after 30 min. This is because the PANI remarkably increases the solar energy absorption of the wood evaporator after modification. The evaporation efficiency of the polyaniline-wood evaporator is 1.66 $kg \cdot m^{-2} \cdot h^{-1}$ under one solar irradiation. In addition, the concentration of main ions in purified water obtained by the evaporator fully meets the WHO requirements for drinking water.

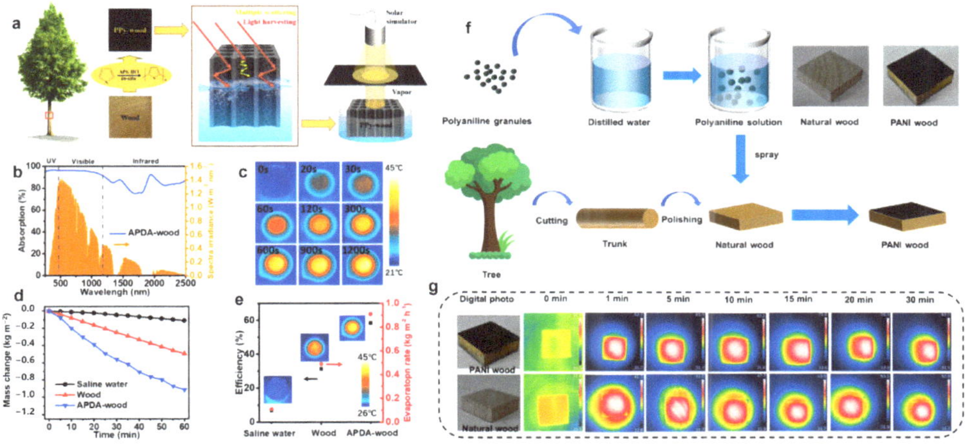

Figure 8. (**a**) Schematic illustration of the PPy-wood for solar steam generation [14]. Reprinted with permission from Ref. [14]. Copyright 2020 Royal Society of Chemistry Ltd. (**b**) Light absorption spectra of APDA-wood ranging from 250 nm to 2500 nm in wet state [90]. (**c**) Time-dependent IR images of the APDA-wood under 635 nm with the irradiation of 1 kW/m^2 [90]. (**d**) Water evaporation rates of saline water, saline water with wood and APDA-wood under 1 kW/m^2 [90]. (**e**) Solar efficiency and evaporation rate of saline water, wood and APDA-wood under visible light [90]. Two differently colored arrows indicate the scales on each side. Reprinted with permission from Ref. [90]. Copyright 2021 Elsevier Ltd. (**f**) Preparation diagram of PANI-wood [93]. (**g**) Infrared image of surface temperature of dry natural wood and PANI-wood under 1 kW/m^2 [93]. Reprinted with permission from Ref. [93]. Copyright 2021 Elsevier Ltd.

Poly (N-phenylglycine) (PNPG) is a conjugated polymer with good light absorption, which can greatly make up for the weak light absorption of wood. In addition, PNPG can effectively improve the photothermal conversion efficiency of wood evaporators. Inspired by the transpiration of trees, Lin et al. [94] developed a PNPG-wood solar evaporator with low thermal conductivity and a special microstructure. Under one solar irradiation, its evaporation rate and conversion efficiency can reach 1.64 kg·m^{-2}·h^{-1} and 90.4%, respectively, which are higher than those of most reported wood solar evaporators. This system has the potential to solve the practical issues of seawater desalination and water purification. Li et al. [95] successfully constructed a polyelectrolyte hydrogel (SCPH) with rapid pumping, enhanced salt discharge, and improved mechanical strength and thermal insulation performance. The wood sponge skeleton of SCPH has a series of characteristics such as high mechanical strength, low thermal conductivity, excellent compressibility and high porosity. In addition, the polyelectrolyte hydrogel has weak mechanical strength, high thermal conductivity, and good pumping properties and salt rejection performance. Therefore, a novel SCPH material was developed by making full use of the advantages of polyelectrolyte hydrogel and balsa sponge. The material has a very low thermal conductivity (0.109 W·m^{-1}·K^{-1}), while its mechanical strength is significantly improved. In addition, it is worth noting that its salt rejection rate reached 88.7% even in a 10% salinity solution. The evaporation rate of the SCPH evaporator reached 2.13 kg m^{-2}·h^{-1}, which greatly exceeded that of a reported polyelectrolyte hydrogel evaporator. It is exciting that the proposed evaporator can run in 20% salt water for 30 days. There is no salt accumulation on its surface, representing an improvement over most reported single-hydrogel evaporators. Sheng et al. [96] used a chemically stable coordination polymer (Ni-DTA) as a hydrophilic photothermal nanomaterial to produce a robust wood-based evaporator (Balsa-NiDTA) with better molecular design performance. Ni-DTA synthesized in situ on the cell wall of balsa wood provides enough light and heat fields, which make the converted energy localize to promote interfacial evaporation. Reasonably controlling the ratio of methanol to dimethylformamide allows 1D nanofibers and 0D nanoparticles to coexist. Therefore, its evaporation rate and energy efficiency under one solar illumination can reach up to 2.75 kg·m^{-2}·h^{-1} and 82%, respectively. The results show that the Ni-DTA polymer with strong hydration ability will reduce the equivalent evaporation enthalpy, which is attributed to the decrease in the H bond density of water molecules near the evaporation interface. The evaporator shows a high chemical stability, which is mainly attributed to the firm Ni-S/Ni-N bond and the excellent cellulose affinity of Ni-DTA. In addition, the evaporator has excellent antibacterial activity and low oil pollution tendency, which are helpful in realizing the efficient and sustainable solar desalination of Balsa-NiDTA under various harsh conditions.

3.5. Wood/Precious Metal Composites

When light is incident on a photothermal conversion layer composed of precious metal nanoparticles, if the incident photon frequency matches the total vibration frequency of precious metal nanoparticles or metal conduction electrons, it will have a strong absorption effect on photon energy and produce the local surface plasmon resonance (LSPR) effect [97]. Due to the LSPR effect, precious metal nanoparticles can show a stronger absorption spectrum in the ultraviolet–visible band. In addition, the LSPR effect also causes near-field enhancement, hot electron generation and photothermal conversion [98]. This is because the electrons are excited from an occupied state to an unoccupied state, forming hot electrons [99], the energy of which is redistributed through electron scattering, thus rapidly increasing the temperature of metal particles [100]. The shape and position of LSPR are closely related to the composition, size, shape, dielectric property and dielectric environment of nanoparticles [101]. In general, reducing the shape symmetry or creating a hollow structure can broaden the spectral LSPR band. However, a variation in particle size or the surrounding media will mainly cause a shift in the LSPR band and broaden the absorption band [102]. Silver is widely known for its high plasma resonance effect and low plasma loss. Gold is famous for its visible–near-infrared plasma resonance effect and chemical stability. Therefore, Au NPs and Ag NPs are the most widely used plasmon metals and photothermal materials for solar evaporation [103]. Goharshadi et al. [104] coated layered porous wood with Ag and Pd nanoparticles (NPs) as a double layer. Then, the coated wood could be used as an effective light absorber in interface solar steam generation (ISSG) (Figure 9a). The results showed that the plasmon resonance effect of Ag and Pd NPs could increase the photothermal conversion efficiency of the wood composite when Ag and Pd NPs were used as the bottom layer and upper layer, respectively. The wood composite's highest evaporation rate was 4.82 kg·m^2·h^{-1} under three solar irradiations. Initially, the salinity, conductivity and pH of seawater were 3138.19 mg·L^{-1}, 6000 μS·cm^{-1} and 8.1, respectively. After treatment, the salinity, conductivity and pH of seawater were reduced to 6.76 mg·L^{-1}, 20.5 μS·cm^{-1} and 6.7, respectively. In addition, the wood composite also presented a stable water generation capability during long-term cycles.

Zhu et al. [105] designed a novel plasmonic wood for efficient steam generation. First, basswood was cut perpendicular to the direction of the tree's growth. Plasma metal nanoparticles were then deposited in the wood microchannels to form plasma wood (Figure 9b). Black plasmonic wood can float on water without additional assistance. The plasma palladium nanoparticles on the surfaces of microchannels convert incident light into heat due to the plasma effect (Figure 9c,d). Therefore, plasmonic wood has a high solar absorptivity (≈99%) in the wide wavelength range of 200~2500 nm. Due to the thin decorative layer, metal nanoparticles will not block the water microchannels, ensuring continuous water transportation. Due to the low thermal conductivity of plasmonic wood, the heat is concentrated on the evaporation surface, resulting in an effective solar steam generator. The hydrophilicity and capillary effect of plasmonic wood can effectively transport water upward to maintain a continuous water supply for steam generation (Figure 9e). The low thermal conductivity of wood microchannels contributes to thermal localization, which can effectively inhibit heat dissipation to bulk water. Therefore, the plasmonic wood has a unique 3D mesoporous arrangement structure. The solar energy conversion efficiency can reach 85% under 10 solar irradiations. The as-obtained plasmonic wood also showed strong stability (over 144 h) and self-cleaning ability. Thanks to the above advantages, plasmonic wood can be applied to many fields such as water sterilization and seawater desalination.

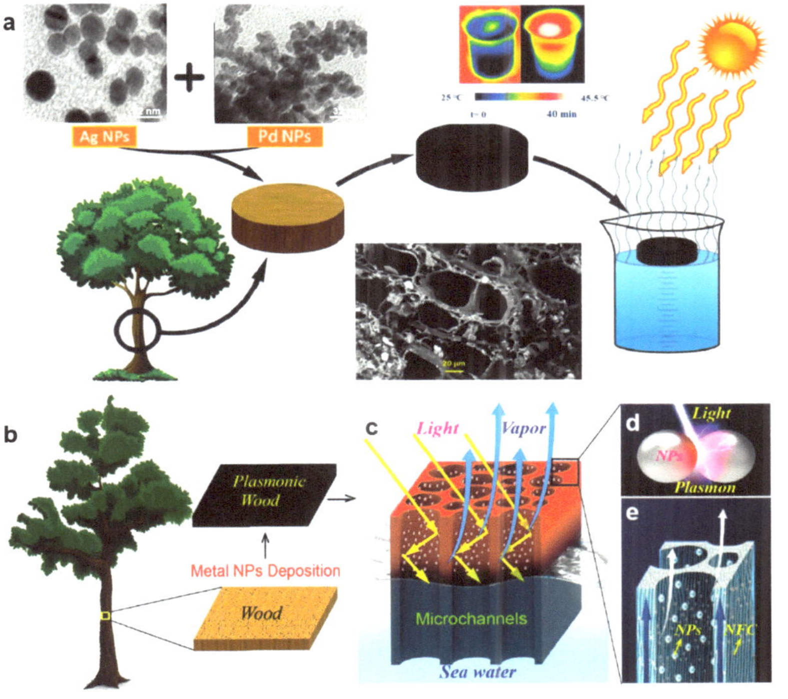

Figure 9. (**a**) Highly efficient plasmonic wood/Ag/Pd photoabsorber in interfacial solar steam generation [104]. Reprinted with permission from Ref. [104]. Copyright 2018 Wiley-Blackwell Ltd. (**b**) Natural wood is cut perpendicular to the growth direction of the tree, and it turns black after nanoparticle decoration due to the plasmonic effect of the metal nanoparticles [105]. (**c**) After metal nanoparticle decoration, light can be guided into the wood lumen and be fully absorbed for steam generation [105]. (**d**) Schematic of plasmonic effect of two adjacent metal nanoparticles (NPs) [105]. (**e**) Zoomed-in schematic illustrating the water transport along microchannels in wood. The cell wall is composed of abundant nanofibrous cellulose (NFC) [105]. Reprinted with permission from Ref. [105]. Copyright 2022 Elsevier Ltd.

4. Summary and Prospects

Wood has a unique natural pore structure and low thermal conductivity. In addition, it has wide sources, large volume, light weight, good toughness, impact resistance, good biocompatibility, reproducibility and biodegradability. These characteristics lay a solid foundation for its application in water pollution purification and solar-driven seawater desalination. The modification of wood is very beneficial for the full contact and interaction of heavy metal ions, aromatic dyes and bacteria with functional groups and nanomaterials in wood nano/micropores. Apparently, such modified wood composites with a high flux and large adsorption capacity are very suitable for industrial wastewater purification. In addition, wood composites combined with precious metals, semiconductors, polymers and carbon nanomaterials and directly carbonized wood exhibit a unique 3D pore structure, low thermal conductivity, good transpiration characteristics, etc. These merits greatly improve the solar-driven seawater desalination rate and efficiency of wood composites.

4.1. Existing Issues

Wood composites are used as photothermal evaporators, filter membranes and adsorbents in the field of water treatment. Many efforts and great breakthroughs have been made in the purification of industrial wastewater containing heavy metal ions, bacteria, aromatic dyes, oil stains, etc. Meanwhile, the catalytic and photothermal properties of nanomaterials and the low thermal conductivity and unique pore structure of wood allow wood composite evaporators to better utilize the full spectrum of sunlight. These outstanding performances accelerate the development of wood composite evaporators in solar-driven seawater desalination. However, there are still many challenges on the road to commercialization; these challenges are listed as follows:

(1) The types of pollutants in actual industrial wastewater are complex. In most reported works, the types of pollutants treated by wood composites are relatively single. The influencing factors and changing rules of industrial wastewater treatment are not investigated deeply enough, and the application scope is still limited.

(2) There are many kinds of woods. Therefore, their pore structures and chemical compositions are quite different. This means that the controllability is relatively low (e.g., uneven distribution of functional groups and nanomaterials and insufficient stability) in the modification of wood (cell walls).

(3) As effective photothermal materials, carbon nanomaterials generally have a high cost, low hydrophilicity, complex production process, etc. These drawbacks cause the uneconomical and poor binding between biomass matrices and carbon nanomaterials.

(4) Although surface carbonization simplifies the preparation of a wood-based interface evaporator, the mechanical strength of carbonized wood is obviously reduced, making structural collapse more likely to occur in harsh environments. In addition, the anisotropic characteristics of wood make it difficult to control its carbonization process and microstructure, which is not conducive to subsequent integration.

(5) Polymers have attracted extensive attention because of their excellent light absorption and mature preparation process. However, the service life and evaporation efficiency of wood/polymer composite solar evaporators are reduced due to the influence of the photodegradation of polymers.

4.2. Further Research

Wood composites with excellent performances in the fields of industrial wastewater purification and solar-driven seawater desalination should be further studied in the following aspects:

(1) Multi-directional improvements and multifunctionalization of wood should be enhanced. The 3D nano/micropore structure of wood (conduits or tracheids along the direction of wood growth and wood rays, pits and nanoholes perpendicular to the direction of wood growth) should be made full use of. With reasonable and optimized structure design, suitable nanomaterial loading and functional group modifications on cell walls and surfaces, wood can be endowed with greater value and more functions. This can greatly enrich the types of pollutant treatments and improve the evaporation performance of wood composites with more flexibility in applied environments in wastewater purification and seawater desalination.

(2) Aside from exploring more effective technologies, more simple, feasible and green technical methods should be studied scientifically for fabricating wood composites. The movement path of fluid in wood microchannels should be explored in greater depth. In addition, the effects of hemicellulose/lignin removal technologies and drying methods on wood's porous structure and its drying shrinkage anisotropy should be studied in depth. The relationship between the hemicellulose and lignin removal and the variation in wood's

porous structure should be investigated carefully. The purpose is to optimize the structure of wood with the appropriate technologies, which should be simple and controllable, green and feasible, low in energy and consumption, and suitable for large-scale production.

(3) The inherent physical and chemical properties of wood have great advantages in the field of polluted water purification and solar-driven seawater desalination, which has built a solid foundation for large-scale and commercial applications. However, research on the application of wood composites in this field is still in the exploratory stage. There is an urgent need to systematically evaluate the multifunctional performance of different woods. The accumulation of data on the practical application and economic effectiveness of wood composites should be enhanced, and a complete development system of wood composites in wastewater purification and seawater desalination should be established.

(4) The advantages and disadvantages of photothermal materials and loading nanomaterials or functional groups should be carefully studied and reasonably combined. The wetting and photothermal properties of a wood composite interface have a great influence on its evaporation performance under solar irradiation. Considering the respective advantages of any raw materials, the development strategy of functional and robust wood composites should be updated and optimized. Wood-based filters and evaporators with low cost, good recombination, high water flux, high photothermal rate, low light reflection, good catalysis, long service life, and simple and feasible characteristics should be prepared more controllably. In addition, it is of great significance to develop wood composites with stimulation response to improve their selectivity and intelligence.

(5) Although the evaporation efficiency of wood composites is high, their solar evaporation is affected by the steam pressure in the limited space. The evaporation result is not ideal in actual seawater desalination applications. Compared with a traditional solar distiller with volume heating, the freshwater productivity of functional wood composites has been improved. However, the service life of wood composites will be reduced in a complex ecological environment with high humidity and high salinity. A similar situation also occurs in the wastewater purification of wood composites. Therefore, future research should not only focus on increasing fresh and clean water production, but also pay more attention to improving wood durability (such as salt resistance, acid and alkali resistance, oil resistance and bacteriostasis).

In summary, there are still many challenges in the fields of industrial wastewater purification and solar-driven seawater desalination. The use of wood composites gives full play to the advantages of renewable resource utilization and effectively solves a series of problems such as the continuous pollution of the ecological environment and the imminent exhaustion of freshwater resources, thus being able to avoid the crisis of water quality and quantity. In addition, it is also a green and sustainable method that provides fresh and clean water for residents' production and life. Therefore, this review provides a favorable solution for the water shortage issue in remote areas and offers a new idea for alleviating energy consumption and environmental pollution.

Author Contributions: D.S.: conceptualization, visualization, writing—original draft preparation; D.Z.: picture collection; Z.L.: copyright application; C.W.: supervision; J.L.: supervision; M.Z.: writing—review and editing, funding acquisition, project administration. All authors have read and agreed to the published version of the manuscript.

Funding: The research was supported by the Natural Science Foundation of Jilin Province (YDZJ2022-01ZYTS441), National Natural Science Foundation of China (32130073), Jilin Province Development and Reform Commission Industrial Innovation Special Fund Project (2023C038-2), State Key Laboratory of Bio-Fibers and Eco-Textiles (KFKT202213) and Fundamental Research Funds for Wood Material Science and Engineering Key Laboratory of Jilin Province.

Institutional Review Board Statement: This manuscript does not contain any studies involving human or animal subjects.

Data Availability Statement: No new data were created or analyzed in this study.

Conflicts of Interest: The authors declare no conflict of interest.

References

1. Chen, C.J.; Kuang, Y.D.; Zhu, S.Z.; Burgert, I.; Keplinger, T.; Gong, A.; Li, T.; Berglund, L.; Eichhorn, S.; Hu, L. Structure-property-function relationships of natural and engineered wood. *Nat. Rev. Mater.* **2020**, *5*, 642–666. [CrossRef]
2. Boutilier, M.; Lee, J.; Chambers, V. Water filtration using plant xylem. *PLoS ONE* **2014**, *9*, e89934. [CrossRef] [PubMed]
3. Wang, M.Z.; Fang, Y.; Li, Q.; Bai, L.; Liu, S.; Li, J.; Chen, W. Nanostructures of plant cell walls and individualization methodology of nanofibrillated cellulose. *Acta Polym. Sin.* **2020**, *51*, 586–597.
4. Zhang, M.; Zhou, N.; Shi, Y.; Ma, Y.; An, C.; Li, J.; Jiang, W.; Wang, C.; Shi, J. Construction of antibacterial, superwetting, catalytic, and porous lignocellulosic nanofibril composites for wastewater purification. *Adv. Mater. Interfaces* **2022**, *9*, 2201388. [CrossRef]
5. Li, Z.; Zhang, M. Progress in the Preparation of Stimulus-Responsive Cellulose Hydrogels and Their Application in Slow-Release Fertilizers. *Polymers* **2023**, *15*, 3643. [CrossRef]
6. Tang, Y.; Chen, L.; Wei, X. Removal of lead ions from aqueous solution by the dried aquatic plant, *Lemna perpusilla* Torr. *J. Hazard. Mater.* **2013**, *244–245*, 603–612. [CrossRef]
7. Liu, C.; Luan, P.C.; Li, Q.; Cheng, Z.; Xiang, P.; Liu, D.; Hou, Y.; Yang, Y.; Zhu, H. Biopolymers derived from trees as sustainable multifunctional materials: A review. *Adv. Mater.* **2020**, *33*, 1654. [CrossRef]
8. Lars, A.B.I. Bioinspired wood nanotechnology for functional materials. *Adv. Mater.* **2018**, *30*, 1–15.
9. Li, J.; Sun, Q.F. Inspirations from nature preliminary discussion of wood bionics. *China Eng. Sci.* **2014**, *16*, 4–12.
10. Yi, J.Y.; Huo, Z.; Asiri, A.M.; Alamry, K.A.; Li, J. Application of agroforestry waste biomass adsorption materials in water pollution treatment. *Chem. Prog.* **2019**, *31*, 760–772.
11. Roccaro, P.; Sgroi, M.; Vagliasind, F.G.A. Removal of xenobiotic compounds from wastewater for environment protection: Treatment processes and costs. *Chem. Eng. Trans.* **2013**, *32*, 505–510.
12. Burakov, A.E.; Galunin, E.V.; Burakova, I.V.; Kucherova, A.E.; Agarwal, S.; Tkachev, A.G.; Gupta, V.K. Adsorption of heavy metals on conventional and nanostructured materials for wastewater treatment purposes: A review. *Ecotoxicol. Environ. Saf.* **2018**, *148*, 702–712. [CrossRef] [PubMed]
13. Santhosh, C.; Velmurugan, V.; Jacob, G.; Jeong, S.K.; Grace, N.A.; Bhatnagar, A. Role of nanomaterials in water treatment applications: A review. *Chem. Eng. J.* **2016**, *306*, 1116–1137. [CrossRef]
14. Zhang, M.; Wang, C.; Ma, Y.; Du, X.; Shi, Y.; Li, J.; Shi, J. Fabrication of superwetting, antimicrobial and conductive fibrous membranes for removing/collecting oil contaminants. *RSC Adv.* **2020**, *10*, 21636–21642. [CrossRef] [PubMed]
15. Jiao, M.L.; Yao, Y.G.; Chen, C.J.; Jiang, B.; Pastel, G.; Lin, Z.; Wu, Q.; Cui, M.; He, S.; Hu, L. Highly efficient water treatment via a wood-based and reusable filter. *ACS Mater. Lett.* **2020**, *2*, 430–437. [CrossRef]
16. Chen, F.J.; Gong, A.S.; Zhu, M.W.; Chen, G.; Lacey, S.D.; Jiang, F.; Li, Y.; Wang, Y.; Dai, J.; Yao, Y.; et al. Mesoporous, three-dimensional wood membrane decorated with nanoparticles for highly efficient water treatment. *ACS Nano* **2017**, *11*, 4275–4282. [CrossRef] [PubMed]
17. An, C.C.; Zhang, M.; Xiao, Z.H.; Yang, Q.; Feng, L.; Li, S.; Shi, M. Lignocellulose/chitosan hybrid aerogel composited with fluorescence molecular probe for simultaneous adsorption and detection of heavy metal pollutants. *J. Environ. Chem. Eng.* **2023**, *11*, 111205. [CrossRef]
18. Tripathi, A.; Ranjanm, R. Heavy metal removal from wastewater using low cost adsorbents. *J. Bioremediat. Biodegrad.* **2015**, *6*, 315. [CrossRef]
19. Zhu, M.W.; Li, Y.J.; Chen, G.; Jiang, F.; Yang, Z.; Luo, X.; Wang, Y.; Lacey, S.D.; Dai, J.; Wang, C.; et al. Tree-inspired design for high-efficiency water extraction. *Adv. Mater.* **2017**, *29*, 1704107. [CrossRef]
20. Ahmad, A.; Rafatullah, M.; Sulaiman, O.; Ibrahim, M.H.; Chii, Y.Y.; Siddique, B.M. Removal of Cu (II) and Pb (II) ions from aqueous solutions by adsorption on sawdust of meranti wood. *Desalination* **2009**, *247*, 636–646. [CrossRef]
21. He, L.; Zhao, L.; Sun, W.; Fang, J.; Liu, X.; Qi, J.; Qian, Y.; Li, H. 3D Wood Microfilter for Fast and Efficient Removal of Heavy Metal Ions from Water. *Langmuir* **2023**, *39*, 15319–15327. [CrossRef] [PubMed]
22. Yang, Z.; Liu, H.; Li, J.; Yang, K.; Zhang, Z.; Chen, F.; Wang, B. High-throughput metal trap: Sulfhydryl-functionalized wood membrane stacks for rapid and highly efficient heavy metal ion removal. *ACS Appl. Mater. Interfaces* **2020**, *12*, 15002–15011. [CrossRef] [PubMed]
23. Pei, L.F.; Jiang, L.L.; Hou, X.G. Applicaton of nanoadsorbent in wastewater treatment of heavy metals and organic matter. *China Metall.* **2018**, *28*, 1–5.
24. Cai, X.H.; Liu, H.W.; Zhi, L.H.; Wen, H.; Yu, A.; Li, L.; Chen, F.; Wang, B. Ag-C_3N_4/rGO nanocomposite as a highly efficient metal-free photocatalyst for direct C-H arylation under visible light irradiation. *RSC Adv.* **2017**, *7*, 46132–46138. [CrossRef]
25. Wang, R.R.; Li, Q.R.; Xiao, H.Y.; Xie, D.; Lu, H. Synthesis of NiO using pine as template and adsorption performance for Pb(II) from aqueous solution. *Appl. Surf. Sci.* **2013**, *279*, 129–136. [CrossRef]
26. Vitas, S.; Keplinger, T.; Reichholf, N.; Figi, R.; Cabane, E. Functional lignocellulosic material for the remediation of copper (II) ions from water: Towards the design of a wood filter. *J. Hazard. Mater.* **2018**, *355*, 119–127. [CrossRef] [PubMed]
27. Liu, M.; Lu, Q.; Yu, W. The improvement of heavy metals removal by wood membrane in drinking water treatment: Comparison with polymer membrane and associated mechanism. *Chemosphere* **2023**, *324*, 138297. [CrossRef]
28. Gu, Y.; Ye, M.X.; Wang, Y.C.; Li, H.M.; Zhang, H.M.; Wang, G.Z.; Zhang, Y.X.; Zhao, H.J. Lignosulfonate functionalized g-C3N4/carbonized wood sponge for highly efficient heavy metal ion scavenging. *J. Mater. Chem. A* **2020**, *8*, 12687–12698. [CrossRef]

29. Guo, Q.; Qiu, Y.Q. Research progress in application of nano-silver in water treatment. *Synth. Mater. Aging Appl.* **2019**, *48*, 125–129.
30. Zhang, M.; Shi, Y.; Wang, R.; Chen, K.; Zhou, N.; Yang, Q.; Shi, J. Triple-functional lignocellulose/chitosan/Ag@TiO_2 nanocomposite membrane for simultaneous sterilization, oil/water emulsion separation, and organic pollutant removal. *J. Environ. Chem. Eng.* **2021**, *9*, 106728. [CrossRef]
31. Macior, A.; Zaborniak, I.; Chmielarz, P.; Smenda, J.; Wolski, K.; Ciszkowicz, E.; Lecka-Szlachta, K. A New Protocol for Ash Wood Modification: Synthesis of Hydrophobic and Antibacterial Brushes from the Wood Surface. *Molecules* **2022**, *27*, 890. [CrossRef]
32. Dai, X.; Qi, Y.; Luo, H.; He, Z.; Wei, L.; Dong, X.; Ma, X.; Yang, D.; Li, Y. Leachability and Anti-Mold Efficiency of Nanosilver on Poplar Wood Surface. *Polymers* **2022**, *14*, 884. [CrossRef]
33. Che, W.; Xiao, Z.; Wang, Z.; Li, J.; Wang, H.; Wang, Y.; Xie, Y. Wood-based mesoporous filter decorated with silver nanoparticles for water purification. *ACS Sustain. Chem. Eng.* **2019**, *7*, 5134–5141. [CrossRef]
34. Wen, J.J.; Tan, X.J.; Hu, Y.Y.; Guo, Q.; Hong, X. Filtration and electrochemical disinfection performance of PAN/PANI/AgNWs-CC composite nanofiber membrane. *Environ. Sci. Technol.* **2017**, *51*, 6395–6403. [CrossRef]
35. Huo, Z.Y.; Luo, Y.F.; Xie, X.; Feng, C.; Jiang, K.; Wang, J.; Hu, H. Carbon-nanotube sponges enabling high-efficiency and reliable cell inactivation by low-voltage electroporation. *Environ. Sci. Nano* **2017**, *4*, 2010–2017. [CrossRef]
36. Yang, Z.; Ni, H.; Liu, P.; Liu, H.; Yang, K.; Zhang, Z.; Wang, B.; Li, X.; Chen, F. Nanofibrils in 3D aligned channel arrays with synergistic effect of Ag/NPs for rapid and highly efficient electric field disinfection. *Chin. Chem. Lett.* **2021**, *32*, 3143–3148. [CrossRef]
37. Du, X.; Shi, L.; Pang, J.; Zheng, H.; Shi, J.; Zhang, M. Fabrication of superwetting and antimicrobial wood-based mesoporous composite decorated with silver nanoparticles for purifying the polluted-water with oils, dyes and bacteria. *J. Environ. Chem. Eng.* **2022**, *10*, 107152. [CrossRef]
38. Cheng, Z.; Guan, H.; Meng, J.; Wang, X. Dual-functional porous wood filter for simultaneous oil/water separation and organic pollutant removal. *ACS Omega* **2020**, *5*, 14096–14103. [CrossRef]
39. Adam, A.M.A.; Saad, H.A.; Atta, A.A.; Alsawat, M.; Hegab, M.S.; Refatm, M.S.; Altalhim, T.A.; Alosaimi, E.H.; Younes, A. Preparation and characterization of new $CrFeO_3$-carbon composite using environmentally friendly methods to remove organic dye pollutants from aqueous solutions. *Crystals* **2021**, *11*, 960. [CrossRef]
40. Habibi, Y.; Lucia, L.A.; Rojas, O.J. Cellulose nanocrystals: Chemistry, self-assembly, and applications. *Chem. Rev.* **2010**, *110*, 3479–3500. [CrossRef]
41. Guo, R.X.; Cai, X.H.; Liu, H.W.; Yang, Z.; Meng, Y.; Chen, F.; Li, Y.; Wang, B. In situ growth of metal-organic frameworks in three-dimensional aligned lumen arrays of wood for rapid and highly efficient organic pollutant removal. *Environ. Sci. Technol.* **2019**, *53*, 2705–2712. [CrossRef]
42. Liu, G.; Chen, D.; Liu, R.; Yu, Z.; Jiang, J.; Liu, Y.; Hu, J.; Chang, S. Antifouling wood matrix with natural water transfer and microreaction channels for water treatment. *ACS Sustain. Chem. Eng.* **2019**, *7*, 6782–6791. [CrossRef]
43. Goodman, S.M.; Bura, R.; Dichiara, A.B. Facile impregnation of graphene into porous wood filters for the dynamic removal and recovery of dyes from aqueous solutions. *ACS Appl. Nano Mater.* **2018**, *1*, 5682–5690. [CrossRef]
44. Sens, M.L.; Emmendoerfer, M.L.; Mueller, L.C. Water filtration through wood with helical cross-flow. *Desalination Water Treat.* **2015**, *53*, 15–26. [CrossRef]
45. Cui, Z.; Wu, J.; Xu, Y.; Wu, T.; Li, H.; Li, J.; Kang, L.; Cai, Y.; Li, J.; Tian, D. In-situ growth of polyoxometalate-based metal-organic frameworks on wood as a promising dual-function filter for effective hazardous dye and iodine capture. *Chem. Eng. J.* **2023**, *451*, 138371. [CrossRef]
46. Liu, Y.; Wang, T. The Wood/MoS_2 Composites with Good Adsorption and Photocatalytic Ability for Highly Efficient Removal of Organic Dye Molecules. *Chem. Sel.* **2023**, *8*, e202300119. [CrossRef]
47. Zhang, M.; Shi, L.; Du, X.; Li, Z.; Shi, Y.; An, C.; Li, J.; Wang, C.; Shi, J. Janus mesoporous wood-based membrane for simultaneous oil/water separation, aromatic dyes removal, and seawater desalination. *Ind. Crops Prod.* **2022**, *188*, 115643. [CrossRef]
48. Li, J.; Zhang, M.; Qiang, T. Research progress in modified PVDF membrane with inorganic particles for oil bearing wastewater treatment in oilfields. *J. For. Environ.* **2016**, *36*, 257–265.
49. Wang, C.Y.; Wang, X.; Yu, Q. New research progress of functional wood. *J. For. Eng.* **2019**, *4*, 10–18.
50. Zhao, M.; Tao, Y.; Wang, J.; He, Y. Facile preparation of superhydrophobic porous wood for continuous oil-water separation. *J. Water Process Eng.* **2020**, *36*, 101279. [CrossRef]
51. Wang, H.; Wu, Y.H.; Wu, P.C.; Chen, S.; Guo, X.; Meng, G.; Peng, B.; Wu, J.; Liu, Z. Environmentally benign chitosan as reductant and supporter for synthesis of Ag/AgCl/chitosan composites by one-step and their photocatalytic degradation performance under visible-light irradiation. *Front. Mater. Sci.* **2017**, *11*, 130–138. [CrossRef]
52. Guan, H.; Cheng, Z.; Wang, X. Highly compressible wood sponges with a spring-like lamellar structure as effective and reusable oil absorbents. *ACS Nano* **2018**, *12*, 10365–10373. [CrossRef]
53. Song, J.W.; Chen, C.J.; Wang, C.W.; Kuang, Y.; Li, Y.; Jiang, F.; Li, Y.; Hitz, E.; Zhang, Y.; Liu, B.; et al. Super flexible wood. *ACS Appl. Mater. Interfaces* **2017**, *9*, 23520–23527. [CrossRef]
54. Chao, W.; Sun, X.; Li, Y.; Cao, G.; Wang, R.; Wang, C.; Ho, S. Enhanced directional seawater desalination using a structure-guided wood aerogel. *ACS Appl. Mater. Interfaces* **2020**, *12*, 22387–22397. [CrossRef]
55. Fu, Q.L.; Ansari, F.; Zhou, Q.; Berglund, L.A. Wood nanotechnology for strong, mesoporous, and hydrophobic biocomposites for selective separation of oil/water mixtures. *ACS Nano* **2018**, *12*, 2222–2230. [CrossRef]

56. Wang, K.; Liu, X.; Tan, Y.; Zhang, W.; Zhang, S.; Li, J.; Huang, A. Highly fluorinated and hierarchical HNTs/SiO_2 Chemical hybrid particles for substrate-independent superamphiphobic coatings. *Chem. Eng. J.* **2019**, *359*, 626–640. [CrossRef]
57. Blanco, M.V.D.; Fscher, E.J.; Cabane, E. Underwater superoleophobic wood cross sections for efficient oil/water separation. *Adv. Mater. Interfaces* **2017**, *4*, 1700584. [CrossRef]
58. Chao, W.X.; Wang, S.B.; Li, Y.D.; Cao, G.; Zhao, Y.; Sun, X.; Wang, C.; Ho, S. Natural sponge-like wood-derived aerogel for solar-assisted adsorption and recovery of high-viscous crude oil. *Chem. Eng. J.* **2020**, *400*, 125865. [CrossRef]
59. Zhu, Y.R.; Li, H.Q.; Huang, W.; Lai, X.; Zeng, X. Facile fabrication of superhydrophobic wood aerogel by vapor deposition method for oil-water separation. *Surf. Interfaces* **2023**, *37*, 102746. [CrossRef]
60. Chen, Z.H.; Su, X.J.; Li, K.Q.; Chen, R.; Lai, Z.; Li, X.; Chen, S.; Wu, W. Superhydrophobic PDMS/PEI-cCNTs wood with Joule heat and demulsification capability for oil-water emulsio70n separation. *Prog. Org. Coat.* **2023**, *185*, 107960. [CrossRef]
61. Wu, J.W.; Tian, J.; Qian, Z.H.; Huang, J.; Sun, D. Highly robust separation for aqueous oils enabled by balsa wood-based cellulose aerogel with intrinsic superior hydrophilicity. *Sep. Purif. Technol.* **2023**, *315*, 123688. [CrossRef]
62. Shi, L.; Zhang, M.; Du, X.; Liu, B.; Li, S.; An, C. In-situ polymerization of pyrrole on elastic wood for high efficiency seawater desalination and oily water purification. *J. Mater. Sci.* **2022**, *57*, 16317–16332. [CrossRef]
63. Zhu, M.; Song, J.; Li, T.; Gong, A.; Wang, Y.; Dai, J.; Yao, Y.; Luo, W.; Henderson, D.; Hu, L. Highly anisotropic, highly transparent wood composites. *Adv. Mater.* **2016**, *28*, 5181–5187. [CrossRef]
64. Grassi, G.; Magnani, F. Stomatal, mesophyll conductance and biochemical limitations to photosynthesis as affected by drought and leaf ontogeny in ash and oak trees. *Plant Cell Environ.* **2005**, *28*, 834–849. [CrossRef]
65. Zhao, X.; Chen, H.; Kong, F.; Zhang, Y.; Wang, S.; Liu, S.; Lucia, L.A.; Fatehi, P.; Pang, H. Fabrication, characteristics and applications of carbon materials with different morphologies and porous structures produced from wood liquefaction: A review. *Chem. Eng. J.* **2019**, *364*, 226–243. [CrossRef]
66. Xue, G.; Liu, K.; Chen, Q.; Yang, P.; Li, J.; Ding, T.; Duan, J.; Qi, B.; Zhou, J. Robust and low-cost flame-treated wood for high-performance solar steam generation. *ACS Appl. Mater. Interfaces* **2017**, *9*, 15052–15057. [CrossRef]
67. Kuang, Y.; Chen, C.; He, S.; Hitz, E.; Wang, Y.; Gan, W.; Mi, R.; Hu, L. A high-performance self-regenerating solar evaporator for continuous water desalination. *Adv. Mater.* **2019**, *31*, 1900498. [CrossRef]
68. Chen, J.; Jian, M.; Yang, X. Highly Effective Multifunctional Solar Evaporator with Scaffolding Structured Carbonized Wood and Biohydrogel. *ACS Appl. Mater. Interfaces* **2022**, *14*, 46491–46501. [CrossRef]
69. Ghafurian, M.M.; Niazmand, H.; Ebrahimnia-Bajestan, E.; Taylor, R.A. Wood surface treatment techniques for enhanced solar steam generation. *Renew. Energy* **2020**, *146*, 2308–2315. [CrossRef]
70. Vélez-Cordero, J.R.; Hernandez-Cordero, J. Heat generation and conduction in PDMS-carbon nanoparticle membranes irradiated with optical fibers. *Int. J. Therm. Sci.* **2015**, *96*, 12–22. [CrossRef]
71. Gao, T.; Li, Y.; Chen, C.; Yang, Z.; Kuang, Y.; Jia, C.; Song, J.; Hitz, E.M.; Liu, B.; Huang, H.; et al. Architecting a floatable, durable, and scalable steam generator: Hydrophobic/hydrophilic bifunctional structure for solar evaporation enhancement. *Small Methods* **2019**, *3*, 1800176. [CrossRef]
72. Zhao, D.L.; Chung, T.S. Applications of carbon quantum dots (CQDs) in membrane technologies: A review. *Water Res.* **2018**, *147*, 43–49. [CrossRef]
73. Ito, Y.; Tanabe, Y.; Han, J.; Fujita, T.; Tanigaki, K.; Chen, M. Multifunctional porous graphene for high-efficiency steam generation by heat localization. *Adv. Mater.* **2015**, *27*, 4302–4307. [CrossRef]
74. Liu, K.K.; Jiang, Q.; Tadepalli, S.; Raliya, R.; Biswas, P.; Naik, R.R.; Singamaneni, S. Wood–graphene oxide composite for highly efficient solar steam generation and desalination. *ACS Appl. Mater. Interfaces* **2017**, *9*, 7675–7681. [CrossRef]
75. Jiang, Q.; Tian, L.; Liu, K.K.; Tadepalli, S.; Raliya, R.; Biswas, P.; Naik, R.R.; Singamaneni, S. Bilayered biofoam for highly efficient solar steam generation. *Adv. Mater.* **2016**, *28*, 9400–9407. [CrossRef]
76. Chao, W.; Li, Y.; Sun, X.; Cao, G.; Wang, C.; Ho, S. Enhanced wood-derived photothermal evaporation system by in-situ incorporated lignin carbon quantum dots. *Chem. Eng. J.* **2021**, *405*, 126703. [CrossRef]
77. Hu, Z.; Hao, L.; Liu, N.; He, P.P.; Bai, H.Y.; Niu, R.; Gong, J. High-performance bilayer solar evaporators constructed by candle-derived carbon nanoparticle/wood hybrid. *Mater. Today Commun.* **2021**, *28*, 102636. [CrossRef]
78. Zhang, X.; Liu, J.; Li, W.; Li, C.; Jiao, F.; Yu, Z.; Li, X. Three-dimensional disposable wooden chopsticks-derived solar-thermal evaporators for environmental energy-enhanced water evaporation and purification. *Compos. Commun.* **2023**, *44*, 101746. [CrossRef]
79. Song, L.; Zhang, X.F.; Wang, Z.; Zheng, T.; Yao, J. Fe_3O_4/polyvinyl alcohol decorated delignified wood evaporator for continuous solar steam generation. *Desalination* **2021**, *507*, 115024. [CrossRef]
80. Fan, D.Q.; Lu, Y.; Xu, X.L.; Tang, Y.; Zhang, H.; Mi, Y.; Yang, X. Multifunctional wood-based hydrogels for wastewater treatment and interfacial solar steam generation. *Chem. Eng. J.* **2023**, *471*, 144421. [CrossRef]
81. He, F.; Han, M.; Zhang, J.; Wang, Z.; Wu, X.; Zhou, Y.; Jiang, L.; Peng, S.; Li, Y. A simple, mild and versatile method for preparation of photothermal woods toward highly efficient solar steam generation. *Nano Energy* **2020**, *71*, 104650. [CrossRef]
82. Yan, C.; Huang, J.; Liang, Z.; Yang, D.; Li, Z. Fabrication of a Highly Efficient Wood-Based Solar Interfacial Evaporator with Self-Desalting and Sterilization Performance. *Langmuir* **2022**, *38*, 12813–12821. [CrossRef]

83. Hu, C.; Li, W.; Zhao, H.Y.; Li, C.; Ma, Z.; Hao, L.; Pang, Y.; Yu, Z.; Li, X. Salt-resistant wood-based solar steam generator with top-down water supply for high-yield and long-term desalination of seawater and brine water. *Chem. Eng. J.* **2023**, *460*, 141622. [CrossRef]
84. He, X.; Zhang, L.; Hu, X.; Zhou, Q. Formation of S defects in MoS_2-coated wood for high-efficiency seawater desalination. *Environ. Sci. Nano* **2021**, *8*, 2069–2080. [CrossRef]
85. Okamoto, T.; Kumagai, S.; Fukuzaki, E.; Ishii, H.; Watanabe, G.; Niitsu, N.; Annaka, Y.; Yamagishi, M.; Tani, Y.; Sugiura, H.; et al. Robust, high-performance n-type organic semiconductors. *Sci. Adv.* **2020**, *6*, eaaz0632. [CrossRef]
86. Long, R.; Li, Y.; Song, L.; Xiong, Y. Coupling solar energy into reactions: Materials design for surface plasmon-mediated catalysis. *Small* **2015**, *11*, 3873–3889. [CrossRef]
87. Brongersma, M.L.; Halas, N.J.; Nordlander, P. Plasmon-induced hot carrier science and technology. *Nat. Nanotechnol.* **2015**, *10*, 25–34. [CrossRef]
88. Zha, Z.; Yue, X.; Ren, Q.; Dai, Z. Uniform polypyrrole nanoparticles with high photothermal conversion efficiency for photothermal ablation of cancer cells. *Adv. Mater.* **2013**, *25*, 777–782. [CrossRef]
89. Zhang, L.; Tang, B.; Wu, J.; Li, R.; Wang, P. Hydrophobic light-to-heat conversion membranes with self-healing ability for interfacial solar heating. *Adv. Mater.* **2015**, *27*, 4889–4894. [CrossRef]
90. Zou, Y.; Yang, P.; Yang, L. Boosting solar steam generation by photothermal enhanced polydopamine/wood composites. *Polymer* **2021**, *217*, 123464. [CrossRef]
91. Huang, W.; Hu, G.; Tian, C.; Wang, X.; Tu, J.; Cao, Y.; Zhang, K. Nature-inspired salt resistant polypyrrole–wood for highly efficient solar steam generation. *Sustain. Energy Fuels* **2019**, *3*, 3000–3008. [CrossRef]
92. Wang, Z.; Yan, Y.; Shen, X.; Jin, C.; Sun, Q.; Li, H. A wood–polypyrrole composite as a photothermal conversion device for solar evaporation enhancement. *J. Mater. Chem. A* **2019**, *7*, 20706–20712. [CrossRef]
93. Qu, M.; Yan, J.; Ge, J.; Zhao, Y.; Xue, Y.; Liu, X.; Liu, H.; Yan, M.; He, J. Nature-inspired wood-based solar evaporation system for efficient desalination and water purification. *J. Mater. Sci.* **2023**, *58*, 6220–6236. [CrossRef]
94. Lin, Z.X.; Wu, T.T.; Jia, B.X.; Shi, J.; Zhou, B.; Zhu, C.; Wang, Y.; Liang, R.; Mizuno, M. Nature-inspired poly (N-phenylglycine)/wood solar evaporation system for high-efficiency desalination and water purification. *Colloids Surf. A Physicochem. Eng. Asp.* **2022**, *637*, 128272. [CrossRef]
95. Li, L.; He, N.; Yang, S.; Zhang, Q.; Zhang, H.; Wang, B.; Dong, T.; Wang, H.; Jiang, B.; Tang, D. Strong tough hydrogel solar evaporator with wood skeleton construction enabling ultra-durable brine desalination. *EcoMat* **2023**, *5*, e12282. [CrossRef]
96. Sheng, K.; Tian, M.; Zhu, J. When Coordination Polymers Meet Wood: From Molecular Design toward Sustainable Solar Desalination. *ACS Nano* **2023**, *17*, 15482–15491. [CrossRef] [PubMed]
97. Ren, H.; Tang, M.; Guan, B.; Wang, K.; Yang, J.; Wang, F.; Wang, M.; Shan, J.; Chen, Z.; Wei, D.; et al. Hierarchical graphene foam for efficient omnidirectional solar–thermal energy conversion. *Adv. Mater.* **2017**, *29*, 1702590. [CrossRef]
98. Gao, M.; Connor, P.K.N.; Ho, G.W. Plasmonic photothermic directed broadband sunlight harnessing for seawater catalysis and desalination. *Energy Environ. Sci.* **2016**, *9*, 3151–3160. [CrossRef]
99. Linic, S.; Aslam, U.; Boerigter, C.; Morabito, M. Photochemical transformations on plasmonic metal nanoparticles. *Nat. Mater.* **2015**, *14*, 567–576. [CrossRef]
100. Gao, M.; Zhu, L.; Peh, C.K.; Ho, G.W. Solar absorber material and system designs for photothermal water vaporization towards clean water and energy production. *Energy Environ. Sci.* **2019**, *12*, 841–864. [CrossRef]
101. Webb, J.A.; Bardhan, R. Emerging advances in nanomedicine with engineered gold nanostructures. *Nanoscale* **2014**, *6*, 2502–2530. [CrossRef] [PubMed]
102. Zhu, L.; Gao, M.; Peh, C.K.N.; Ho, G.W. Solar-driven photothermal nanostructured materials designs and prerequisites for evaporation and catalysis applications. *Mater. Horiz.* **2018**, *5*, 323–343. [CrossRef]
103. Zhou, L.; Tan, Y.; Ji, D.; Zhu, B.; Zhang, P.; Xu, J.; Gan, Q.; Yu, Z.; Zhu, J. Self-assembly of highly efficient, broadband plasmonic absorbers for solar steam generation. *Sci. Adv.* **2016**, *2*, e1501227. [CrossRef] [PubMed]
104. Goharshadi, K.; Sajjadi, S.A.; Goharshadi, E.K.; Mehrkhah, R. Highly efficient plasmonic wood/Ag/Pd photoabsorber in interfacial solar steam generation. *Mater. Res. Bull.* **2022**, *154*, 111916. [CrossRef]
105. Zhu, M.; Li, Y.; Chen, F.; Zhu, X.; Dai, Y.; Li, Y.; Yang, Z.; Yan, X.; Song, J.; Wang, Y.; et al. Plasmonic wood for high-efficiency solar steam generation. *Adv. Energy Mater.* **2018**, *8*, 1701028. [CrossRef]

polymers

MDPI

Article

Comparison of Effects of Plasma Surface Modifications of Bamboo and Hemp Fibers on Mechanical Properties of Fiber-Reinforced Epoxy Composites

Pornchai Rachtanapun [1,2,3,†], Choncharoen Sawangrat [4,†], Thidarat Kanthiya [5], Kannikar Kaewpai [6], Parichat Thipchai [7], Nuttapol Tanadchangsaeng [8], Patnarin Worajittiphon [2,9], Jonghwan Suhr [10], Pitiwat Wattanachai [11,*] and Kittisak Jantanasakulwong [1,2,3,*]

1 Faculty of Agro-Industry, Chiang Mai University, Chiang Mai 50100, Thailand; pornchai.r@cmu.ac.th
2 Center of Excellence in Materials Science and Technology, Chiang Mai University, Chiang Mai 50200, Thailand; patnarin.w@cmu.ac.th
3 Center of Excellence in Agro Bio-Circular-Green Industry, Faculty of Agro-Industry, Chiang Mai University, Chiang Mai 50100, Thailand
4 Department of Industrial Engineering, Faculty of Engineering, Chiang Mai University, Chiang Mai 50200, Thailand; choncharoen@step.cmu.ac.th
5 Office of Research Administration, Chiang Mai University, Chiang Mai 50200, Thailand; thidaratkanthiya05@gmail.com
6 Science and Technology Park (STeP), Chiang Mai University, Chiang Mai 50100, Thailand; kannikar@step.cmu.ac.th
7 Nanoscience and Nanotechnology, Faculty of Science, Chiang Mai University, Chiang Mai 50200, Thailand; parichat_thi@cmu.ac.th
8 College of Biomedical Engineering, Rangsit University, Pathumthani 12000, Thailand; nuttapol.t@rsu.ac.th
9 Department of Chemistry, Faculty of Science, Chiang Mai University, Chiang Mai 50200, Thailand
10 School of Mechanical Engineering, Sungkyunkwan University, Suwon-si 16419, Republic of Korea; suhr@skku.edu
11 Department of Civil Engineering, Faculty of Engineering, Chiang Mai University, Chiang Mai 50200, Thailand
* Correspondence: pitiwat@step.cmu.ac.th (P.W.); kittisak.jan@cmu.ac.th (K.J.); Tel.: +66-(0)53948274 (P.W. & K.J.); Fax: +66-(0)53948230 (P.W. & K.J.)
† These authors contributed equally to this work.

Citation: Rachtanapun, P.; Sawangrat, C.; Kanthiya, T.; Kaewpai, K.; Thipchai, P.; Tanadchangsaeng, N.; Worajittiphon, P.; Suhr, J.; Wattanachai, P.; Jantanasakulwong, K. Comparison of Effects of Plasma Surface Modifications of Bamboo and Hemp Fibers on Mechanical Properties of Fiber-Reinforced Epoxy Composites. *Polymers* **2024**, *16*, 3394. https://doi.org/10.3390/polym16233394

Academic Editor: Ming Zhang

Received: 7 November 2024
Revised: 25 November 2024
Accepted: 27 November 2024
Published: 30 November 2024

Abstract: In this study, we investigated the behaviors of epoxy composites reinforced with bamboo (BF) and hemp (HF) fibers. Both fibers were treated using dielectric barrier discharge (DBD) plasma for various durations (2.5 to 20 min). Epoxy resin (ER) was mixed with BF or HF with and without plasma treatment. The Fourier-transform infrared spectra of the plasma-treated fibers showed an enhanced peak intensity of carboxyl groups. ER/BF treated for 20 min exhibited a high tensile strength (up to 56.5 MPa), while ER/HF treated for 20 min exhibited a more significant increase in elongation at break (6.4%). Flexural tests indicated that the plasma treatment significantly improved the flexural strength of the hemp composites (up to 62.2 MPa) compared to the bamboo composites. The plasma treatment increased the fiber surface roughness and interfacial bonding in both composites. The thermal stability and wettability were improved by the DBD plasma treatment. The DBD plasma treatment enhanced the interfacial adhesion between fibers and ER matrix, which improved the mechanical, thermal, and wettability properties of the composites.

Keywords: plasma; surface modification; bamboo; hemp; epoxy composites

1. Introduction

In recent years, biocomposites have emerged as versatile materials in various industries [1,2]. They are environmentally friendly material alternatives for the replacement of petroleum-based polymers due to their ability to combine the strengths of polymers with reinforcing materials. The reinforced polymers by biomaterials exhibit enhanced mechanical, thermal, and chemical properties. Composites are produced by embedding a reinforcing phase within a polymer matrix [3], which provides lightweight structures, strength, and durability. The use of cellulose fibers, such as bamboo, flax, sisal, jute, and kenaf, in polymer composites as reinforcement gained popularity in several engineering applications due to the low cost, density, favorable mechanical properties, and recyclability [4,5]. Cellulose fibers are environmentally friendly, nontoxic, and renewable materials. Manufacturing industries, such as the packaging, building construction, automotive, and furniture industries, have been encouraged to use plant fibers instead of reinforcing materials [6–8].

Natural fibers are a biomaterial used to reinforce a polymer matrix. Bamboo belongs to the grass family poaceae, which is known for its strength and ability to thrive in diverse climates [9]. Bamboo is composed of a series of nodes and internodes, which yields a hollow cylindrical structure. The cellulose and lignin contents of bamboo fibers are up to 40–50% and 20–30%, respectively, which provides rigidity, high tensile strength, and high stiffness [10,11]. This composition makes bamboo a strong and flexible material, which can be used for various construction and industrial applications. In the context of fiber reinforcement, bamboo is processed into bamboo fibers (BFs) for use in polymer composites. These fibers exhibit a high tensile strength, comparable to those of some synthetic fibers [12,13]. Bamboo fiber composites are increasingly used in construction, automotive components, and packaging due to their lightweight structure, strength, and biodegradability. Hemp is another natural alternative for composite reinforcement. Hemp fibers (HFs) are strong and lightweight, and exhibit favorable thermal properties, making them ideal for composites requiring enhanced toughness and thermal stability. HF contains cellulose (60–70%), hemicellulose (15–25%), and lignin (5–10%), which provides excellent mechanical properties. Hemp has been used in textiles for centuries. Its potential as a reinforcing fiber in polymer composites is particularly recognized in sustainable manufacturing sectors [14,15]. However, the hydrophilic property of plant fibers, hydrophobic property of resins, and compatibility problems limit the application of composites [16]. Surface treatments with chemical and physical methods are employed to enhance the interfacial adhesion between natural fibers and matrix resins. Substantial studies have been carried out to improve interfacial adhesion by chemical, physical, or other modification methods. Common surface modification methods include alkali, silane, plasma, and enzymatic treatments [17–21]. These treatments improve the fiber surface by increasing the roughness or introducing functional groups that enhance the adhesion, which leads to transfer load between the fiber and matrix [22,23]. Natural fiber composites are increasingly used in construction, automotive components, and packaging [24,25] due to their lightweight structure, strength, and biodegradability.

Plasma treatment involves exposure of a material surface to a partially ionized gas (plasma), electrons, ions, and neutral particles. The interaction between the plasma and surface induces chemical, physical, and structural modifications of materials [26]. The use of dielectric barrier discharge (DBD) plasma is a specific form of atmospheric plasma treatment with a dielectric barrier that separates two electrodes. This setup provides uniformity of nonthermal plasma generation at atmospheric pressure. The DBD plasma is frequently used to treat natural fibers due to its energy efficiency and scalability. In the DBD plasma, various gasses can be used, such as O_2, N_2, or Ar, depending on the desired surface modification [27–29]. Natural fiber surface improvement by DBD plasma is a novel process to increase roughness and reactive functional groups on fiber surface to form physical and chemical crosslinking with the polymer matrix, respectively. The synergistic crosslinking power may increase properties of polymer composite which can be applied for construction, electrical, coating, packaging, agriculture, and medical applications.

Epoxy resin (ER) is the most used thermoset structure in engineering fields owing to its chemical resistance, excellent mechanical properties, and electrical insulation [30]. ERs are created through reactions between epoxides (three-membered cyclic ethers) and curing agents (hardener agents), such as amines, anhydrides, or phenols. Upon curing, epoxy forms a rigid crosslinked network structure that provides the characteristic strength and durability [31]. Fiber-reinforced epoxy composite is a material obtained by embedding high-strength fibers into an ER matrix. This combination creates a composite material with superior mechanical properties compared to those of the individual components. The fibers act as a reinforcing material to provide strength and stiffness, while the ER acts as a matrix that binds the fibers together, distributes stress, and protects the fibers from damage [32,33]. Some researchers attempted to develop natural fiber-reinforced ERs without surface modification, which led to weak mechanical properties and connecting phase between fibers and the ER matrix. The use of chemical and physical processes is a target approach to improve the fiber surface before blending with ER. When the surface of the fiber is improved by chemical and physical roughness, this improves the reaction, interfacial adhesion, and properties of the blend.

Therefore, in this study, a combined surface modification, with chemical and physical processes, by the plasma technology was used to improve the surface of BF and HF. BF and HF were plasma-treated using an argon + oxygen gas (Ar + O_2) for different times of 0 to 20 min, followed by plating in an NH_4OH solution to improve the chemical structure of the fibers. NH_4OH was used to provide grafting $-NH_2$ groups onto the surface structure of the fibers, which can react with epoxy groups of ER. The DBD plasma technique was used to improve the roughness surface and polarity of the fibers. The properties of both fiber composites were investigated. Fourier-transform infrared (FTIR) spectroscopy was utilized to evaluate the chemical bonding and reaction mechanisms within the composites, while tensile strength and flexural properties were evaluated to assess the impact of the plasma treatment. Morphological and thermal stability studies were carried out to explain the impact of the plasma treatment on the composite structure and performance. This research provides insights into the potential of plasma-treated natural fiber composites in creating stronger and more durable ecofriendly materials for industrial applications.

2. Materials and Methods

2.1. Materials

BFs were obtained from South Samoeng, Chiang Mai, Thailand. HFs were purchased from Royal Project Foundation, Chiang Mai, Thailand. ER was of grade A302. A hardener of grade A301 was purchased from Easy Resin, Co., Ltd., Nonthaburi, Thailand. All chemicals for the surface modification, including sodium hydroxide (NaOH), sodium chlorite ($NaClO_2$), and ammonium hydroxide (NH_4OH), were purchased from Merck & Co. Inc., Darmstadt, Germany.

2.2. Surface Treatment of Fibers

Before the plasma treatment, bamboo and hemp fibers were mildly alkaline treated in an aqueous NaOH solution (20% w/v) at 80 °C for 5 h. The pulps were bleached with $NaClO_2$ to remove lignin and hemicellulose, as described in our previous report [28]. The fibers were sieved through a 180 μm sieve. The bamboo and hemp fibers were treated for surface modification using a DBD plasma approach. The DBD plasma machine is shown in Figure 1. The fibers were put on the tray, while DBD was generated through two parallel high-power electrodes covered by a thin quartz foil. The grounded electrode can be adjusted to change the discharge gap for the fibers. The DBD plasma was generated by a high-voltage power of 180 W (3.45 W/cm^2) with a constant frequency of 13.56 kHz. Ar + O_2 gasses were used for plasma flow rates of 8 and 10 L/min, respectively. The DBD plasma was applied on the bamboo and hemp fibers for 2.5, 5, 10, 15, and 20 min. The treated fibers under each plasma condition were modified on the surface by mixing into an NH_4OH solution (1:10% w/v). The sample was heated by a hot plate at 60 °C and stirred continuously for 1 h. Afterward, the NH_4OH solution was evaporated from the final product by heating at 60 ± 2 °C for 48 h.

Figure 1. Dielectric barrier discharge (DBD) plasma machine.

2.3. Fiber Composite Material Fabrication

The fiber composite materials were fabricated by the hand layup technique followed by the mixing of ER and hardener (2:1% *w*/*w*), as indicated in Table 1. The fiber composites were separated into untreated (F_{untr}) and treated (F_{tr}) plasma groups for the different natural fiber types (bamboo and hemp), and then the fiber samples were mixed with ER at room temperature. The mixed samples were stirred to enhance the dispersion of fibers, while air bubbles were removed in the ER with an aspirated vacuum. The mixture was then cast into a silicone mold of a bone shape followed by drying at 80 °C for 5 h.

Table 1. Notation of the composite material samples (% *w*/*w*).

Sample	Type of Fiber		Epoxy:Hardener (2:1)	Fiber
	Bamboo	Hemp	(%)	(%)
ER/BF_{untr}	✓		95	5
ER/BF_{tr} 2.5	✓		95	5
ER/BF_{tr} 5	✓		95	5
ER/BF_{tr} 10	✓		95	5
ER/BF_{tr} 15	✓		95	5
ER/BF_{tr} 20	✓		95	5
ER/HF_{untr}		✓	95	5
ER/HF_{tr} 2.5		✓	95	5
ER/HF_{tr} 5		✓	95	5
ER/HF_{tr} 10		✓	95	5
ER/HF_{tr} 15		✓	95	5
ER/HF_{tr} 20		✓	95	5

2.4. FTIR Spectroscopy

The chemical functional groups of the samples were examined using FTIR spectroscopy (Thermo Nicolet 6700 FTIR spectrometer, Thermo Fisher Scientific, Woodland, CA, USA) with the ATR mode. FTIR spectra were acquired in the range of 500–4000 cm^{-1} (32 scans, scan resolution of 4 cm^{-1}).

2.5. Tensile Properties

The bone-shaped configuration samples were prepared with dimensions of 2 mm × 5 mm × 1 mm (width × length × thickness). The test was carried out according to the JISK-6251-7 standard (Model MCT1150, Tokyo, Japan) at room temperature. The elongation at break (EB) and tensile strength (TS) were measured at a crosshead speed of 50 mm/min. Ten replicates were performed on each sample.

2.6. Flexural Test

The samples were cast into the silicone mold with dimensions of 13 mm × 65 mm × 3 mm (width × length × thickness) according to the American Society for Testing and Materials (ASTM) D790 standard. The testing was conducted using a universal testing machine (H1KS, Hounsfield Test Co., Ltd., Surrey, UK) in a three-point bending configuration with a force of 1 kN at room temperature. The average from five samples was recorded for the analysis. The flexural strength and modulus were calculated by the following [28]:

$$Flexural\ strength = \frac{3FL}{2bd^2}, \tag{1}$$

$$Flexural\ modulus = \frac{mL^3}{4bd^3} \tag{2}$$

where F, L, b, d, and m are the maximum failure load (N), length of span (mm), width, thickness, and slope of the load–displacement curve tangent to the initial line, respectively.

2.7. Morphological Analysis

The fractured morphologies of the samples were studied by scanning electron microscopy (SEM; JSM-IT300LV, JEOL Co., Ltd., Tokyo, Japan). The impact fracture surface broken in liquid nitrogen was evaporated in a hot air oven at 60 °C for 2 h. The samples were sputter-coated with gold and observed at 15 kV under a vacuum.

2.8. Thermal Stability

The thermal stabilities of the epoxy and fiber-reinforced epoxy composites were evaluated by a thermogravimetric analysis (TGA; Mettler Toledo STARe TGA/DSC3+, Greifensee, Switzerland) under a nitrogen atmosphere. The testing temperature range was 25–600 °C, while the rate was 10 °C/min.

2.9. Contact Angle

The water contact angle was measured using a water droplet instrument (DSA30E, Krüss Co., Ltd., Hamburg, Germany). Samples were formed by casting them into a silicone mold and pasting them on glass slides. The water drop shape was recorded at 0–10 min. At least five measurements of each sample were performed to average the water contact angles.

2.10. Statistical Analysis

A statistical comparison was carried out using a one-way analysis of variance (ANOVA) with the SPSS software (SPSS version 17, Armonk, NY, USA). The statistical significance ($p < 0.05$) was estimated using the Duncan test.

3. Results and Discussion

3.1. FTIR Spectroscopy

The chemical characteristics of the fiber-reinforced epoxy composites evaluated by FTIR spectroscopy for bamboo and hemp fibers with DBD plasma treatment are shown in Figure 2. The untreated and treated fibers exhibited hydroxyl (O−H) stretching vibrations at 3330 to 3400 cm^{-1} [34,35]. The band at 2890 cm^{-1} was assigned to the C−H stretching vibration of the cellulose fiber containing a functional group of alkanes [36], while the band at 1635 cm^{-1} was attributed to the aromatic of lignin [37,38]. The peak around 1730 cm^{-1} was indicated to the carbonyl (C=O) stretching of acetyl groups of hemicellulose [39]. Both plasma-treated and untreated fibers exhibited peak disappearance of lignin at 1635 cm^{-1} partly due to the alkaline treatment and blenching before the plasma treatment. In the ER/BF and ER/HF composites, ER exhibited C−C stretching peaks at 1610 and 1585 cm^{-1} and C=C stretching vibrations of the ER aromatic ring at 1508 and 1454 cm^{-1} [40]. The bands at 1241, 913, and 827 cm^{-1} were C−O symmetrical stretching, C−O asymmetrical stretching, and C−O−C stretching in the oxirane ring, respectively [41,42]. For the plasma-treated composites, an increase in peak intensity at 3200−3600 cm^{-1} was observed due to O−H stretching of hydroxyl groups of the fibers. The treatment with the Ar + O_2 gas generated the polar groups of carbonyls (C=O) and carboxyl (−COOH) at 1680−1720 cm^{-1}. In addition, carbon hydrogen group (CH_2, CH_3) stretching was observed at 2850–3000 cm^{-1} [28,43]. The amount of oxygen-containing groups on the surface increased with the time of plasma treatment owing to bonding between the fiber and resin [44]. However, the epoxy groups of ER (912 cm^{-1}) were not observed on the ER/BF_{tr} composite (Figure 2a). The ER/BF_{tr} composites exhibited a new shoulder peak at 1700 cm^{-1}, which disappeared upon the treatment for 20 min. This new peak indicates a new C−O formation from the reaction between the ER and BF surface, which changed the form with the 20 min treatment. HF_{untr} exhibited –COOH at 1730 cm^{-1} and C−O at 1360 cm^{-1}. The intensities of these two peaks increased with the plasma treatment due to the increasing contents of the −COOH and C−O groups on the surface of HF. The results indicate that the −COOH groups react with NH_4OH and epoxy groups of ER. −COOH groups disappeared after mixing with ER. The results also indicate that $-NH_2$ groups graft on the fiber surface and react with epoxy groups of ER. For the ER/HF composites, the intensity of epoxy groups of ER at 912 cm^{-1} decreased with the time of plasma treatment owing to the high reaction rate between the surface of HF and ER. The plasma treatment improves the reactive functional groups of the fibers surface, which react with epoxy groups of the ER main matrix [10,42,45]. The plasma treatment caused changes in the chemical and physical structures of the bamboo and hemp fibers [27,46,47], resulting in the formation of functional groups that induced bonds and interfacial adhesion between EB matrix and the fiber. These reactions could improve the properties of composites.

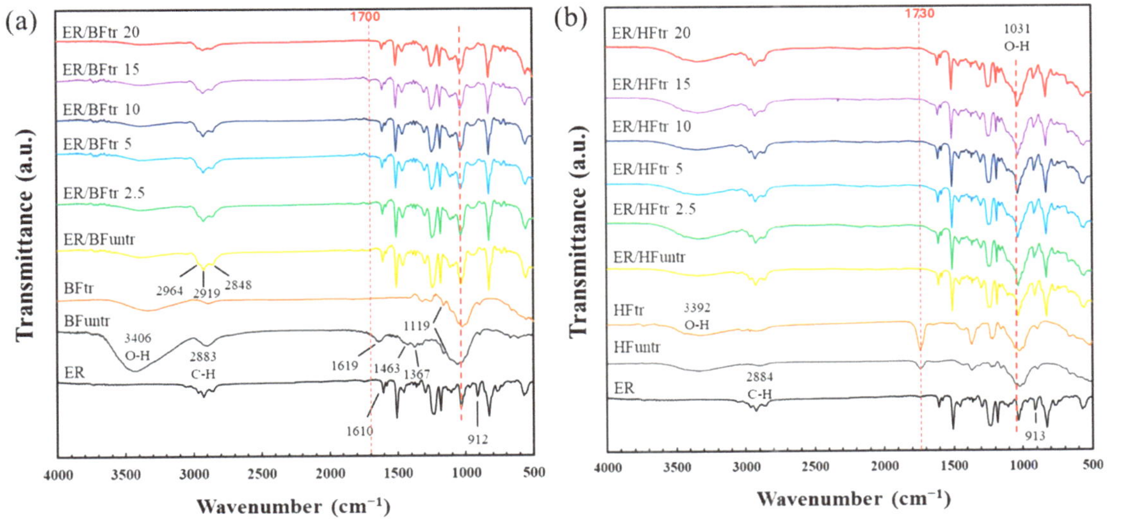

Figure 2. FTIR spectra of (**a**) bamboo and (**b**) hemp fiber-reinforced epoxy composites (untreated and treated with Ar + O_2 for 2.5–20 min + NH_4OH).

3.2. Tensile Properties

Figure 3 shows the tensile properties of ER composites with plasma treatment (0–20 min) on BF and HF. The tensile strength and elongation at break of the ER/BF_{untr}

composite using bamboo untreated fibers were 42 MPa and 4.7%, respectively. The EP/BF_{tr} composites treated with the Ar + O_2 gas for 2.5–20 min exhibited increasing tensile strength and elongation at break with the treatment time. They exhibited the highest tensile strength at 20 min (56.5 MPa). The tensile strength and elongation at break of the hemp untreated composite (ER/HF_{untr}) were 43.4 MPa and 3.6%, respectively. The EP/HF_{tr} composites exhibited higher tensile strengths than that of the untreated composite (highest value of 53.2 MPa at 5 min), which slightly decreased with the increase in treatment time. The elongation at break of the EP/HF_{tr} composite significantly increased with the treatment time from 3.6% to 6.4%. The improved tensile properties of the bamboo and hemp fiber-reinforced epoxy composites were attributed to the adhesion between the fibers and ER matrix. The plasma treatment of the fibers significantly affected the adhesion performance of the interface between the reinforced fibers and ER [48]. The samples using the hemp-treated fiber reinforcement were improved in terms of elongation at break, while the bamboo-treated fibers provided a larger improvement in tensile strength. The surface roughness of the fibers increased due to plasma etching, which enhanced the penetration and diffusion at the fiber–matrix interface and generated interlocking bonds [27,49]. The high degree of interfacial chemical reaction between epoxy groups of ER and –COOH groups of fibers with physical crosslinking with a high fiber roughness could improve the mechanical properties of the composites. Optimal plasma processing parameters relate to mechanical properties of the composites [50].

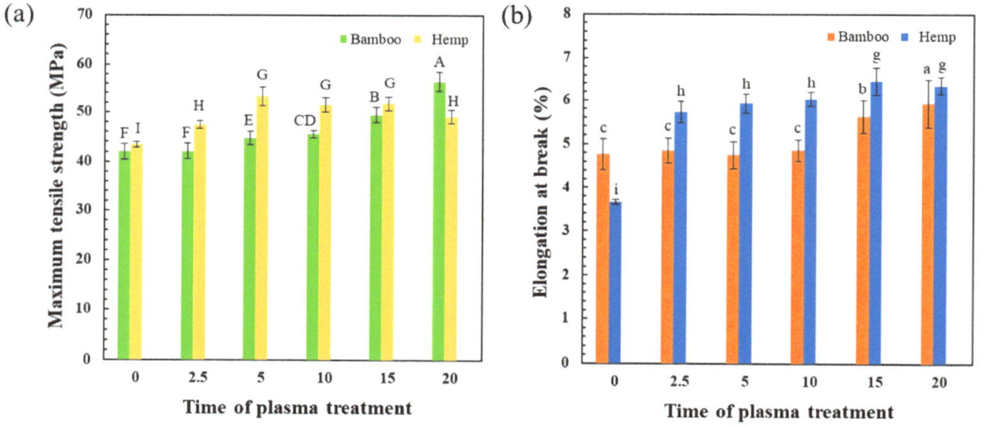

Figure 3. Comparison of tensile properties between bamboo and hemp fiber-reinforced epoxy composites, untreated (0) and treated with Ar + O_2 for 2.5–20 min + NH_4OH. (**a**) Maximum tensile strength (MPa) and (**b**) elongation at break (%). The mean values of the tensile strength (uppercase letter) and elongation at break (lowercase letter) differ significantly ($p < 0.05$).

3.3. Flexural Test

We evaluated the flexural strengths of the composites according to the plasma treatment of the bamboo and hemp fibers, as shown in Table 2. The flexural strength and modulus of the bamboo untreated composite (ER/BF_{untr}) were 55.9 MPa and 7.6 GPa, respectively. The flexural strengths of ER/BF_{tr} treated for 2.5, 5, 10, 15, and 20 min were 54.8, 50.5, 56.8, 58.1, and 57.5 MPa, while the flexural moduli were 6.9, 6.6, 8.2, 8.1, and 7.7 GPa, respectively. The ER/HF_{untr} composite exhibited a flexural strength and modulus of 54.7 MPa and 7.6 GPa, respectively. The flexural strengths (53.9, 47.9, 56.7, 54.4, and 62.2 MPa) and moduli (7.7, 6.7, 7.8, 7.5, and 8.6 GPa) of the hemp fiber-reinforced composites after the plasma treatment exhibited increasing trends with the treatment time due to the improvements in HF_{tr} roughness and interfacial adhesion of the ER/HF_{tr} composites. The DBD plasma treatment with O_2 gas improved the mechanical properties of both bamboo and hemp [51]. Both flexural stress and modulus of hemp were significantly improved compared to those of bamboo due to the different chemical structure of the hemp fibers with a responsive structural basis to the plasma treatment compared to the bamboo [52,53].

Table 2. Comparison of flexural strengths and flexural moduli of the composites with untreated and plasma-treated fibers.

Sample	Bamboo (B)		Hemp (H)	
	Flexural Stress (MPa)	Flexural Modulus (GPa)	Flexural Stress (MPa)	Flexural Modulus (GPa)
ER/F_{untr}	55.9 ± 4.6 ^{BC}	7.6 ± 0.6 ^{ab}	54.7 ± 0.8 ^{I}	7.6 ± 0.1 ^{j}
ER/F_{tr} 2.5	54.8 ± 6.8 ^{AB}	6.9 ± 0.5 ^{bc}	53.9 ± 0.7 ^{I}	7.7 ± 0.1 ^{i}
ER/F_{tr} 5	50.5 ± 7.8 ^{A}	6.6 ± 0.3 ^{c}	47.9 ± 0.8 ^{J}	6.7 ± 0.1 ^{k}
ER/F_{tr} 10	56.8 ± 5.1 ^{C}	8.2 ± 0.7 ^{a}	56.7 ± 0.9 ^{H}	7.8 ± 0.1 ^{i}
ER/F_{tr} 15	58.1 ± 5.4 ^{C}	8.1 ± 0.7 ^{a}	54.4 ± 1.6 ^{I}	7.5 ± 0.2 ^{j}
ER/F_{tr} 20	57.5 ± 5.9 ^{C}	7.7 ± 0.8 ^{ab}	62.2 ± 1.1 ^{G}	8.6 ± 0.2 ^{h}

Ignificantly different mean values of the flexural stress (uppercase letter) and flexural strain (lowercase letter) ($p < 0.05$).

3.4. Morphological Analysis

SEM was employed to acquire fractured surface images of the composites. Figure 4 shows the morphologies of the fiber-reinforced epoxy composites with bamboo and hemp structures, untreated and plasma-treated for different times. The ER/BF_{untr} composite with the untreated fibers exhibited a smooth surface of the bamboo fiber similar to that of the ER/BF_{tr} 2.5 sample. The increase in the time of plasma treatment (5–20 min) resulted in a high fiber roughness due to the oxygen in the discharge gasses inducing oxygen groups onto the fiber structure. The plasma treatment induces etching and formation of chemical species on the surface, which corresponds to the roughness [54]. Fine BF_{tr} and HF_{tr} fibers distribution on ER matrix were observed, which provided the properties improvement of composite [55,56]. The bamboo fiber-reinforced composites exhibited obvious gaps between fibers and matrix due to the plasma treatment with O_2 gas resulting in surface etching and reducing the polarity of the fiber by the formation of oxygen bonding [28,57,58]. Reduction in gaps between the BF surface and ER matrix were observed for the plasma treatment of 20 min, which related to the increased tensile strength. However, ER/HF_{tr}20 showed low gaps between HF and ER matrix with reduction in tensile strength due to the occurred degradation of HF fiber which was weaker structure than BF fiber. Combination between interfacial crosslinking and fiber degradation provided slightly decreased tensile strength. The SEM images of the HF-reinforced composites showed different characteristics from those of the bamboo fibers. The morphology of the ER/HF_{untr} composite with untreated fibers clearly exhibited hemp stem sections [59] with characteristics of smooth fibers, large pores, and irregular structure [22,52]. The EP/HF_{tr} composites exhibited low gaps between the HF and BR matrix for all treatment times owing to the high interfacial adhesion between HF and BR via chemical and physical HF surface improvements. The ER/HF_{tr} composites exhibited surface improvements without removed fibers due to the excellent adhesion between the HF and EP [28,60]. The DBD plasma affected the etching of the fiber surface, enhanced the roughness, and induced –COOH groups on the HF structure, due to the surface ion bombardment, etching, and ablation of surface layers [61]. Long-term treatment provided crosslinking reaction which improved interfacial adhesion and mechanical properties of the composites.

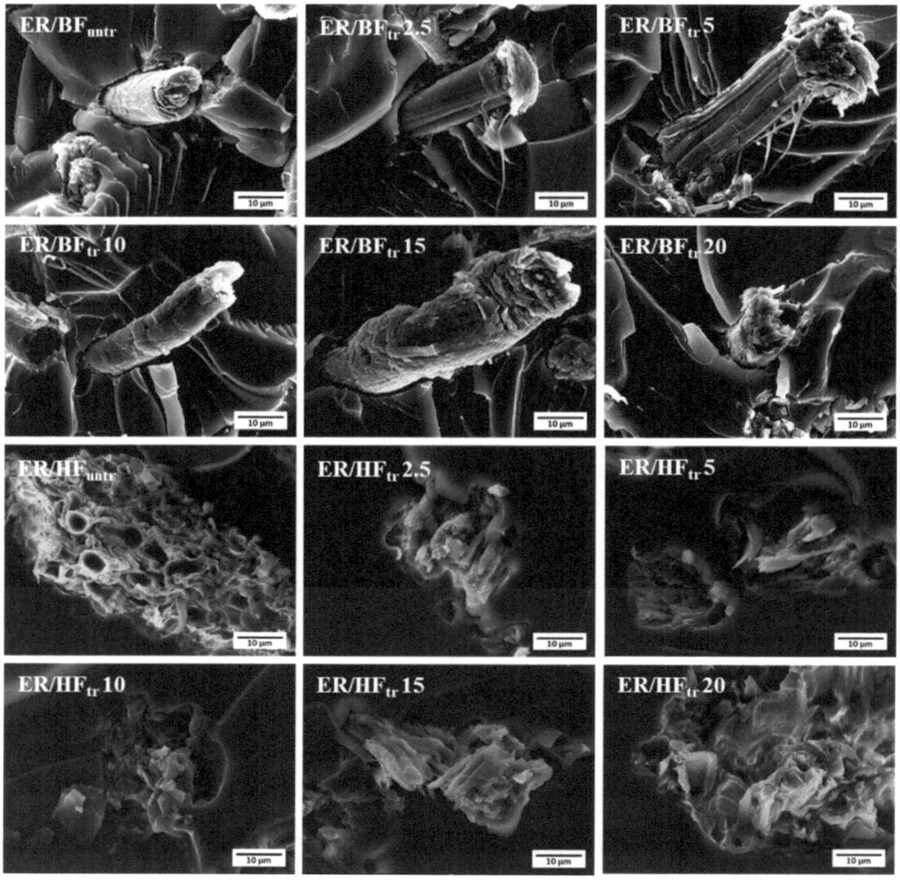

Figure 4. SEM images of BF- and HF-reinforced epoxy composites, untreated and treated with Ar + O_2 gas for different times at 1000×.

3.5. Thermal Stability

The results of the thermal analysis of the pure epoxy and fiber-reinforced ER composites with untreated fibers and plasma-treated fibers with Ar + O_2 gas for different times are shown in Figure 5. Weight loss of the epoxy is observed in the first stage at approximately 150 °C due to the released humidity of the composites [62,63]. The ER/BF and ER/HF composites exhibited similar behaviors with the treatment time. In the temperature range of 200–350 °C, the ER blend with fibers exhibited a higher weight loss than the pure ER, which increased with the treatment time. All composites exhibited fiber degradation at 370 °C owing to the degradation of cellulose and hemicellulose composition in the fibers [64]. The phase of lignin was decomposed in the range of 400 to 460 °C with a weight loss percentage of 6%. A higher temperature is needed to decompose hemicellulose, cellulose, and lignin. The remaining component at 500 to 600 °C had an ash content, increasing with the treatment time due to the high degree of crosslinking reaction around the interface of the fibers with ER [65]. The crosslinking improved the interfacial adhesion and mechanical properties of the composites.

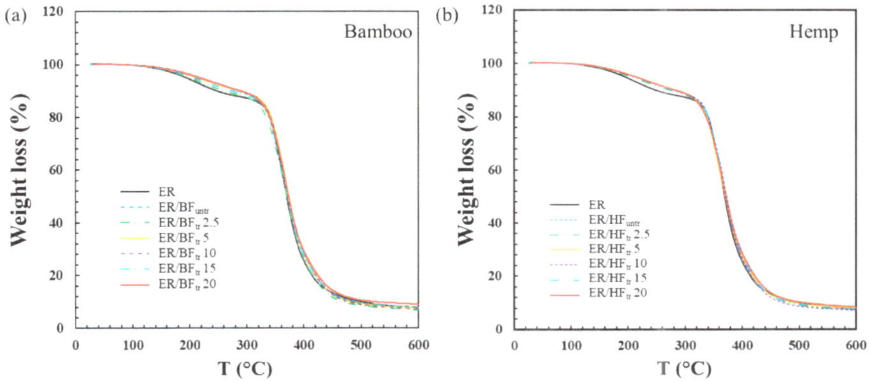

Figure 5. TGA curves of (**a**) BF- and (**b**) HF-reinforced epoxy composites with untreated fibers and fibers treated with Ar + O_2 gas for different times.

3.6. Water Contact Angle

The wettability properties of the composites were calculated using a contact angle measurement, as shown in Figure 6. Water droplets were dropped onto the composite surfaces and automatically recorded from 0 to 10 min. The untreated BF composite (EP/BF$_{untr}$) exhibited a contact angle of 72° after 10 min (Figure 6a), which implies a low surface wettability. The treated samples exhibited decreased water contact angles after 10 min with the increase in treatment time (65, 62, 62, and 57°) due to the high polarity of the plasma treatment, which increased with the 20 min treatment (67°) because of the high crosslinking degree and surface roughness effect of BF$_{tr}$. The ER/HF$_{untr}$ composite exhibited a contact angle of 85° after 10 min (Figure 6b), which decreased with the enhance in the plasma treatment time (82, 82, 83, 81, 77°). The decreasing water droplet contact angle of ER/HF$_{tr}$ indicates a high amount of generated −COOH and −OH groups on the HF$_{tr}$ structure, which reduced the contact angle by its hydrophilicity. However, for BR/BF$_{tr}$ treated for 20 min, the contact angle was increased due to the high crosslinking density between the BF$_{tr}$ and ER matrix. Crosslinking reaction and the enhanced crosslinking amount with time treatment were confirmed by FTIR and TGA results, respectively. Both composite fibers exhibited decreasing contact angles with the increase in the DBD plasma treatment time due to the increases in the surface roughness, porosity, and amount of hydrophilic groups of the fibers subjected to the Ar + O_2 gas treatment, which led to etching of the surface fibers [48,66]. The long-term plasma treatment affected the water penetration into the fibers and decreased the water resistance of the composite [28,67].

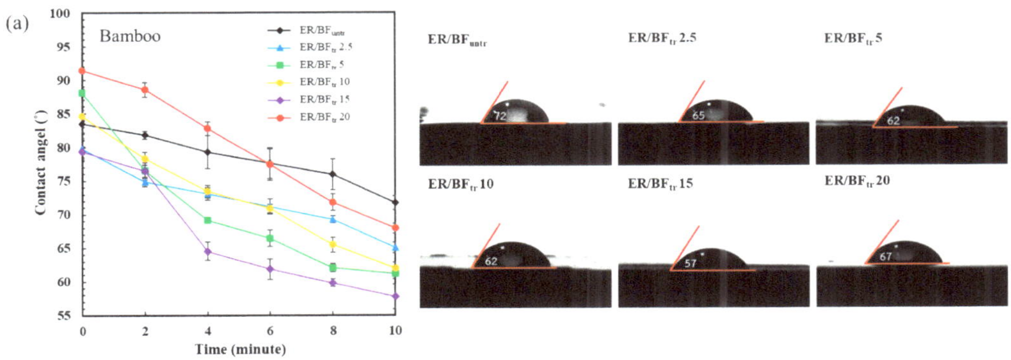

Figure 6. *Cont.*

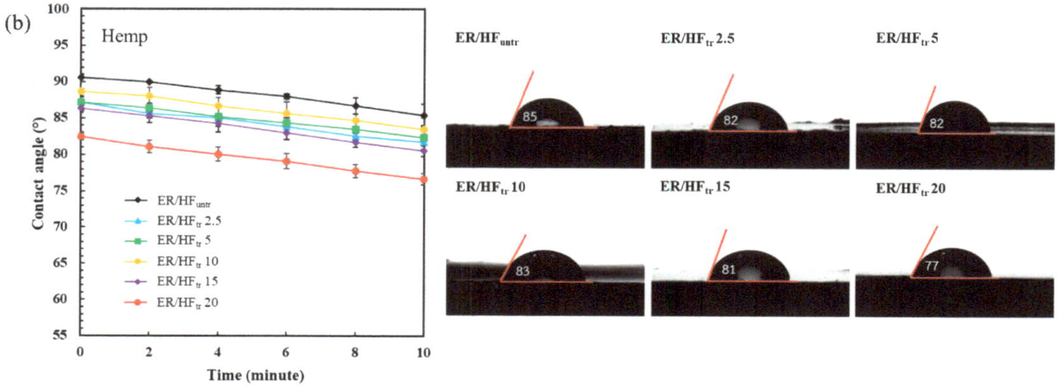

Figure 6. Water droplet contact angles and images of water contact angles at 10 min for the fiber-reinforced epoxy composites with untreated and plasma-treated structures: (**a**) BF and (**b**) HF.

4. Conclusions

ER/BF and ER/HF were successfully developed with DBD Ar + O_2 plasma treatment. The Ar + O_2 gas plasma treatment significantly enhanced the properties of the bamboo and hemp fiber-reinforced epoxy composites. ER/BFtr was better in terms of tensile strength, while ER/HFtr exhibited a larger improvement in elongation at break. The FTIR spectroscopy confirmed the reaction between epoxy groups of ER with C–O and $-NH_2$ groups of the treated fibers, which improved the interfacial adhesion between the fiber surface and ER matrix. The ER/BF$_{tr}$ and ER/HF$_{tr}$ composites exhibited improved flexural strengths and flexural moduli after the Ar + O_2 plasma treatment due to the improved polarity and roughness of the fiber surface. The SEM analysis showed the fiber surface roughness with a low gap between fibers and ER matrix. The BFs exhibited a gradual increase in surface roughness with the treatment time, while the HFs exhibited a good interfacial adhesion with a low gap. The TGA also confirmed the improved thermal stability and reactions by the plasma treatment with a high weight loss for the long-term Ar + O_2 plasma treatment. The plasma treatment decreased the water contact angle owing to the increased content of hydrophilic groups, etching, porosity, and roughness of the fiber surface. The ER/BF$_{tr}$ and ER/HF$_{tr}$ composites can be employed in packaging, coating, construction, and electrical applications.

Author Contributions: K.J., P.W. (Pitiwat Wattanachai), C.S., and P.R. designed the research study; K.J., P.W. (Pitiwat Wattanachai), C.S., and P.R. methodology; P.R., P.W. (Pitiwat Wattanachai), C.S., P.W. (Patnarin Worajittiphon), J.S., N.T., and K.J. visualization, data analysis, and supervision; P.R., P.W. (Pitiwat Wattanachai), C.S., J.S., N.T., K.J., P.W. (Patnarin Worajittiphon), T.K., K.K., and P.T. prepared and investigated the samples; P.R., P.W. (Pitiwat Wattanachai), C.S., J.S., N.T., P.W. (Patnarin Worajittiphon), and K.J. discussed the results; K.J., T.K., P.W. (Pitiwat Wattanachai), C.S., and P.R. writing—original draft preparation. All authors have read and agreed to the published version of the manuscript.

Funding: This research has received funding support from the NSRF via the Program Management Unit for Human Resources and Institutional Development, Research, and Innovation (grant number B16F640206). This research work was partially supported by Chiang Mai University. The present study was partially supported by the Thailand Research Fund (TRF) Research Team Promotion Grant, RTA, Senior Research Scholar (N42A671052). This research and innovation activity was funded by the National Research Council of Thailand (NRCT) under the "Hub of Talents in Bioplastics for Use in Medical Applications" (Grant No. N34E670071).

Institutional Review Board Statement: Not applicable.

Informed Consent Statement: Not applicable.

Data Availability Statement: The data presented in this study are available upon request from the corresponding author.

Acknowledgments: This research work was partially supported by Chiang Mai University. The authors wish to express their sincere thanks to the Agriculture and Bio Plasma Technology Center (ABPlas), Science and Technology Park (STeP), and Faculty of Agro-industry, Chiang Mai University for laboratory equipment and resources support and the Center of Excellence in Materials Science and Technology, Chiang Mai University.

Conflicts of Interest: The authors declare no conflicts of interest.

Abbreviations and Symbols

ER, epoxy resin; BF, bamboo fibers; HF, hemp fibers; DBD, dielectric barrier discharge; FTIR, Fourier-transform infrared spectroscopy; SEM, scanning electron microscopy; TGA, thermogravimetric analysis.

References

1. Low, I.M.; McGrath, M.; Lawrence, D.; Schmidt, P.; Lane, J.; Latella, B.A.; Sim, K.S. Mechanical and fracture properties of cellulose-fibre-reinforced epoxy laminates. *Compos. A Appl. Sci. Manuf.* **2007**, *38*, 963–974. [CrossRef]
2. Shih, Y.-F. Mechanical and thermal properties of waste water bamboo husk fiber reinforced epoxy composites. *Mater. Sci. Eng. A* **2007**, *445–446*, 289–295. [CrossRef]
3. Gassan, J. Effects of corona discharge and UV treatment on the properties of jute-fibre epoxy composites. *Compos. Sci. Technol.* **2000**, *60*, 2857–2863. [CrossRef]
4. Radoor, S.; Karayil, J.; Rangappa, S.M.; Siengchin, S.; Parameswaranpillai, J. A review on the extraction of pineapple, sisal and abaca fibers and their use as reinforcement in polymer matrix. *Express Polym. Lett.* **2020**, *14*, 309–335. [CrossRef]
5. Ramadan, N.; Taha, M.; Elsabbagh, A. Processing and characterization of flame-retardant natural fibre-reinforced epoxy composites and construction of selection charts for engineering applications. *Express Polym. Lett.* **2024**, *18*, 779–795. [CrossRef]
6. Sukwijit, C.; Seubsai, A.; Charoenchaitrakool, M.; Sudsakorn, K.; Niamnuy, C.; Roddecha, S.; Prapainainar, P. Production of PLA/cellulose derived from pineapple leaves as bio-degradable mulch film. *Int. J. Biol. Macromol.* **2024**, *270*, 132299. [CrossRef] [PubMed]
7. Bachtiar, D.; Sapuan, S.M.; Hamdan, M.M. The effect of alkaline treatment on tensile properties of sugar palm fibre reinforced epoxy composites. *Mater. Des.* **2008**, *29*, 1285–1290. [CrossRef]
8. Alamri, H.; Low, I.M. Characterization of epoxy hybrid composites filled with cellulose fibers and nano-SiC. *J. Appl. Polym. Sci.* **2012**, *126*, E222–E232. [CrossRef]
9. Emamverdian, A.; Ding, Y.; Ranaei, F.; Ahmad, Z. Application of bamboo plants in nine aspects. *Sci. World J.* **2020**, *2020*, 7284203. [CrossRef]
10. Kiattipornpithak, K.; Rachtanapun, P.; Thanakkasaranee, S.; Jantrawut, P.; Ruksiriwanich, W.; Sommano, S.R.; Leksawasdi, N.; Kittikorn, T.; Jantanasakulwong, K. Bamboo pulp toughening poly (lactic acid) composite using reactive epoxy resin. *Polymers* **2023**, *15*, 3789. [CrossRef]
11. Huang, J.; Zhang, L.; Zhou, Y.; Huang, M.; Sha, Y. Study on the suitability of bamboo fiber for manufacturing insulating presspaper. *IEEE Trans. Dielectr. Electr. Insul.* **2016**, *23*, 3641–3651. [CrossRef]
12. Liu, D.; Song, J.; Anderson, D.P.; Chang, P.R.; Hua, Y. Bamboo fiber and its reinforced composites: Structure and properties. *Cellulose* **2012**, *19*, 1449–1480. [CrossRef]
13. Lokesh, P.; Surya Kumari, T.S.A.; Gopi, R.; Babu Loganathan, G. A study on mechanical properties of bamboo fiber reinforced polymer composite. *Mater. Today Proc.* **2020**, *22*, 897–903. [CrossRef]
14. Rohen, L.A.; Neves, A.C.C.; dos Santos, J.L.; Nascimento, L.F.C.; Monteiro, S.N.; de Assis, F.S.; Simonassi, N.T.; da Silva, L.C. Comparative analysis of the tensile properties of polyester and epoxy composites reinforced with hemp fibers. *Mater. Sci. Forum* **2018**, *930*, 201–206. [CrossRef]
15. Islam, M.S.; Pickering, K.L.; Foreman, N.J. Influence of alkali fiber treatment and fiber processing on the mechanical properties of hemp/epoxy composites. *J. Appl. Polym. Sci.* **2010**, *119*, 3696–3707. [CrossRef]
16. Valadez-Gonzalez, A.; Cervantes-Uc, J.M.; Olayo, R.; Herrera-Franco, P.J. Effect of fiber surface treatment on the fiber–matrix bond strength of natural fiber reinforced composites. *Compos. B Eng.* **1999**, *30*, 309–320. [CrossRef]
17. Thipchai, P.; Punyodom, W.; Jantanasakulwong, K.; Thanakkasaranee, S.; Hinmo, S.; Pratinthong, K.; Kasi, G.; Rachtanapun, P. Preparation and characterization of cellulose nanocrystals from bamboos and their application in cassava starch-based film. *Polymers* **2023**, *15*, 2622. [CrossRef]
18. Suriyatem, R.; Noikang, N.; Kankam, T.; Jantanasakulwong, K.; Leksawasdi, N.; Phimolsiripol, Y.; Insomphun, C.; Seesuriyachan, P.; Chaiyaso, T.; Jantrawut, P.; et al. Physical properties of carboxymethyl cellulose from palm bunch and bagasse agricultural wastes: Effect of delignification with hydrogen peroxide. *Polymers* **2020**, *12*, 1505. [CrossRef] [PubMed]

19. Zhou, F.; Cheng, G.; Jiang, B. Effect of silane treatment on microstructure of sisal fibers. *Appl. Surf. Sci.* **2014**, *292*, 806–812. [CrossRef]
20. Xie, J.; Xin, D.; Cao, H.; Wang, C.; Zhao, Y.; Yao, L.; Ji, F.; Qiu, Y. Improving carbon fiber adhesion to polyimide with atmospheric pressure plasma treatment. *Surf. Coat. Technol.* **2011**, *206*, 191–201. [CrossRef]
21. Karaduman, Y.; Gokcan, D.; Onal, L. Effect of enzymatic pretreatment on the mechanical properties of jute fiber-reinforced polyester composites. *J. Compos. Mater.* **2012**, *47*, 1293–1302. [CrossRef]
22. Oliveira, F.R.; Erkens, L.; Fangueiro, R.; Souto, A.P. Surface modification of banana fibers by DBD plasma treatment. *Plasma Chem. Plasma Process.* **2012**, *32*, 259–273. [CrossRef]
23. Phan, K.T.K.; Phan, H.T.; Brennan, C.S.; Regenstein, J.M.; Jantanasakulwong, K.; Boonyawan, D.; Phimolsiripol, Y. Gliding arc discharge non-thermal plasma for retardation of mango anthracnose. *LWT* **2019**, *105*, 142–148. [CrossRef]
24. Faruk, O.; Bledzki, A.K.; Fink, H.-P.; Sain, M. Progress report on natural fiber reinforced composites. *Macromol. Mater. Eng.* **2014**, *299*, 9–26. [CrossRef]
25. Peças, P.; Carvalho, H.; Salman, H.; Leite, M. Natural fibre composites and their applications: A Review. *J. Compos. Sci.* **2018**, *2*, 66. [CrossRef]
26. Liston, E.M. Plasma treatment for improved bonding: A Review. *J. Adhes.* **1989**, *30*, 199–218. [CrossRef]
27. Sawangrat, C.; Thipchai, P.; Kaewapai, K.; Jantanasakulwong, K.; Suhr, J.; Wattanachai, P.; Rachtanapun, P. Surface modification and mechanical properties improvement of bamboo fibers using dielectric barrier discharge plasma treatment. *Polymers* **2023**, *15*, 1711. [CrossRef]
28. Rachtanapun, P.; Sawangrat, C.; Kanthiya, T.; Thipchai, P.; Kaewapai, K.; Suhr, J.; Worajittiphon, P.; Tanadchangsaeng, N.; Wattanachai, P.; Jantanasakulwong, K. Effect of plasma treatment on bamboo fiber-reinforced epoxy composites. *Polymers* **2024**, *16*, 938. [CrossRef] [PubMed]
29. Lee, D.; Hong, S.H.; Paek, K.-H.; Ju, W.-T. Adsorbability enhancement of activated carbon by dielectric barrier discharge plasma treatment. *Surf. Coat. Technol.* **2005**, *200*, 2277–2282. [CrossRef]
30. Biswas, S.; Shahinur, S.; Hasan, M.; Ahsan, Q. Physical, mechanical and thermal properties of jute and bamboo fiber reinforced unidirectional epoxy composites. *Procedia Eng.* **2015**, *105*, 933–939. [CrossRef]
31. Mashouf Roudsari, G.; Mohanty, A.K.; Misra, M. Study of the curing kinetics of epoxy resins with biobased hardener and epoxidized soybean oil. *ACS Sustain. Chem. Eng.* **2014**, *2*, 2111–2116. [CrossRef]
32. Uflyand, I.E.; Irzhak, V.I. Recent advances in the study of structure and properties of fiber composites with an epoxy matrix. *J. Polym. Res.* **2021**, *28*, 440. [CrossRef]
33. Mohan, P. A Critical Review: The modification, properties, and applications of epoxy resins. *Polym. Plast. Technol. Eng.* **2013**, *52*, 107–125. [CrossRef]
34. Kodsangma, A.; Homsaard, N.; Nadon, S.; Rachtanapun, P.; Leksawasdi, N.; Phimolsiripol, Y.; Insomphun, C.; Seesuriyachan, P.; Chaiyaso, T.; Jantrawut, P.; et al. Effect of sodium benzoate and chlorhexidine gluconate on a bio-thermoplastic elastomer made from thermoplastic starch-chitosan blended with epoxidized natural rubber. *Carbohydr. Polym.* **2020**, *242*, 116421. [CrossRef]
35. Ma, H.; Whan Joo, C. Influence of surface treatments on structural and mechanical properties of bamboo fiber-reinforced poly(lactic acid) biocomposites. *J. Compos. Mater.* **2011**, *45*, 2455–2463. [CrossRef]
36. Sgriccia, N.; Hawley, M.C.; Misra, M. Characterization of natural fiber surfaces and natural fiber composites. *Compos. A Appl. Sci. Manuf.* **2008**, *39*, 1632–1637. [CrossRef]
37. Bozaci, E.; Sever, K.; Sarikanat, M.; Seki, Y.; Demir, A.; Ozdogan, E.; Tavman, I. Effects of the atmospheric plasma treatments on surface and mechanical properties of flax fiber and adhesion between fiber–matrix for composite materials. *Compos. B Eng.* **2013**, *45*, 565–572. [CrossRef]
38. Rayung, M.; Ibrahim, N.A.; Zainuddin, N.; Saad, W.Z.; Razak, N.I.; Chieng, B.W. The effect of fiber bleaching treatment on the properties of poly(lactic acid)/oil palm empty fruit bunch fiber composites. *Int. J. Mol. Sci.* **2014**, *15*, 14728–14742. [CrossRef]
39. Liu, W.; Mohanty, A.K.; Drzal, L.T.; Askel, P.; Misra, M. Effects of alkali treatment on the structure, morphology and thermal properties of native grass fibers as reinforcements for polymer matrix composites. *J. Mater. Sci.* **2004**, *39*, 1051–1054. [CrossRef]
40. Kiatiporntipthak, K.; Thajai, N.; Kanthiya, T.; Rachtanapun, P.; Leksawasdi, N.; Phimolsiripol, Y.; Rohindra, D.; Ruksiriwanich, W.; Sommano, S.R.; Jantanasakulwong, K. Reaction mechanism and mechanical property improvement of poly(lactic acid) reactive blending with epoxy resin. *Polymers* **2021**, *13*, 2429. [CrossRef]
41. Dimitrakellis, P.; Faubert, F.; Wartel, M.; Gogolides, E.; Pellerin, S. Plasma surface modification of epoxy polymer in air DBD and gliding arc. *Processes* **2022**, *10*, 104. [CrossRef]
42. Kanthiya, T.; Kiattipornpithak, K.; Thajai, N.; Phimolsiripol, Y.; Rachtanapun, P.; Thanakkasaranee, S.; Leksawasdi, N.; Tanadchangsaeng, N.; Sawangrat, C.; Wattanachai, P.; et al. Modified poly(lactic acid) epoxy resin using chitosan for reactive blending with epoxidized natural rubber: Analysis of annealing time. *Polymers* **2022**, *14*, 1085. [CrossRef] [PubMed]
43. Baghery Borooj, M.; Mousavi Shoushtari, A.; Nosratian Sabet, E.; Haji, A. Influence of oxygen plasma treatment parameters on the properties of carbon fiber. *J. Adhes. Sci. Technol.* **2016**, *30*, 2372–2382. [CrossRef]
44. Pitto, M.; Fiedler, H.; Kim, N.K.; Verbeek, C.J.R.; Allen, T.D.; Bickerton, S. Carbon fibre surface modification by plasma for enhanced polymeric composite performance: A review. *Compos. Part A* **2024**, *180*, 108087. [CrossRef]

45. Kanthiya, T.; Thajai, N.; Chaiyaso, T.; Rachtanapun, P.; Thanakkasaranee, S.; Kumar, A.; Boonrasri, S.; Kittikorn, T.; Phimolsiripol, Y.; Leksawasdi, N.; et al. Enhancement in mechanical and antimicrobial properties of epoxidized natural rubber via reactive blending with chlorhexidine gluconate. *Sci. Rep.* **2023**, *13*, 9974. [CrossRef]
46. Manjula, S.; Shanmugasundaram, O.L.; Ponappa, K. Optimization of plasma process parameters for surface modification of bamboo spunlace nonwoven fabric using glow discharge oxygen plasma. *J. Ind. Text.* **2019**, *51*, 225–245. [CrossRef]
47. Sánchez, M.L.; Patiño, W.; Cárdenas, J. Physical-mechanical properties of bamboo fibers-reinforced biocomposites: Influence of surface treatment of fibers. *J. Build. Eng.* **2020**, *28*, 101058. [CrossRef]
48. Li, S.; Han, K.; Rong, H.; Li, X.; Yu, M. Surface modification of aramid fibers via ammonia-plasma treatment. *J. Appl. Polym. Sci.* **2013**, *131*, 40250. [CrossRef]
49. Mohit, H.; Arul Mozhi Selvan, V. A comprehensive review on surface modification, structure interface and bonding mechanism of plant cellulose fiber reinforced polymer-based composites. *Compos. Interfaces* **2018**, *25*, 629–667. [CrossRef]
50. Chen, Y.; Xu, C.; Wang, C.H.; Bilek, M.M.M.; Cheng, X. An effective method to optimise plasma immersion ion implantation: Sensitivity analysis and design based on low-density polyethylene. *Plasma Process. Polym.* **2022**, *19*, e2100199. [CrossRef]
51. Kim, S.Y.; Hong, K.; Kim, K.; Yu, H.K.; Kim, W.-K.; Lee, J.-L. Effect of N_2, Ar, and O_2 plasma treatments on surface properties of metals. *J. Appl. Phys.* **2008**, *103*, 076101. [CrossRef]
52. Pejic, B.M.; Kramar, A.D.; Obradovic, B.M.; Kuraica, M.M.; Zekic, A.A.; Kostic, M.M. Effect of plasma treatment on chemical composition, structure and sorption properties of lignocellulosic hemp fibers (*Cannabis sativa* L.). *Carbohydr. Polym.* **2020**, *236*, 116000. [CrossRef]
53. Ivanovska, A.; Milošević, M.; Obradović, B.; Svirčev, Z.; Kostić, M. Plasma treatment as a sustainable method for enhancing the wettability of jute fabrics. *Sustainability* **2023**, *15*, 2125. [CrossRef]
54. Ren, Z.; Tang, X.; Wang, H.; Qiu, G. Continuous modification treatment of polyester fabric by Ar-O_2(10:1) Discharge at atmospheric pressure. *J. Ind. Text.* **2016**, *37*, 43–53.
55. Zhang, D.; Huang, Y. Dispersion characterizations and adhesion properties of epoxy composites reinforced by carboxymethyl cellulose surface treated carbon nanotubes. *Powder Technol.* **2022**, *404*, 117505. [CrossRef]
56. Zhang, D.; Huang, Y.; Xia, W.; Xu, L.; Wang, X. Dispersion characteristics and mechanical properties of epoxy nanocomposites reinforced with carboxymethyl cellulose functionalized nanodiamond, carbon nanotube, and graphene. *Polym. Compos.* **2023**, *45*, 398–412. [CrossRef]
57. Movahed, S.; Nguyen, A.K.; Goering, P.L.; Skoog, S.A.; Narayan, R.J. Argon and oxygen plasma treatment increases hydrophilicity and reduces adhesion of silicon-incorporated diamond-like coatings. *Biointerphases* **2020**, *15*, 041007. [CrossRef] [PubMed]
58. Kan, C.W.; Chan, K.; Yuen, C.W.M. Surface characterization of low temperature plasma treated wool fiber. *Fibers Polym.* **2004**, *5*, 52–58. [CrossRef]
59. Marrot, L.; Lefeuvre, A.; Pontoire, B.; Bourmaud, A.; Baley, C. Analysis of the hemp fiber mechanical properties and their scattering (Fedora 17). *Ind. Crops Prod.* **2013**, *51*, 317–327. [CrossRef]
60. Relvas, C.; Castro, G.; Rana, S.; Fangueiro, R. characterization of physical, mechanical and chemical properties of Quiscal Fibres: The influence of atmospheric DBD plasma treatment. *Plasma Chem. Plasma Process.* **2015**, *35*, 863–878. [CrossRef]
61. Alonso-Montemayor, F.J.; López-Badillo, C.M.; Aguilar, C.N.; Ávalos-Belmontes, F.; Castañeda-Facio, A.O.; Reyna-Martínez, R.; Neira-Velázquez, M.G.; Soria-Argüello, G.; Navarro-Rodríguez, D.; Delgado-Aguilar, M.; et al. Effect of cold air plasmas on the morphology and thermal stability of bleached hemp fibers. *Rev. Mex. Ing. Quim.* **2020**, *19*, 457–467. [CrossRef]
62. Silva, T.T.D.; Silveira, P.; Ribeiro, M.P.; Lemos, M.F.; da Silva, A.P.; Monteiro, S.N.; Nascimento, L.F.C. Thermal and chemical characterization of kenaf fiber (*Hibiscus cannabinus*) reinforced epoxy matrix composites. *Polymers* **2021**, *13*, 2016. [CrossRef] [PubMed]
63. Misnon, M.I.; Islam, M.M.; Epaarachchi, J.A.; Lau, K.T. Analyses of woven hemp fabric characteristics for composite reinforcement. *Mater. Des.* **2015**, *66*, 82–92. [CrossRef]
64. Scalici, T.; Fiore, V.; Valenza, A. Effect of plasma treatment on the properties of *Arundo Donax* L. leaf fibres and its bio-based epoxy composites: A preliminary study. *Compos. B Eng.* **2016**, *94*, 167–175. [CrossRef]
65. Mohammed, Z.; Jeelani, S.; Rangari, V.K. Effect of low-temperature plasma treatment on starch-based biochar and its reinforcement for three-dimensional printed polypropylene biocomposites. *ACS Omega* **2022**, *7*, 39636–39647. [CrossRef]
66. Lieberman, M.A.; Lichtenberg, A.J. Principles of Plasma Discharges and Materials Processing. *MRS Bull.* **2005**, *30*, 899.
67. Tan, S.H.; Nguyen, N.T.; Chua, Y.C.; Kang, T.G. Oxygen plasma treatment for reducing hydrophobicity of a sealed polydimethylsiloxane microchannel. *Biomicrofluidics* **2010**, *4*, 32204. [CrossRef]

polymers

MDPI

Article

A Survey on the Effect of the Chemical Composition on the Thermal, Physical, Mechanical, and Dynamic Mechanical Thermal Analysis of Three Brazilian Wood Species

Matheus de Prá Andrade [1], Heitor Luiz Ornaghi, Jr. [1], Francisco Maciel Monticeli [2,*], Matheus Poletto [1] and Ademir José Zattera [1]

[1] Postgraduate Program in Engineering of Processes and Technologies, University of Caxias do Sul, Caxias do Sul 95070-560, Brazil; mpandrade@ucs.br (M.d.P.A.); ornaghijr.heitor@gmail.com (H.L.O.J.); mpolett1@ucs.br (M.P.); ajzatter@ucs.br (A.J.Z.)

[2] Department of Aerospace Structures and Materials, Faculty of Aerospace Engineering, Delft University of Technology, 2629 HS Delft, The Netherlands

* Correspondence: f.m.monticeli@tudelft.nl

Abstract: Wood is a versatile material extensively utilized across industries due to its low density, favorable mechanical properties, and environmental benefits. However, despite considerable research, the diversity in species with varying compositions and properties remains insufficiently explored, particularly for native woods. A deeper understanding of these differences is crucial for optimizing their industrial applications. This study investigated the composition, tensile strength, flexural strength, Young's modulus, bending stiffness and elongation at break, thermal behavior, and viscoelastic properties of three Brazilian native wood species: *Araucaria angustifolia* (ARA), *Dipterix odorata* (DOD), and *Tabeuia ochracea* (TOC). The density of these woods showed a linear correlation with mechanical properties such as Young's modulus (0.9) and flexural modulus (0.9). The research revealed a linear correlation between the woods' density and mechanical properties, with lignin content emerging as a key determinant of thermal stability. This study highlights the importance of understanding wood species' composition and physical properties, and provides valuable insights into their behavior.

Keywords: wood; correlation; thermal properties; mechanical properties

Citation: de Prá Andrade, M.; Ornaghi, H.L., Jr.; Monticeli, F.M.; Poletto, M.; Zattera, A.J. A Survey on the Effect of the Chemical Composition on the Thermal, Physical, Mechanical, and Dynamic Mechanical Thermal Analysis of Three Brazilian Wood Species. *Polymers* **2024**, *16*, 2651. https://doi.org/10.3390/polym16182651

Academic Editor: Ming Zhang

Received: 26 August 2024
Revised: 17 September 2024
Accepted: 18 September 2024
Published: 20 September 2024

1. Introduction

Wood remains a highly competitive material today, with one of its key advantages being its beneficial impact on CO_2 emissions. Wood generally captures more CO_2 than it emits; for instance, 600 kg of wood can sequester approximately 1.5 tons of CO_2 [1]. Its composition primarily includes cellulose, hemicellulose, lignin, and low molecular weight components called extractives [2–4]. Cellulose is generally responsible for the mechanical properties, while lignin is responsible for thermal stability (the end of the degradation curve) due to its antioxidant capacity and the presence of aromatic rings [3,5,6]. A high ratio of hemicellulose to extractives may be preferred in applications where stiffness and thermal stability (in the initial plateau of the degradation curve) are required, but at the cost of higher moisture absorption, which also acts as an accelerator of thermal degradation [2,3]. Wood's chemical composition, climate, and region also contribute to the diversity of wood species worldwide. Commonly studied species, such as *Pinus elliot* and *Eucalyptus grandis*, are widely used in industry, resulting in numerous investigations of their mechanical and thermal properties [3]. Figure 1 shows a schematic representation of the cellular structure of wood species [7].

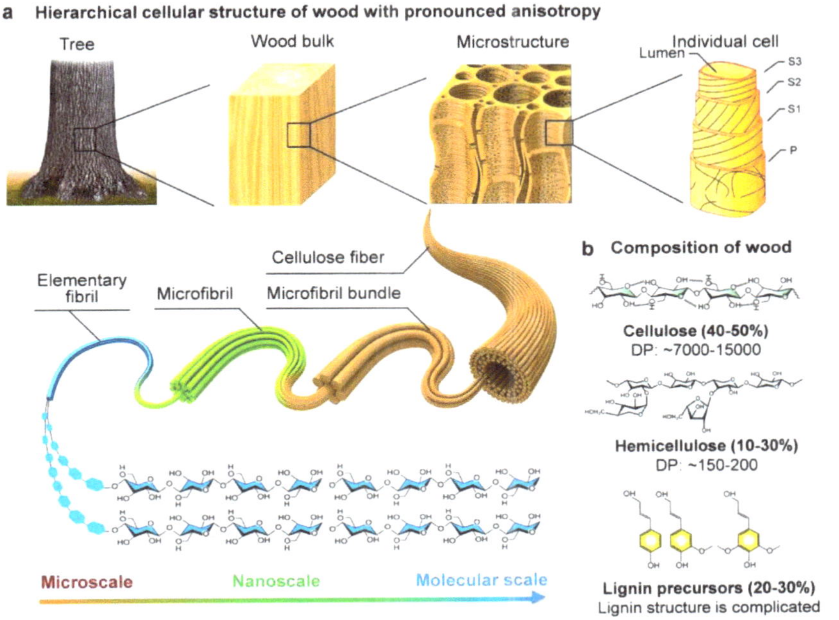

Figure 1. Schematic representation of the cellular structure of wood species showing the (**a**) Hierarchical cellular structure of wood and (**b**) composition of wood (figure used under the terms and conditions of creative commons attribution [7]).

The extensive tropical native forests in Brazil, which account for approximately 12% of the world's forest area, are home to diverse wood species with considerable variety in their composition and properties. On the other hand, only a limited number of studies have focused on elucidating the properties of these lesser-known species [8], such as their mechanical, physical, and thermal properties. The study of these species is essential for determining their optimal applications and identifying the industries where they can be most effectively utilized. By understanding their mechanical properties, it becomes possible to assess whether their use in specific applications is reliable and suitable for long-term performance. This knowledge also makes it possible to compare the woods' properties between the wood classes or with other material classes, such as polymers and metals [1].

Wood is a natural composite and an orthotropic material, meaning its properties vary along three distinct directions: longitudinal, radial, and tangential [1]. Additionally, moisture content plays a significant role in affecting its mechanical properties. As the moisture content increases, there is typically a reduction in properties such as the modulus of elasticity. This is because water can act as a plasticizer, weakening the bonds between cellulose fibers and thereby reducing the overall strength of the material [1,9,10].

The mechanical properties of wood are mainly influenced by the density and cellulose content [11,12]. Pertuzzati et al. (2018) [13] studied the effect of mechanical densification on two fast-growing species, pine and eucalyptus. The authors claimed an increase in the mechanical properties with density (from 0.53 to 0.99 $g{\cdot}cm^{-3}$), with a 294% increase in the modulus of rupture and an 85% increase in Young's modulus for the pine species. To confirm these results, Li et al. [14] investigated the effect of woods' properties on the fire performance of glulam columns made of six different wood species in China: poplar, Chinese fir, Douglas fir, hemlock, larch, and spruce. The densities ranged from 0.36 to 0.645 $g{\cdot}cm^{-3}$, and the fire tests showed lower charring rates and slower heat penetration rates for the higher-density glulam columns.

Yue et al. [15] studied thermally treated Chinese poplar wood, with a focus on structural applications, and claimed that the variation in mechanical properties is directly related to the thermal degradation of the chemical composition. Also, from a structural point of view, the best temperature tested in the treatment of poplar wood was 170–180 °C. Qian et al. [16] increased bamboo's strength and thermal stability for sustainable construction. The authors produced an environmentally sustainable composite of modified glued bamboo and obtained an improvement in all mechanical properties from 40 to 120%. The composite also showed excellent flame-retardant properties. Pereira et al. [17] investigated the effect of chemical treatment of pineapple crown fiber on the production, chemical composition, crystalline structure, thermal stability, and kinetics of thermal degradation. The authors used different methods to extract cellulose from pineapple crown fiber. The authors claimed that stability, thermal stability, and some physical and chemical properties were improved, depending on the chemical treatment used. Pereira et al. [18] obtained bleached cellulose from orange bagasse using different sequential chemical treatments. Three different pretreatments were used: alkaline treatment, organosolv and residual insoluble alcohol. All pretreated samples were bleached. The FTIR and XRD results indicated that hemicellulose and lignin were largely removed after the first pretreatment. The thermal degradation curves also showed the presence of two degradation steps instead of three.

Thermal properties are critical, especially when the wood species is incorporated into polymeric matrices [19,20]. To understand the thermal properties of wood and other lignocellulosic materials, the most common technique is thermogravimetric analysis (TGA), from which it is possible to verify the degradation steps, kinetics, and thermodynamic parameters, as many studies have already demonstrated [21–25]. Tabal et al. [26] evaluated the thermal degradation of *Ficus nitida* wood using isoconversion methods such as the Friedman, Flynn–Wall–Ozawa (FWO), and Vyazovkin methods. The authors claimed that the degradation of *Ficus nitida* wood can proceed through three-stage degradation, with activation energies ranging from 171.4 to 248.23 kJ·mol^{-1}. Ornaghi Jr. [27] studied the thermal degradation effect of each component of biomass (cellulose, hemicellulose, and lignin, with their respective chars, among others) using a statistical kinetic approach. The main results indicated that cellulose plays a major role in the activation energy, while the reaction order is determined by hemicellulose and lignin. In a review, Hill et al. [28] investigated the effect of chemical modification and hygroscopicity on the thermal properties of various woods. The authors reviewed the current status of thermally modified wood under dry and wet conditions and the role of some polysaccharides in lignocellulosic materials on the thermal properties, dimensional stability, and mass loss. The main information is on the role of OH groups in the sorption behavior and how this affects other properties.

The properties of native woods in Brazil are not well-established, possibly due to the dominance of exotic species in the market, such as *Pinus elliottii* and *Eucalyptus grandis*, which have been the subject of far more research [13,29–31]. In this study, three native Brazilian wood species were investigated: *Araucaria angustifolia* (ARA), *Dipteryx odorata* (DOD), and *Tabebuia ochracea* (TOC). *Araucaria angustifolia*, a key species in southern Brazil, is known for its relatively low density and well-balanced mechanical properties [32–34]. DOD is present in south of Brazil and spreads over the north and midwest regions; it is known for its high density (>950 kg·m^{-3}) and mechanical resistance [35,36]. These species were selected in part because they are native to the southern region of Brazil, making them representative choices for this investigation.

Many other studies of wood properties may focus on the composition, dynamic mechanical thermal analysis, or physical properties, which are generally analyzed without considering other aspects of wood, and correlations among the properties studied are scarce. The aim of this work was to evaluate and correlate the composition and thermal, physical, mechanical, and dynamic properties of three Brazilian wood species: *Araucaria angustifolia* (ARA), *Dipterix odorata* (DOD), and *Tabeuia ochracea* (TOC).

2. Materials and Methods

2.1. Materials

Three Brazilian native woods were studied in panel format without any prior treatment: *Araucaria angustifolia* (ARA), *Dipterix odorata* (DOD), and *Tabebuia ochracea* (TOC). These wood samples were exclusively extracted from the heartwood, with no portions from the outer sections of the tree utilized. Figure 2 schematically shows the process of obtaining samples for tensile and bending tests and the corresponding characterizations. Wood was obtained from a local provider, Madeireira Bianchi (Bento Gonçalves, RS, Brazil), in board format. Samples for the flexural and tensile tests were machined from the wood board using a computer numerical control (CNC) machine (Jaraguá, SC, Brazil) with the machining aligned along the fibers' direction. Prior to analysis, the samples were oven-dried at 105 °C for 4 h to eliminate residual moisture.

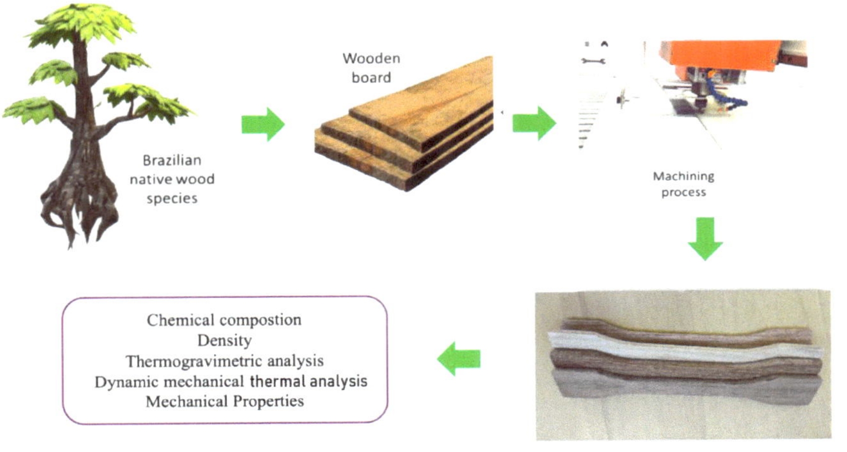

Figure 2. Schematic representation of the process of obtaining the specimens and their characterization.

2.2. Composition

The composition of three wood samples was measured using established methods. Wood extractives were determined according to TAPPI T204 [37] using a 1:2 ethanol/benzene solution. The Klason lignin content was determined according to TAPPI T222 [38]. The Van Soest method was used for factors such as moisture, cellulose, and hemicellulose. All tests were performed in triplicate to ensure the reliability of the results.

2.3. Mechanical Tests and Density

Two tests were used to measure the mechanical properties of wood. The bending tests were carried out on an EMIC (Caxias do Sul, Brazil) universal testing machine, model DL 3000 (Instron, Norwood, MA, USA), with a 200 kg load cell. The tests were carried out following ASTM D790 [39], with rectangular samples measuring 125 × 12.7 × 3.2 mm. The tensile tests were performed again on an EMIC DL 3000 (Caxias do Sul, RS, Brazil) universal testing machine under ASTM D638 [40], with base dimension of 165 × 13 × 3.2 mm. The standards were used without any adjustment.

In order to understand the relationship between the physical and mechanical properties, the densities of the wood species were also measured. This was determined by the volumetric method according to ASTM D792 [41], using the machined specimens from the bending tests (25 × 13 × 4 mm), and the samples were oven dried for 4 h at the temperature of 105 °C. Equation 1 was used to determine the density of the wood.

$$\rho = \frac{(a * b)}{(a - c)} \tag{1}$$

where ρ is the density ($g \cdot cm^{-3}$), a is the sample's mass (g), b is the density of water ($g \cdot cm^{-3}$), and c is the sample's mass underwater ($g \cdot cm^{-3}$). Five samples were used for each of the three wood species for each measurement.

2.4. Statistical Analysis

The coefficient of variation (CV) is a statistical measure that represents the ratio of the standard deviation to the mean of a dataset, which represents the standard deviation divided by the average (Equation (2)). It is expressed as a percentage and used to compare the relative variability of data between different datasets or measures. A higher CV indicates greater variability relative to the mean, while a lower value indicates greater consistency.

$$CV\ (\%) = 100 \cdot \frac{\sigma}{\mu} \tag{2}$$

where σ is the standard deviation and μ represents the average.

A Weibull model (Equations (3) and (4)) was used to assess the reliability of the mechanical test data. In this model, the variable x represents the measured physical and mechanical properties, i.e., density, tensile strength, Young's modulus, elongation at breaking, flexural strength, and modulus; β represents the shape parameter indicating whether the distribution follows an exponential trend ($\beta = 1$) or a polynomial trend ($\beta > 1$). The α is the scale parameter, reflecting the scale of the measured values. The function $F(x)$ is the probability density function describing the probability of each variable taking a particular value, which is directly related to the confidence level, where $R(x) = 1 - F(x)$. OriginLab 2021 software was used to fit the reliability curves for each parameter. The purpose of this method was to activate the 95% reliability level for all physical and mechanical parameters.

$$F(x) = 1 - exp[-(x/\alpha)]^{\beta} \tag{3}$$

$$ln\{ln[1/R(x)]\} = \beta \cdot ln(x) + \beta \cdot ln(\alpha) \tag{4}$$

2.5. Thermal Analysis

The samples were subjected to the thermogravimetric analysis (TGA) for thermal analysis. TGA was performed using Shimadzu (Kyoto, Japan) TGA-50 equipment, using a nitrogen atmosphere with a flow of 50 $mL \cdot min^{-1}$, starting from 25 up to 800 °C, a heating rate of 10 $°C \cdot min^{-1}$, and a platinum crucible. Prior to testing, the samples were oven-dried at 105 °C for 4 h to minimize the residual moisture content.

2.6. Dynamic Mechanical Thermal Analysis (DMTA)

All samples were subjected to DMTA in a dynamic mechanical analyzer from TA Instruments (Newcastle, WA, USA), Model Q800, with a dual cantilever mode. The storage modulus (E'), loss modulus (E''), and damping factor (tan δ) were obtained. The analysis was conducted from −130 °C to 150 °C with a heating rate of 5 $°C \cdot min^{-1}$. The deformation was set at 0.1%, and the frequency was 1 Hz. The samples were dried in an oven at 105 °C for 4 h before testing to reduce any remaining moisture content.

3. Results and Discussion

3.1. Composition

The composition of the materials tested is shown in Table 1. This table shows the average results of three measurements for each sample and the standard deviation. Cellulose presented lower variation and lower CV as a consequence. In terms of the results, it can be seen that for constituents with a higher proportion, there was less variability due to the greater sensitivity of the measurements and the reliability of their values, whereas for those with reduced proportions, there was less sensitivity and greater variability in the results.

Table 1. Chemical composition of the three wood species.

Sample	Cellulose (%)	*CV (%)	Hemicellulose (%)	*CV (%)	Lignin (%)	*CV (%)	Extractives (%)	*CV (%)	Ash (%)	*CV (%)
ARA	49.9 ± 0.3	0.7	9.1 ± 1.8	20.2	26.8 ± 3.3	12.4	3.48 ± 1.0	28.0	0.33 ± 0.1	35.8
DOD	45.9 ± 3.3	7.1	7.4 ± 0.1	1.3	31.8 ± 2.5	8.0	5.33 ± 0.6	11.3	0.30 ± 0.1	6.9
TOC	27.9 ± 2.9	9.4	9.7 ± 2.3	23.8	48.9 ± 0.9	1.9	4.98 ± 0.1	8.6	0.28 ± 0.1	15.1

*CV, coefficient of variation.

ARA and DOD had a higher cellulose content than TOC. In theory, the higher the cellulose content, the better the mechanical properties, because cellulose is made up of long, linear chains of glucose molecules linked together by strong hydrogen bonds, forming very stable and rigid structures that can improve the tensile strength of the wood. Another important aspect of these materials is their lignin content. TOC had the highest lignin content of the woods studied, reaching a total of 48.9%. Lignin is an amorphous biopolymer with a very complex structure compared with the other components. With its high molecular weight and aromatic rings, lignin is a natural material with high thermal stability, decomposing over a wide temperature range (200 °C to 500 °C). However, the degradation of some components of lignin can also occur at temperatures below 200 °C. Notably, lignin is often selected as a natural antioxidant in polymer applications, as reported by several authors [42–45]. DOD showed the highest extractive content, reaching 5.33%, but was similar to TOC, considering the standard deviation. The higher the concentration of extractives, the lower the thermal stability. The degradation of low molecular weight compounds accelerates the thermal degradation of the material. Finally, the ash content of all samples ranged from 0.28% to 0.33%.

The hemicellulose content was very similar for all woods, with ARA and DOD showing almost identical results of 9.1% and 9.7%, respectively. DOD showed a slightly lower hemicellulose content of 7.4%. Hemicellulose is considered to be the primary substance in wood cell walls, as it is the material that coats and shapes the cellulose structure and is responsible for the flexibility of lignocellulosic materials. Therefore, wood with a high hemicellulose content can have a more flexible structure and absorb more moisture. However, hemicellulose can act as an accelerator of degradation, reducing the thermal stability of these materials [2,46,47].

The results found for ARA were consistent with the literature, where Barros et al. (2021) [48] found that *Araucaria angustifolia* contained a total of 7% extractives, 34% lignin, 9% hemicellulose, and 46% α-cellulose.

3.2. Mechanical Tests and Density

The wood samples were subjected to mechanical tensile and bending tests, and density measurements. The results are presented in Table 2, where the average results of five sample measurements are shown.

Table 2. Tensile, flexural, and density properties of wood species.

Sample	Density ($g{\cdot}cm^{-3}$)	Tensile Strength (MPa)	Young's Modulus (GPa)	Elongation at Breaking (%)	Flexural Strength (MPa)	Flexural Modulus (GPa)
ARA	0.486 ± 0.031	67.67 ± 14.6	0.63 ± 0.5	12.9 ± 1.9	134.0 ± 18.7	9.85 ± 3.7
DOD	1.059 ± 0.017	65.05 ± 31.7	1.49 ± 0.3	6.8 ± 1.8	294.6 ± 38.5	30.46 ± 2.8
TOC	0.906 ± 0.012	112.70 ± 32.2	1.57 ± 0.2	10.0 ± 2.1	184.0 ± 25.4	13.64 ± 5.4

Considering that the variability of each result was high, Weibull methodology was applied to activate the values, with a reliability of 95%. The Weibull equation followed the procedure in [49]. The results presented in Figure 3 represent the values found at a 95% confidence level. In general, they were lower than the averages due to the influence of the variability of the replicates, in addition to the high level of rigor applied. The averages presented a reliability close to 60%, which was less conservative. However, the trends were the same as those observed for the average.

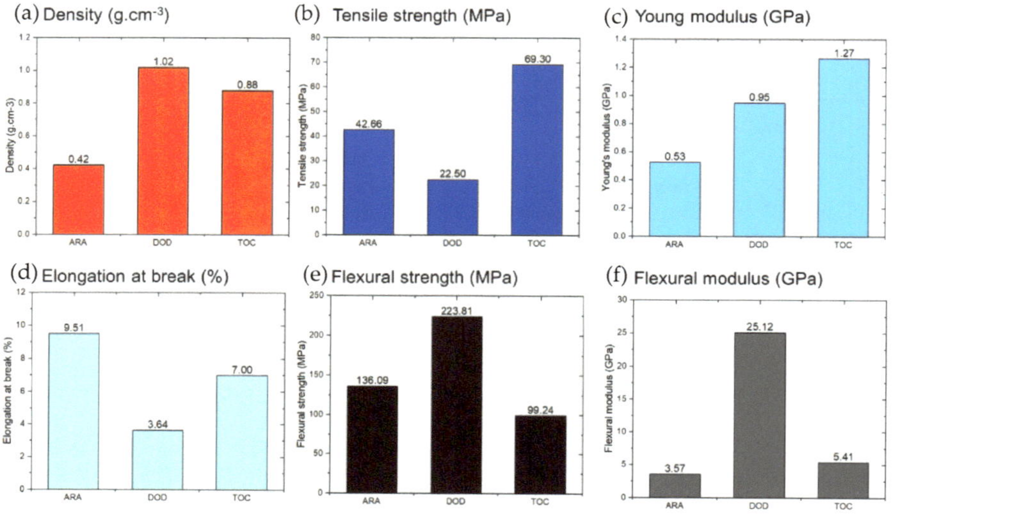

Figure 3. Weibull method reliability (95%) for (**a**) density, (**b**) tensile strength, (**c**) Young modulus, (**d**) elongation at break, (**e**) flexural strength, and (**f**) flexural modulus for different wood species.

TOC showed the highest values for tensile properties, reaching 69.30 MPa for tensile strength and 1.27 GPa for Young's modulus. DOD also showed high values for both properties, with a very similar value for Young's modulus. In the flexural tests, ARA still showed the lowest values of the species analyzed, with the most notable result being the flexural strength, which was 45% lower than DOD. These properties were probably most affected by the density of these materials. Table 3 shows that DOD and TOC had significantly higher densities than ASU, with 1.02 g·cm^{-3} and 0.88 g·cm^{-3}, respectively. In general, the cellulose content is critical in terms of the mechanical properties. However, TOC outperformed DOD in some of these properties, even though DOD had a higher cellulose content.

Table 3. Relaxation temperatures for ARA, DOD, and TOC woods.

Sample	Molecular Transitions		
	γ (°C)	β (°C)	α (°C)
ARA	−89.3	−23.9	103.7
DOD	−92.7	12.9	128.1
TOC	−95.0	2.2	135.9

The most notable difference was flexural strength, where DOD had a value of 223.81 MPa compared with ARA, which had 136.09 MPa, indicating a more rigid behavior. It can be observed that density has a more significant effect on mechanical performance than cellulose content in the case of the properties of TOC and ARA. Although ARA had a higher cellulose content, the density of TOC was almost twice that of ARA, suggesting a significant influence on the mechanical properties. The oven-dry density of ARA was up to twice that of DOD. Correspondingly, the flexural strength of ARA followed this trend and was significantly higher than that of DOD. However, the tensile strength of both materials was almost identical. This study did not focus on the fibers' orientation or grain structure, but these factors may account for the observed differences in the mechanical properties. In addition, the presence and variability of defects may have played a significant role [50,51]. In particular, the tensile strength of DOD had twice the standard deviation of ARA, suggesting that the variability of the DOD samples may have been greater than those of ARA.

The Pearson correlations between some of the woods' composition and mechanical properties is shown in Figure 4. Three wood samples were used to calculate the correlations among these properties.

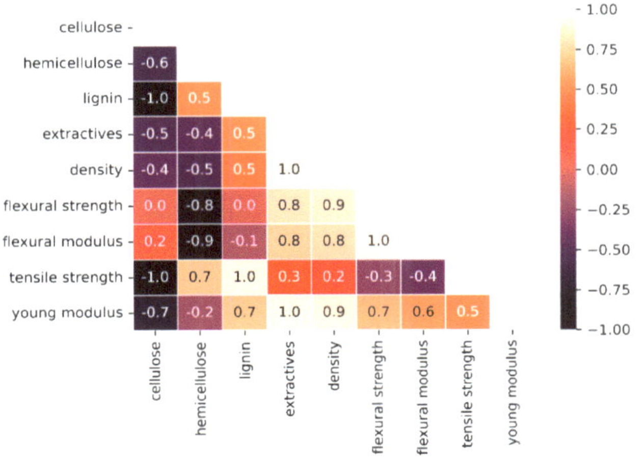

Figure 4. Heatmap of Pearson's correlations for woods' composition and mechanical properties.

Pearson's correlation coefficient measures the linear correlation between two variables and helps to identify variations in the woods' composition correlated with the mechanical properties. A correlation coefficient close to −1 or 1 indicates a high linear correlation between two variables, whereas a value close to −1 indicates a negative correlation. The graphic shows the correlation of each mechanical property with the thermal and other properties. The heat map shows a clear correlation between the density and mechanical properties of wood, particularly Young's modulus (0.9), flexural strength (0.8), and flexural modulus (0.8). In addition, hemicellulose showed a negative correlation with most of the mechanical properties, suggesting that hemicellulose plays a negative role in wood's properties when it comes to the mechanical properties investigated in our study (tensile strength, flexural strength, modulus of elasticity, bending stiffness, and elongation at breaking).

In contrast, in a study by Yu et al. [15], the authors claimed that hemicellulose determined the longitudinal shear strength of wood. This effect could be attributed to the chemical properties of hemicellulose. As a heteropolymer composed of sugar monomers that form a matrix around cellulose microfibrils, hemicellulose can act as a plasticizer, especially at higher temperatures [52].

3.3. Thermal Analysis

Thermal analysis was performed on all samples, and the TGA and differential thermogravimetric analysis (DTG) curves are shown in Figure 5.

Figure 5 shows the TGA and DTG of ARA samples, and three main degradation steps can be identified. An initial mass loss was observed around 100 °C, associated with releasing chemically bound water and some low molecular weight components. The first degradation step was observed around 200 °C and was mainly attributed to hemicellulose. Hemicellulose, being the component with the lowest thermal stability among the main elements of wood, contributed to this degradation step [2]. As a comparison of the thermal stability among the species, the temperature at which 10% mass loss occurred (T10) was measured at a heating rate of 10 °C·min^{-1}. For ARA, T10 was 275.5 °C and the temperature at which the sample reached its maximum degradation rate (Tmax) was 348.4 °C.

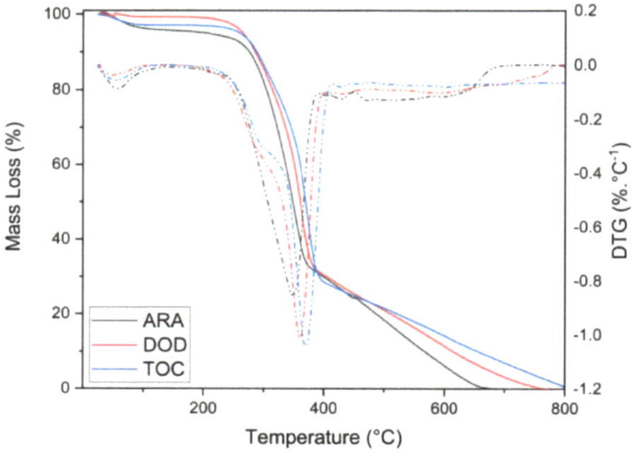

Figure 5. TGA (full lines) and DTG (dashed lines) curves for ARA (black lines), DOD (red lines), and TOC (blue lines).

A second stage occurred at around 370 °C and was closely linked to the first. This stage was mainly concerned with the degradation of cellulose. The thermal degradation of cellulose started in the amorphous phase and progressed to the crystalline phase as the temperature rose, due to its more stable structure. In the case of ARA, a third step was unclear, but it is very common for wood materials to present a third degradation step related to lignin. Lignin undergoes thermal degradation over a wide temperature range, initiating the degradation of less stable structures. Due to its complex structure of benzene–propane units, which are highly cross-linked and have a high molecular weight, lignin requires significantly more energy to degrade than hemicellulose and cellulose. Consequently, the degradation of lignin continues until the wood sample is entirely degraded.

Figure 5 shows the TGA and DTG for DOD. TGA clearly showed three degradation steps for DOD compared with ARA, a behavior that is more commonly observed in wood species. T10 occurred at 285.4 °C and Tmax at 360 °C. The first step, which started at around 200 °C, involved the degradation of low molecular weight compounds such as extractives and hemicelluloses. In the case of DOD, the degradation of hemicelluloses could be verified by DTG, where a shoulder was observed in the range of 210–315 °C. In the second stage, starting at around 320 °C and continuing up to 400 °C, cellulose degradation took place. Starting at 400 °C, the third stage is generally associated with lignin degradation. DOD showed a slightly higher thermal stability compared with ARA, with a shift in T10 and a slightly more pronounced difference in Tmax, with a variation of 12 °C. Although the composition of these two species was very similar, DOD had a higher lignin content, suggesting that the antioxidant behavior of lignin may have improved its thermal stability. However, the main difference between these species may be their density. As the density of DOD was twice that of ARA, its more compact cell structure may make the diffusion of degradation more difficult. Another relevant aspect is that ARA showed more pronounced water retention, although the samples were subjected to the same drying process before TGA. The water content may have facilitated the initiation of degradation, potentially reducing the thermal stability [27,53].

Similar to what was observed for ARA, the TGA curve of TOC in Figure 5 shows a mass loss below 100 °C, which is related to residual moisture that was not completely removed during the drying process. The general behavior observed was the same as that of DOD. For TOC, T10 occurred at 288.2 °C and Tmax at 370 °C. The first degradation step started at around 212 °C, where a shoulder attributed to hemicellulose could be observed for DTG. The second degradation step began at around 326 °C and extended to 415 °C, where the third and final step occurred. Compared with ARA and DOD, all temperatures observed for TOC were higher, indicating a higher thermal stability compared with ARA and DOD.

In brief, the three main components of wood fiber are hemicellulose, cellulose, and lignin. Hemicellulose consists of xylose, mannose, glucose, galactose, and other saccharides in an amorphous and branched structure. Cellulose is a glucose chain without branches and has higher thermal stability than hemicellulose. This property is achieved because cellulose has a higher chain order, resulting in a higher packing of the glucose chains. Finally, lignin also has branches with aromatic rings, resulting in degradation over a wide temperature range. In other words, hemicellulose degrades first (200–320 °C), followed by cellulose (320–400 °C) and lignin (150–900 °C). It is worth noting that the degradation events overlap at specific temperatures. Chemically, hemicellulose has a higher CO and CO2 yield, while lignin has a higher CH4 yield. The organic compounds are mainly released at lower temperatures (<500 °C) from hemicellulose and cellulose. Lignin has a low release of organic compounds below this temperature [54,55].

Several aspects of wood and lignocellulosic materials contribute to their thermal stability. The mass fraction ratio of the components plays an important role, with higher ratios of extractives and hemicelluloses leading to earlier degradation. In addition, the complex structure of lignin requires more energy to degrade. Therefore, if a lignocellulosic material has a higher lignin content, it may also have a higher thermal stability [27,53,56]. In the case of TOC, it had the highest lignin content, combined with its relatively high density. These properties likely contributed significantly to the observed thermal stability of the TGA results.

3.4. *Dynamic Mechanical Thermal Analysis (DMTA)*

The DMTA of the wood samples is presented in Figure 6. This technique was employed to investigate the viscoelastic properties of the studied woods. The graphic shows the storage (E′) (Figure 6a) and loss modulus (E″) (Figure 6b), which represent the elastic and viscous responses, respectively.

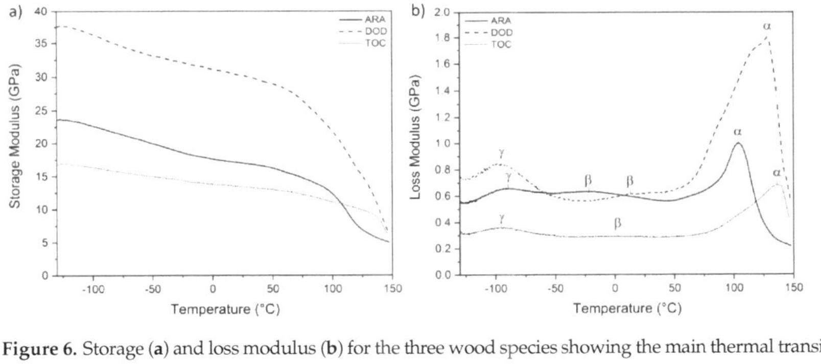

Figure 6. Storage (**a**) and loss modulus (**b**) for the three wood species showing the main thermal transitions.

Within the loss modulus (Figure 6b), three relaxation events could be observed for each wood sample. The first observed event was visualized in the range of −95 to −89 °C. The specific temperatures are given in Table 3 and are labelled γ. This third-order relaxation could be attributed to the methylol and hydroxymethyl groups in non-crystalline regions of hemicellulose or cellulose, which agrees with the observation of several authors [31,56,57].

At a higher temperature range, from −23.9 to 12.9 °C, the β relaxation, also a secondary event, was observed. The interpretation and location of the peaks for this event can be challenging, as they may overlap with the γ relaxation event. This event can be attributed to the loss of hemicellulose or the moisture content in the hemicellulose regions, as suggested by previous studies [57]. The observation of this peak may be more complicated, depending on the moisture content of the wood [31].

The last set of peaks, observed above 100 °C, can be related to the micro-Brownian motion of the polymer chains during the glassy to viscous transition. This phenomenon is generally associated with the glass transition (T_g) of lignin [56]. This was only detectable via DMTA because lignin and hemicellulose are primarily amorphous polymers and behave as thermoplastic polymers. Although cellulose is a semi-crystalline material, detecting the relaxation of hemicellulose and lignin is much easier than that of cellulose, and DMTA plays a crucial role in investigating the viscoelastic behavior of wood.

Only lignin T_g could be identified in the temperature range studied, as hemicellulose and amorphous cellulose showed relaxation at higher temperatures, above 200 °C. In Table 3, the α-relaxation was shifted to higher temperatures for DOD and TOC compared with ARA. This shift can be attributed to the higher lignin content in these two species, and it is possible that this result is correlated with the thermal behavior of the wood species. This finding may explain why these species exhibited higher thermal stability, as the lignin-softening temperature was higher, requiring more energy than the lignin T_g of ARA.

Figure 6a plots the storage modulus (E′), showing the differences between the three wood species analyzed, with DOD showing the highest stiffness. The numerical results are tabulated in Table 4, and DOD presented an E′ at 25 °C of 30.16 GPa, which was 44% and 55% higher than that of ARA and TOC, respectively. This result was consistent with the mechanical results, especially the flexural strength and modulus, where DOD showed the highest values compared with ARA and TOC. From the curve, it can be seen that the E values decreased with temperature. This was due to the increase in chains' mobility in the amorphous phase of the wood constituents [58].

Table 4. E′, E″, and tan δ measurements for ARA, DOD, and TOC woods.

Sample	E′ at 25 °C (GPa)	E″ at 25 °C (GPa)	Tan δ at 25 °C	Tan δ Peak Height	E′ at 25 °C (GPa)
ARA	16.97	0.57	0.034	0.089	16.97
DOD	30.16	0.62	0.020	0.130	30.16
TOC	13.43	0.29	0.022	0.082	13.43

The loss factor (tan δ) was also higher for DOD, as seen in Table 4 and Figure 7, indicating a more dissipative and elastic behavior, consistent with the observed E′ results [56].

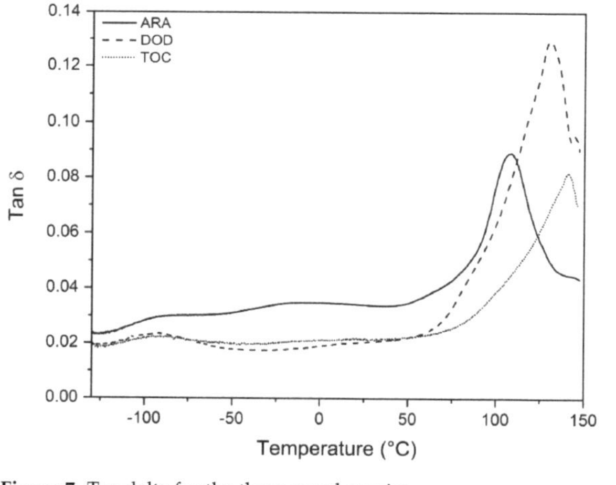

Figure 7. Tan delta for the three wood species.

The effect of wood's fiber components, hemicellulose, cellulose, and lignin is that they are more difficult to separate in dynamic mechanical thermal analysis. Different chemical and physical interactions in the wood fibers can affect the dynamic mechanical properties. In addition to the intrinsic properties of cellulose, hemicellulose, and lignin, other interactions can affect mechanical resistance, such as ester bonds to hemicellulose for ferulic and p-coumaric acid, ester and ether bonds between p-coumaric acid and lignin, ester and ether bonds between hemicellulose and lignin, hemicellulose cross-linking by diferulic acid, diferulic acid cross-linking (ester to polysaccharides and ether to lignin), and ferulic acid cross-linking (ester to polysaccharides and ether to lignin), among others. Knots, cross-grains, checks, splits, moisture, soil conditions, and growing space also affect the properties of wood [55].

4. Conclusions

The study investigated the composition, mechanical properties, thermal behavior, and viscoelastic properties of three Brazilian wood species: *Araucaria angustifolia* (ARA), *Dipterix odorata* (DOD), and *Tabeuia ochracea* (TOC). The species analyzed showed different compositions, with ARA and DOD having a higher cellulose content compared with TOC, while TOC had a higher lignin content. Although wood with a higher cellulose content typically has higher mechanical strength, this was not the case in this study. TOC presented the lower limit of cellulose content and exhibited higher strength and modulus than ARA, suggesting that wood density plays a critical role in mechanical properties, a conclusion supported by Pearson analysis. Thermal analysis indicated that ARA may undergo a two-step degradation process, whereas DOD and TOC undergo a three-step process. TOC showed the highest thermal stability due to its high lignin content and density.

Regarding the viscoelastic properties, DOD showed the highest E' and $\tan \delta$. Relaxation temperatures α, β, and γ were observed. The γ relaxations ranged from −95 to −89, β relaxations were from −23.9 to 12.9, and α relaxations were from 103.7 to 135.9. The storage modulus values that were corroborated by the observed mechanical properties were for DOD, which had the highest values for both. This study highlights the importance of understanding wood's composition and physical properties when assessing its mechanical and thermal properties. Density and composition can have a direct effect on these properties. Thus, this research contributes to selecting and using wood materials in various applications.

Author Contributions: Conceptualization, M.d.P.A., M.P. and A.J.Z.; methodology, M.d.P.A., M.P. and A.J.Z.; software, M.d.P.A.; writing—original draft preparation, M.d.P.A., H.L.O.J. and F.M.M.; writing—review and editing, M.P., H.L.O.J. and F.M.M.; supervision, M.P. and A.J.Z.; project administration, A.J.Z. All authors have read and agreed to the published version of the manuscript.

Funding: This research received no external funding.

Institutional Review Board Statement: Not applicable.

Data Availability Statement: The original contributions presented in the study are included in the article, further inquiries can be directed to the corresponding author.

Acknowledgments: The authors would like to thank Madereira Bianchi (Garibaldi, Brazil) for the kind donation of the *Araucaria angustifolia* (ARA), *Dipterix odorata* (DOD), and *Tabeuia ochracea* (TOC) woods. The authors would also like to thank CAPES and CNPq.

Conflicts of Interest: The authors declare no conflicts of interest.

References

1. Arriaga, F.; Wang, X.; Íñiguez-González, G.; Llana, D.F.; Esteban, M.; Niemz, P. Mechanical Properties of Wood: A Review. *Forests* **2023**, *14*, 1202. [CrossRef]
2. de Prá Andrade, M.; Piazza, D.; Poletto, M. Pecan Nutshell: Morphological, Chemical and Thermal Characterization. *J. Mater. Res. Technol.* **2021**, *13*, 2229–2238. [CrossRef]
3. Poletto, M.; Zattera, A.J.; Forte, M.M.C.; Santana, R.M.C. Thermal Decomposition of Wood: Influence of Wood Components and Cellulose Crystallite Size. *Bioresour. Technol.* **2012**, *109*, 148–153. [CrossRef] [PubMed]
4. Dong, X.; Gan, W.; Shang, Y.; Tang, J.; Wang, Y.; Cao, Z.; Xie, Y.; Liu, J.; Bai, L.; Li, J.; et al. Low-Value Wood for Sustainable High-Performance Structural Materials. *Nat. Sustain.* **2022**, *5*, 628–635. [CrossRef]
5. Anushikha; Gaikwad, K.K. Lignin as a UV Blocking, Antioxidant, and Antimicrobial Agent for Food Packaging Applications. *Biomass Convers. Biorefin* **2023**, *14*, 16755–16767.
6. Lu, X.; Gu, X.; Shi, Y. A Review on Lignin Antioxidants: Their Sources, Isolations, Antioxidant Activities and Various Applications. *Int. J. Biol. Macromol.* **2022**, *210*, 716–741. [CrossRef]

7. Mariani, A.; Malucelli, G. Transparent Wood-Based Materials: Current State-of-the-Art and Future Perspectives. *Materials* **2022**, *15*, 9069. [CrossRef]
8. Lahr, F.A.R.; Arroyo, F.N.; Rodrigues, E.F.C.; Almeida, J.P.B.; de Moura Aquino, V.B.; Wolenski, A.R.V.; Dos Santos, H.F.; Ferraz, A.L.N.; Chahud, E.; Molina, J.C.; et al. Models to Estimate Longitudinal Compressive Strength of Brazilian Hardwood Based on Apparent Density. *Bioresources* **2021**, *16*, 1373–1381. [CrossRef]
9. Báder, M.; Németh, R. Moisture-Dependent Mechanical Properties of Longitudinally Compressed Wood. *Eur. J. Wood Wood Prod.* **2019**, *77*, 1009–1019. [CrossRef]
10. Startsev, O.V.; Makhonkov, A.; Erofeev, V.; Gudojnikov, S. Impact of Moisture Content on Dynamic Mechanical Properties and Transition Temperatures of Wood. *Wood Mater. Sci. Eng.* **2017**, *12*, 55–62. [CrossRef]
11. Yang, X.; Berglund, L.A.; Yang, X.; Berglund, A. Structural and Ecofriendly Holocellulose Materials from Wood: Microscale Fibers and Nanoscale Fibrils. *Adv. Mater.* **2021**, *33*, 2001118. [CrossRef] [PubMed]
12. Furtos, G.; Molnar, L.; Silaghi-Dumitrescu, L.; Pascuta, P.; Korniejenko, K. Mechanical and Thermal Properties of Wood Fiber Reinforced Geopolymer Composites. *J. Nat. Fibers* **2022**, *19*, 6676–6691. [CrossRef]
13. Pertuzzatti, A.; Missio, A.L.; de Cademartori, P.H.G.; Santini, E.J.; Haselein, C.R.; Berger, C.; Gatto, D.A.; Tondi, G. Effect of Process Parameters in the Thermomechanical Densification of Pinus Elliottii and Eucalyptus Grandis Fast-Growing Wood. *Bioresources* **2018**, *13*, 1576–1590. [CrossRef]
14. Li, X.; Yue, K.; Zhu, L.; Lv, C.; Wu, J.; Wu, P.; Li, Q.; Xu, C.; Sun, K. Relationships between Wood Properties and Fire Performance of Glulam Columns Made from Six Wood Species Commonly Used in China. *Case Stud. Therm. Eng.* **2024**, *54*, 104029. [CrossRef]
15. Yue, K.; Qian, J.; Wu, P.; Jiao, X.; Lu, D.; Song, X. Experimental Analysis of Thermally-Treated Chinese Poplar Wood with Focus on Structural Application. *Ind. Crops Prod.* **2023**, *197*, 116612. [CrossRef]
16. Qian, J.; Yue, K.; Liu, S.; Lu, D.; Wu, P.; Li, Q. Augmenting Bamboo Strength and Thermal Stability for Sustainable Construction. *J. Clean. Prod.* **2024**, *451*, 142073. [CrossRef]
17. Pereira, P.H.F.; Ornaghi, H.L.; Arantes, V.; Cioffi, M.O.H. Effect of Chemical Treatment of Pineapple Crown Fiber in the Production, Chemical Composition, Crystalline Structure, Thermal Stability and Thermal Degradation Kinetic Properties of Cellulosic Materials. *Carbohydr. Res.* **2021**, *499*, 108227. [CrossRef]
18. Pereira, P.H.F.; Fernandes, L.L.; Santos, L.B.U.; Ornaghi, H.L.; Cioffi, M.O.H. Different Sequential Chemical Treatments Used to Obtain Bleached Cellulose from Orange Bagasse. *J. Nat. Fibers* **2022**, *19*, 12849–12861. [CrossRef]
19. Asim, M.; Paridah, M.T.; Chandrasekar, M.; Shahroze, R.M.; Jawaid, M.; Nasir, M.; Siakeng, R. Thermal Stability of Natural Fibers and Their Polymer Composites. *Iran. Polym. J.* **2020**, *29*, 625–648. [CrossRef]
20. Khan, M.Z.R.; Srivastava, S.K.; Gupta, M.K. A State-of-the-Art Review on Particulate Wood Polymer Composites: Processing, Properties and Applications. *Polym. Test.* **2020**, *89*, 106721. [CrossRef]
21. Sobek, S.; Werle, S. Kinetic Modelling of Waste Wood Devolatilization during Pyrolysis Based on Thermogravimetric Data and Solar Pyrolysis Reactor Performance. *Fuel* **2020**, *261*, 116459. [CrossRef]
22. Wang, W.; Luo, G.; Zhao, Y.; Tang, Y.; Wang, K.; Li, X.; Xu, Y. Kinetic and Thermodynamic Analyses of Co-Pyrolysis of Pine Wood and Polyethylene Plastic Based on Fraser-Suzuki Deconvolution Procedure. *Fuel* **2022**, *322*, 124200. [CrossRef]
23. Sharma, A.; Aravind Kumar, A.; Mohanty, B.; Sawarkar, A.N. Critical Insights into Pyrolysis and Co-Pyrolysis of Poplar and Eucalyptus Wood Sawdust: Physico-Chemical Characterization, Kinetic Triplets, Reaction Mechanism, and Thermodynamic Analysis. *Renew. Energy* **2023**, *210*, 321–334. [CrossRef]
24. Ali, S.; Hussain, S.A.; Mohd Tohir, M.Z.; Nuruddin, A.A. Thermogravimetric Analysis and Pyrolysis Kinetics of Malaysian Wood Species. *IOP Conf. Ser. Mater. Sci. Eng.* **2020**, *811*, 012003. [CrossRef]
25. Liu, H.; Li, M.; Zhao, S.; Mensah, R.A.; Das, O.; Jiang, L.; Xu, Q. Insights into Wood Species and Aging Effects on Pyrolysis Characteristics and Combustion Model by Multi Kinetics Methods and Model Constructions. *Renew Energy* **2023**, *206*, 784–794. [CrossRef]
26. Tabal, A.; Barakat, A.; Aboulkas, A.; El harfi, K. Pyrolysis of Ficus Nitida Wood: Determination of Kinetic and Thermodynamic Parameters. *Fuel* **2021**, *283*, 119253. [CrossRef]
27. Ornaghi, H.L.; Ornaghi, F.G.; Neves, R.M.; Monticeli, F.; Bianchi, O. Mechanisms Involved in Thermal Degradation of Lignocellulosic Fibers: A Survey Based on Chemical Composition. *Cellulose* **2020**, *27*, 4949–4961. [CrossRef]
28. Hill, C.; Altgen, M.; Rautkari, L. Thermal Modification of Wood—a Review: Chemical Changes and Hygroscopicity. *J. Mater. Sci.* **2021**, *56*, 6581–6614. [CrossRef]
29. Bal, B.C.; Bektaş, I. The Effects of Heat Treatment on Some Mechanical Properties of Juvenile Wood and Mature Wood of Eucalyptus Grandis. *Drying Technology* **2013**, *31*, 479–485. [CrossRef]
30. de Jesus, D.S.; Castro, J.P.; Guedes, T.; da Silva, J.R.M.; Latorraca, J.V.d.F.; Nascimento, A.M.d. Assessment of the Acoustic Change in Heat-Treated Pinus Elliottii and Bertholletia Excelsa Wood from Homogeneous Plantations in Brazil. *South. For. A J. For. Sci.* **2022**, *84*, 320–324. [CrossRef]
31. Sharma, M.; Brennan, M.; Chauhan, S.S.; Entwistle, K.M.; Altaner, C.M.; Harris, P.J. Wood Quality Assessment of Pinus Radiata (Radiata Pine) Saplings by Dynamic Mechanical Analysis. *Wood Sci Technol* **2015**, *49*, 1239–1250. [CrossRef]
32. Trevisan, R.; Zanella, A.; Da Silva, F.M.; Rosa, M.; Fioresi, T.; De Oliveira Fortes, F. Axial Variation of Basic Density of Araucaria Angustifolia Wood in Different Diameter Classes. *Ciência Rural* **2016**, *46*, 1969–1972. [CrossRef]
33. De Oliveiraa, R.M.; Brisolaria, A.; Salesb, A.; Gonçalvesa, D. Wettability, Shrinkage and Color Changes of Araucaria Angustifolia After Heating Treatment. *Materials Research* **2010**, *13*, 351–354. [CrossRef]

34. Medina, M.; Flores, M.P.; Ritter, L.J.; Goya, J.F.; Campanello, P.I.; Arturi, M.F. Wood Density and Leaf Traits Independently Relate to Growth Rate of Naturally Regenerated Tree Species in Araucaria Angustifolia Plantations in the Atlantic Forest, Argentina. *Can. J. For. Res.* **2023**, *54*, 1–11. [CrossRef]
35. de Sousa, B.C.M.; de Castro, S.P.; Lourido, K.A.; Kasper, A.A.M.; Paulino, G.d.S.; Delarmelina, C.; Duarte, M.C.T.; Sartoratto, A.; Vieira, T.A.; Lustosa, D.C.; et al. Identification of Coumarins and Antimicrobial Potential of Ethanolic Extracts of Dipteryx Odorata and Dipteryx Punctata. *Molecules* **2022**, *27*, 5837. [CrossRef]
36. de Oliveira Araüjo, S.; Vital, B.R.; Oliveira, B.; de Cássia Oliveira Carneiro, A.; Lourenço, A.; Pereira, H. Physical and Mechanical Properties of Heat Treated Wood from Aspidosperma Populifolium, Dipteryx Odorata and Mimosa Scabrella. Maderas. *Cienc. Y Tecnol.* **2016**, *18*, 143–156. [CrossRef]
37. TAPPI. *204 Solvent Extractives of Wood and Pulp (Proposed Revision of T 204 cm-97)*; TAPPI: Atlanta, GA, USA, 2007.
38. TAPPI. 222 om-02: Acid-insoluble lignin in wood and pulp. In *2002–2003 TAPPI Test Methods*; TAPPI: Atlanta, GA, USA, 2002.
39. *ASTM D790-17*; Standard Test Methods for Flexural Properties of Unreinforced and Reinforced Plastics and Electrical Insulating Materials. ASTM: West Conshohocken, PA, USA, 2017.
40. *ASTM D638-22*; Standard Test Method for Tensile Properties of Plastics. American Society for Testing and Materials: West Conshohocken, PA, USA, 2022.
41. *ASTM D792*; Standard Test Methods for Density and Specific Gravity (Relative Density) of Plastics by Displacement. American Society For Testing Materials: West Conshohocken, PA, USA, 2020.
42. Abdullah, T.; Colombani, T.; Alade, T.; Bencherif, S.A.; Memić, A.M. Injectable Lignin-Co-Gelatin Cryogels with Antioxidant and Antibacterial Properties for Biomedical Applications. *Biomacromolecules* **2021**, *22*, 4110–4121. [CrossRef]
43. Mahmood, Z.; Yameen, M.; Jahangeer, M.; Riaz, M.; Ghaffar, A.; Javid, I. Lignin as Natural Antioxidant Capacity. *Lignin—Trends Appl.* **2018**, *10*, 181–205.
44. Gadioli, R.; Waldman, W.R.; De Paoli, M.A. Lignin as a Green Primary Antioxidant for Polypropylene. *J. Appl. Polym. Sci.* **2016**, *133*, 43558. [CrossRef]
45. Trevisan, H.; Rezende, C.A. Pure, Stable and Highly Antioxidant Lignin Nanoparticles from Elephant Grass. *Ind. Crops Prod.* **2020**, *145*, 112105. [CrossRef]
46. Hosseinaei, O.; Wang, S.; Enayati, A.A.; Rials, T.G. Effects of Hemicellulose Extraction on Properties of Wood Flour and Wood–Plastic Composites. *Compos. Part. A Appl. Sci. Manuf.* **2012**, *43*, 686–694. [CrossRef]
47. Salmén, L. On the Organization of Hemicelluloses in the Wood Cell Wall. *Cellulose* **2022**, *29*, 1349–1355. [CrossRef]
48. de Sá Barros, S.; Pessoa, W.A., Jr.; Júnior, A.C.; Borges, Z.V.; Poffo, C.M.; Regis, D.M.; de Freitas, F.A.; Manzato, L. Value Aggregation of Pine (Araucaria Angustifolia) Nuts Agro-Industrial Waste by Cellulose Extraction. *Res. Soc. Dev.* **2021**, *10*, e270101018836. [CrossRef]
49. Monticeli, F.M.; Almeida, J.H.S., Jr.; Neves, R.M.; Ornaghi, F.G.; Ornaghi, H.L. On the 3D Void Formation of Hybrid Carbon/Glass Fiber Composite Laminates: A Statistical Approach. *Compos. Part. A Appl. Sci. Manuf.* **2020**, *137*, 106036. [CrossRef]
50. Xie, Y.; Wang, Q.; Fu, Q.; Wang, Q.; Xiao, Z.; Militz, H. Effects of Chemical Modification on the Mechanical Properties of Wood. *Eur. J. Wood Wood Prod.* **2013**, *71*, 401–416. [CrossRef]
51. Jakob, M.; Mahendran, A.R.; Gindl-Altmutter, W.; Bliem, P.; Konnerth, J.; Müller, U.; Veigel, S. The Strength and Stiffness of Oriented Wood and Cellulose-Fibre Materials: A Review. *Prog. Mater. Sci.* **2022**, *125*, 100916. [CrossRef]
52. Agustin-Salazar, S.; Cerruti, P.; Medina-Juárez, L.Á.; Scarinzi, G.; Malinconico, M.; Soto-Valdez, H.; Gamez-Meza, N. Lignin and Holocellulose from Pecan Nutshell as Reinforcing Fillers in Poly (Lactic Acid) Biocomposites. *Int. J. Biol. Macromol.* **2018**, *115*, 727–736. [CrossRef]
53. Cavinato, C.D.; Poletto, M.; Cavinato, C.D.; Poletto, M. Kinetic Analysis of Thermal Degradation of Cedrela Odorata, Marmaroxylon Racemosum and Tectona Grandis from Timber Industry. *Maderas. Ciencia tecnología* **2021**, *23*, 1–10. [CrossRef]
54. Yang, H.; Yan, R.; Chen, H.; Lee, D.H.; Zheng, C. Characteristics of Hemicellulose, Cellulose and Lignin Pyrolysis. *Fuel* **2007**, *86*, 1781–1788. [CrossRef]
55. Zhang, J.; Choi, Y.S.; Yoo, C.G.; Kim, T.H.; Brown, R.C.; Shanks, B.H. Cellulose-Hemicellulose and Cellulose-Lignin Interactions during Fast Pyrolysis. *ACS Sustain. Chem. Eng.* **2015**, *3*, 293–301. [CrossRef]
56. Broda, M.; Spear, M.J.; Curling, S.F.; Dimitriou, A. Effects of Biological and Chemical Degradation on the Properties of Scots Pine—Part II: Wood-Moisture Relations and Viscoelastic Behaviour. *Forests* **2022**, *13*, 1390. [CrossRef]
57. Ashaduzzaman, M.; Hale, M.D.; Ormondroyd, G.A.; Spear, M.J. Dynamic Mechanical Analysis of Scots Pine and Three Tropical Hardwoods. *Int. Wood Prod. J.* **2020**, *11*, 189–203. [CrossRef]
58. Jambreković, B.; Bajsić, E.G.; Španić, N.; Sedlar, T.; Sinković, T. Viscoelastic and Thermal Properties of Styrene Modified Fir Wood. *Polymers* **2022**, *14*, 786. [CrossRef] [PubMed]

polymers

MDPI

Article

Surface Roughness, Dynamic Wettability, and Interphase of Modified Melamine Formaldehyde-Based Adhesives on Jabon Wood

Yusup Amin [1,2,*], Naresworo Nugroho [2], Effendi Tri Bahtiar [2], Wahyu Dwianto [1], Muhammad Adly Rahandi Lubis [1], Ulfa Adzkia [2] and Lina Karlinasari [2,*]

1 Research Center for Biomass and Bioproducts, National Research and Innovation Agency, Cibinong 16911, Indonesia; wahy002@brin.go.id (W.D.); muha142@brin.go.id (M.A.R.L.)

2 Department of Forest Products, Faculty of Forestry and Environment, IPB University, Bogor 16680, Indonesia; nares@apps.ipb.ac.id (N.N.); bahtiar_et@apps.ipb.ac.id (E.T.B.); ulfaadzkia@apps.ipb.ac.id (U.A.)

* Correspondence: yusu007@brin.go.id (Y.A.); karlinasari@apps.ipb.ac.id (L.K.)

Abstract: The surface roughness and wettability of wood are critical aspects to consider when producing laminated wood products with adhesive applications. This study aims to investigate the surface roughness and dynamic wettability of Jabon wood in the presence of melamine formaldehyde (MF)-based adhesives. Commercial MF adhesives (MF-0) and modified MF adhesives (MF-1) were applied to Jabon wood, which includes tangential (T), radial (R), and semi-radial (T/R) surfaces. The surface roughness of Jabon wood was assessed using a portable stylus-type profilometer. The low-bond axisymmetric drop shape analysis (LB-ADSA) method was employed to identify the contact angle (θ) of the MF-based adhesives on Jabon wood. The wettability was determined by evaluating the constant contact angle change rate (K value) using the Shi and Gardner (S/G) model. Dynamic mechanical analysis (DMA) was employed to investigate the viscoelastic characteristics of the interphase analysis of the wood and MF-based adhesives. The roughness level (Ra) of the Jabon board ranged from 5.62 to 6.94 μm, with the T/R having a higher level of roughness than the R and T. MF-0 exhibited a higher K value (0.262–0.331) than MF-1 (0.136–0.212), indicating that MF-0 wets the surface of Jabon wood more easily than MF-1. The wood–MF-0 interphase reached a maximum stiffness of 957 N/m at 123.0 °C, while the wood–MF-1 had a maximum stiffness of 2734 N/m at 110.5 °C. In addition, the wood–MF-0 had a maximum storage modulus of 12,650 MPa at a temperature of 128.9 °C, while the wood–MF-1 had a maximum storage modulus of 22,950 MPa at 113.5 °C.

Keywords: contact angle analysis; dynamic mechanical analysis; melamine formaldehyde; Jabon wood; wood–adhesive interphase

Citation: Amin, Y.; Nugroho, N.; Bahtiar, E.T.; Dwianto, W.; Lubis, M.A.R.; Adzkia, U.; Karlinasari, L. Surface Roughness, Dynamic Wettability, and Interphase of Modified Melamine Formaldehyde-Based Adhesives on Jabon Wood. *Polymers* **2024**, *16*, 1084. https://doi.org/10.3390/polym16081084

Academic Editor: Ming Zhang

Received: 23 January 2024
Revised: 24 March 2024
Accepted: 26 March 2024
Published: 12 April 2024

1. Introduction

Adhesive quality is a critical component that must be meticulously evaluated during the production of laminated timber products [1]. In the manufacturing of wood products, surface preparation is a critical step preceding the gluing or finishing procedures [2–5]. Sanding is a method that is frequently employed in this process. The wettability of wood is directly influenced by the texture of its surface, which is itself altered by sanding [6,7]. The properties of the liquid being applied and the characteristics of the wood itself are the two primary determinants of liquid penetration into wood [8,9]. Furthermore, Walinder (2000) [9] states that investigations into wetting phenomena on wood surfaces could contributefundamental knowledge regarding a wood material and its intricacies, in addition to being essential for comprehending the interaction between wood and substances such as adhesives or coatings. Wetting refers to the phenomenon that occurs when a liquid makes contact with a solid surface [10,11]. In other terms, wetting is an intermolecular interaction that occurs when liquids and solids come into direct contact at their interfaces [9]. So, the roughness and molecular physicochemical properties of solid surfaces reatlyaffect wetting phenomena and the transmission of liquids [12,13].

The penetration of a liquid into wood pores is a crucial aspect of bond formation since wood is a porous material [10]. The principles of adhesive wetting on a wood surface include the creation of a contact angle at the solid–adhesive interface, the spread of the adhesive over a solid surface, and the penetration of the adhesive into the porous solid substrate [10,14]. In several recent investigations [3,11,15–20], the wettability qualities of wood were determined by measuring the contact angle between a liquid and the surface of the wood. As an anisotropic material, the adhesive wetting properties of wood may be different along and across the wood grain direction [8,10]. Numerous factors influence wood wettability, including surface roughness and machining conditions [14,21], wood species and the location of wood [10,22], and treatment and drying methods [23]. In addition, Yuan and Lee [24] have reported that the factors of wood characteristics and a liquid's properties affect wettability.

Numerous adhesives can be used to fabricate laminated wood products such as urea formaldehyde (UF), phenol formaldehyde (PF), melamine formaldehyde (MF), melamine–urea formaldehyde (MUF), emulsion polymer isocyanate (EPI), and polyurethane [25–29]. Among them, MF adhesives are one type of adhesives ordinarily used to produce laminated wood products [1,26,27,29], such as cross-laminated timber (CLT), glued laminated timber (GLT), and laminated veneer lumber (LVL). Alongside PF adhesives [25], MF adhesives are extensively used in the production of semi-exterior- and exterior-grade wood-based panels [29]. These adhesives are widely selected by glulam or CLT producers because of their relatively low costs, transparent bonding lines, excellent durability, heat resistance, and resistance to water and moisture [30,31]. MF, being a thermosetting glue, is typically utilized in conjunction with hot presses, resulting in higher costs associated with heat energy consumption. Modifying formaldehyde-based adhesives is a remarkably effective strategy for increasing the economic advantages of these adhesives and preserving the environment [25]. A recent innovation concerning the modification of MF-based adhesives for CLT production from Jabon wood by means of cold-pressing application has been successfully developed [1]. Nevertheless, the specific attributes of the bonding mechanism between Jabon wood and MF-based adhesives have yet to be discovered. The bonding quality of a laminated wood product is influenced by the surface roughness and wettability qualities of the wood. This study aims to investigate the surface roughness and wettability of Jabon wood (*Anthocephalus cadamba* Roxb. Miq.) in the presence of an MF-based adhesive.

2. Materials and Methods

2.1. Materials

The two main materials used in this study are Jabon wood and MF resin as an adhesive. Jabon wood (*Anthocephalus cadamba* Roxb. Miq.) with a density of 0.44 g/cm^3 and a moisture content of 12 ± 2% was obtained from a community forest located in the Bogor Region, West Java. Meanwhile, the commercial melamine adhesive (MA-204) was supplied by PT Pamolite Adhesive Industry, Probolinggo, East Java, Indonesia; Polymeric 4,4-methylene diphenyl diisocyanate (pMDI) was supplied by PT Anugerah Raya Kencana, Tangerang, Banten, Indonesia. Additional materials such as wheat flour and technical-grade citric acid powder were either obtained from a commercial market or provided by the Integrated Laboratory of Bioproducts (iLab), National Research and Innovation Agency (BRIN), Cibinong, Indonesia.

2.2. Methods

2.2.1. Surface Roughness Test

Jabon lumbers with a dimension of 90 mm × 40 mm × 15 mm in length, width, and thickness were used for measuring surface roughness. In conformance with ISO 1997, the surface roughness of Jabon wood lumbers was determined using a portable stylus-type profilometer (Mitutoyo Surftest® SJ-210, Mitutoyo Corporation, Kanagawa, Japan), after sanding with a 100-grit belt sander (P100). The surface roughness measurement of the wood specimens was performed perpendicular to the fiber direction in three different positions (Figure 1) by a diamond tip with a radius of 5 mm, a tracing length of 6 mm, a cut off of 0.8 mm, and a speed of 0.5 mm/s [3,15,32]. The arithmetical mean roughness value (Ra) was used to assess the surface roughness level [3,4,11,14,15]. Ra measurements were conducted on tangential (T), radial (R), and semi-radial (T/R) cross-sections of Jabon lumbers, with each kind of lumber being tested over three replicas. The Keyence VHX 6000 Digital Microscope (Keyence Corporation, Osaka, Japan) was used for observing the surface topography profile of Jabon wood on three different surfaces. This microscope can generate high-quality digital images [33]. The ultrasmall zoom lens models VH-Z20T and VH-250T dual-light enable high-magnification recording, with magnification capabilities of 20–200× and 250–2500×, respectively.

Figure 1. The surface roughness measurement of Jabon wood specimens in three different positions using a portable stylus-type profilometer (Mitutoyo Surftest® SJ-210).

2.2.2. Contact Angle Measurement

Commercial melamine formaldehyde adhesives (MF-0) and modified MF adhesives (MF-1) were chosen to investigate the wettability of adhesives on Jabon wood. In this research, MF-1 was composed, starting from MF-0, by adding citric acid (20 wt%) as a catalyst, pMDI (96.2 wt%) as a cross-linker, and wheat flour as a filler [1]. Citric acid and pMDI were added based on the solids content of the control adhesive: 5% and 3%, respectively. Wheat flour made up as much as 10% of the total mixture. The resulting mixture was then manually stirred for 1–2 min at 27 ± 2 °C, yielding a cold-setting melamine-based adhesive. According to a previously published study, the fundamental properties of the adhesives used in this study are shown in Table 1 [1].

Table 1. Basic properties of MF-based adhesives [1].

Type of Adhesive	Properties			
	Solids Content (%)	Gelation Time (min, T = 100 °C)	Viscosity (cPs, T = 25 °C)	pH
MF-0	51.89 ± 0.36	30.27 ± 0.15	484.51 ± 3.69	7.67 ± 0.58
MF-1	48.56 ± 0.08	6.90 ± 0.10	570.75 ± 14.80	6.67 ± 0.58

The evaluation of wettability was performed through the measurement of the dynamic contact angle between the MF-based adhesives and the surface of Jabon wood. Distilled water (aquades) was employed as a control. Each of these liquids, i.e., MF-0, MF-1, and aquades, were dripped using a micropipette and a syringe onto the Jabon wood surface, employing the same specimen and point location as the surface roughness test. The sessile droplet volume used for the experiment was 0.02 mL for aquades [3,32] and about 0.03 mL for the MF-based adhesives [11,15].

The experiment involved using a Dino-Lite Digital Microscope Basic Am 2111 series (AnMo Electronics Corporation, New Taipei City, Taiwan) with a USB camera connected to a personal computer (PC) to record videos (Figure 2). A Dino-Capture tool with a magnification of 40× was utilized to record the process of droplet distribution and absorption from the initial placing until the water penetrated the wood surface. The video data in a Windows media video (WMV) format were further processed using the GOM Player software (GOM Player 2.3.90.5360) to extract visual segments at intervals of 3 s, resulting in seven individual images, for a total duration of 18 s. As in the surface roughness test, the contact angle measurements were conducted on T, R, and T/R surfaces of Jabon wood in three replicas, respectively. The contact angle (θ) of an individual drop image was determined using the low-bond axisymmetric drop shape analysis (LB-ADSA) method, provided by the Image-J software (ImageJ 1.46r) [34]. In contrast to predominant droplet shape analyses, such as the drop-snakes analysis method [3,4,8], the LB-ADSA method fails to distinguish between left and right droplet image contours and contact angles. The approach applied by LB-ADSA is that the contact angles on the left and right sides of an object are identical (Figure 3), resulting in a reflection resembling a mirror on both sides [34].

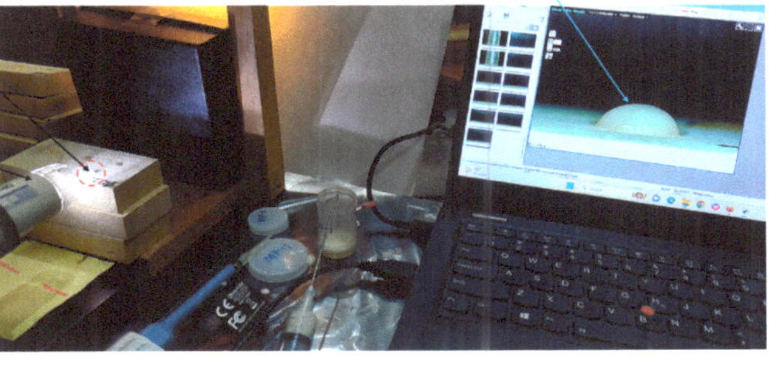

Figure 2. Capturing droplet distribution and change patterns with a Dino-Lite Digital Microscope (Am 2111 series) adjusted for 40× magnification.

2.2.3. Determination of Equilibrium Contact Angle and Constant Contact Angle Change Rate

When the liquid was first poured on the wood surface, the droplet was normally circular and gradually flattened over time [15]. With time, the drop shape tended to stabilize, resulting in an equilibrium contact angle (θ_e). The change in the contact angle as a function of time was determined using a segmented regression model between the time (x) and the contact angle (y), using the SAS STAT's PROC NLIN [3,11,15]. The rate of contact angle change is proportional to the rate of liquid penetration and spread across a solid surface [10,11]. In this study, wettability was quantitatively assessed by examining the constant contact angle change rate (*K* value) on the S/G model [10]. The *K* value indicates the rate at which a liquid distributes and permeates into the porous structure of wood. Increasing values of K correspond to a reduced time needed for the contact angle to reach a state of relative equilibrium and for the liquid to spread and permeate. The S/G model is explained by Equation (1):

$$\theta = \frac{\theta_i \times \theta_e}{\theta_i + (\theta_e - \theta_i)\exp\left[K\left(\frac{\theta_e}{\theta_e - \theta_i}\right)t\right]} \quad (1)$$

where θ represents the contact angle at a certain time, θ_i is the initial contact angle, θ_e is equilibrium contact angle, t is the wetting time, and K represents the constant contact angle change rate. A non-linear regression model was used to calculate the K-value using the defined function to fit the S/G equation via XLSTAT Addinsoft (XLSTAT 2014.5.03) [3,11].

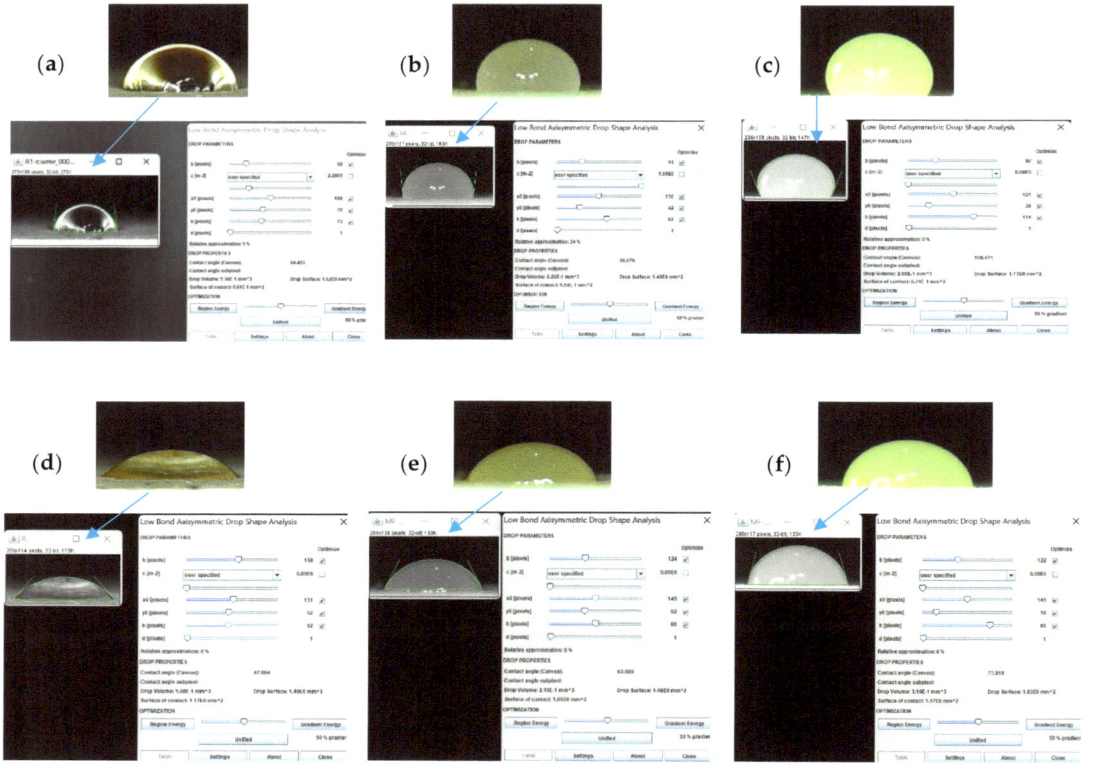

Figure 3. Contact angle measurements by LB-ADSA method: (**a**) aquades at 0 s; (**b**) MF-0 adhesive at 0 s; (**c**) MF-1 adhesive at 0 s; (**d**) aquades at 3 s; (**e**) MF-0 adhesive at 3 s; and (**f**) MF-1 adhesive at 3 s.

2.2.4. Wood–Adhesives Interphase Analysis Using Dynamic Mechanical Analyzer

Dynamic mechanical analysis (DMA) can be employed to investigate the viscoelastic characteristics of the interphase analysis of wood and adhesives. Most polymeric materials or polymeric interfacial phases exhibit a viscoelastic behavior, which combines the characteristics of both solids (elasticity) and liquids (viscosity) [35–37]. Dynamic mechanical analysis (DMA 8000, Perkin Elmer Inc., Waltham, MA, USA) was utilized to investigate the MF resin and Jabon wood interphase. Each MF adhesive was used to bond two thin Jabon wood veneers (50 mm × 10 mm × 0.5 mm) with a glue spread of 300 g/m^2. All the specimens were precured in an oven at 50 °C for 10 min before the DMA analysis. The storage modulus (E′), loss modulus (E″), and tan delta of each specimen were determined at a frequency of 1 Hz, a strain level of 0.01%, and a heating rate of 1 °C/minin the scanning range of 25–300 °C in the dual-cantilever mode [38].

2.2.5. Mechanical Properties of Wood–MF Composites

Block shear strength samples were prepared by gluing Jabon wood with MF-0 and MF-1 resins. Two-ply composites were made via cold-pressing at 1 MPa for 2 h at different glue spreads of 250, 280, and 300 g/m^2. Furthermore, the composites were conditioned for a week prior to block shear testing. A block shear strength analysis was undertaken to determine the bonding strength. The block shear samples were tested using a 50 kN universal testing machine (UTM AG-IS 50 kN, Shimadzu, Kyoto, Japan) with a crosshead speed of 2 mm/min.

3. Results and Discussion

3.1. Surface Roughness of Jabon Wood

The result in Table 2 shows the surface roughness value of Jabon wood after sanding with a 100-grit belt sander (P-100) on different surface types, determined using a portable stylus-type profilometer. The surface roughness of solid materials is generally measured using three parameters: average surface roughness (Ra), root mean square roughness (Rq), and ten-point mean roughness (Rz) [39]. However, Ra is the most used metric for determining wood surface roughness. Consequently, the Ra parameter was employed in this investigation to quantify the surface irregularity of Jabon wood. An approach methodology utilizing Ra values has been implemented in prior studies [3–5,14,15,39,40]. The average roughness (Ra) of a profile is calculated by averaging the individual depths and heights (irregularities) of its arithmetic mean elevation. In addition, Ra represents the mean deviation of the profile from the mean line, determined across the entire length of the assessment [41]. Jabon wood treated with a P-100-grit sander had an Ra of 5.62 μm on the tangential surface (T), 5.77 μm on the radial surface (R), and 6.94 μm on the semi-radial surface (T/R). The tangential surface had a lower Ra value than the radial or semi-radial surfaces, indicating that the roughness level of Jabon wood on a tangential surface was better than that of the radial and semi-radial surfaces. The findings of this study aligned with prior studies [41] indicating that the radial surface of rubberwood exhibited a rougher appearance compared to its tangential surface. The average Ra value of the three types of Jabon wood surfaces investigated in this study (6.11 μm) is comparable to prior research [15], which found that the average Ra value of Jabon wood before heat treatment was 6.22 m.

Table 2. Surface roughness of Jabon wood after being sanded with a 100-grit belt sander.

Roughness Parameter	Tangential Surface (T)	Radial Surface (R)	Semi-Radial Surface (T/R)
Ra (mm)	5.62 ± 1.54	5.77 ± 1.65	6.94 ± 1.63
Rq (mm)	7.27 ± 2.13	7.14 ± 2.13	8.68 ± 1.89
Rz (mm)	33.21 ± 9.52	32.09 ± 10.64	39.09 ± 7.07

The evaluation of surface quality can be accomplished through the utilization of topographic measurements [42]. Figure 4 shows the 3D surface topographical profile of Jabon wood after it was sanded with a 100-grit belt sander (P100) and analyzed with a digital microscope (Keyence VHX 6000, Keyence Corporation, Osaka, Japan). The tangential surface of Jabon wood (Figure 4a) had a lower maximum peak height value than the radial (Figure 4b) and semi-radial (Figure 4c) surfaces. This finding is comparable with the results obtained using the stylus method mentioned previously and in line with the results of previous studies [43]. Visually, Figure 4b,c demonstrate that the radial surface has greater amounts of pattern (color) changes than the semi-radial surface, rendering it rougher. However, it is crucial to highlight that significant fluctuations in the surface roughness profile do not imply a rougher surface when compared with a surface roughness profile with slight variations [43]. Based on a 3D analysis of the surface topographical profiles, the maximum value of the peak height has a greater impact on wood surface roughness than the frequency of color change patterns.

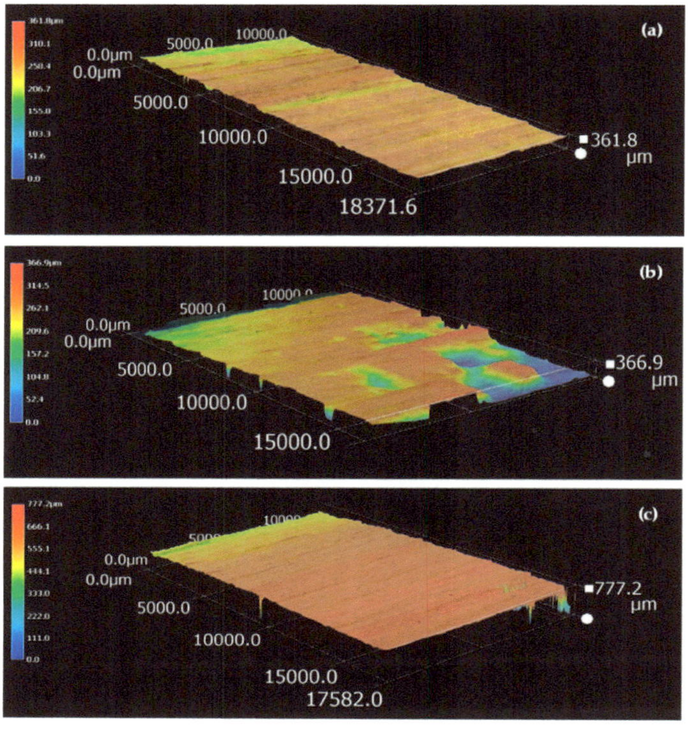

Figure 4. Three-dimensional surface topography profile of Jabon wood after being sanded with a 100-grit belt sander (P100) at a 100× magnification: (**a**) tangential surface; (**b**) radial surface; and (**c**) semi-radial surface.

The surface properties of wood are essential in the manufacturing processes of wood products, such as adhesive bonding or finishing [5,41]. Wetting analyses, topography measurements, and cell damage evaluations can all help to determine surface quality. Numerous processing techniques used on wood can influence the structure, morphology, and chemical composition of the wood surface, hence altering the ability of liquids to wet the wood [14]. Different wood machining techniques result in various levels of liquid wettability on wood. Previous research [21] investigated the effect of surface conditions caused by various machining methods on the wettability properties of Mediterranean wood species. The same authors reported that sanded surfaces exhibited a significantly higher wettability compared to planed or disc-sawn surfaces. Sanding is a necessary and time-consuming procedure in the woodworking industry [6]. Sanded wooden surfaces are often varnished or glued in the manufacturing of furniture products. Several studies have been conducted to investigate the effect of sander grit size on the level of roughness of a wood surface [4,7,11,41]. Generally, the surface roughness of wood decreases as the grit number of the abrasive paper increases [11,41]. The sanding technique, whether applied tangentially, radially, or semi-radially, serves to refine the surface cells and make them smoother. The sanded surface disintegrates, creating fine particles (dust) which will occupy the pores on the surface [41]. A wood structure reveals variations in its constituents across its tangential, radial, and longitudinal surfaces [8]. An example of this distinction is the variation in the orientation of ray cells. The tangential surface is considerably smoother (lower Ra) than the radial surface. Possibly, the radial surface presents a greater obstacle during measurements than the tangential surface; this indicates that the radial surface is considerably coarser in texture [41]. It has been reported in other studies that surface roughness does not invariably diminish as sandpaper grain size increases [4]. Wood surface roughness parameters do not indicate that sanding with an abrasive of a smaller grain size results in a smoother surface [7]. Therefore, in our study, it was determined that 100-grain sandpaper (P-100) would be adequate to produce Jabon lumber with a suitable surface for applying MF-based adhesives in the manufacturing of laminated wood products. Previous investigations [44,45] verified a substantial rise in surface roughness parameters when P-100-grit sandpaper was utilized. The abrasive grains formed deep grooves, peak heights, and cell wall fibrillation on the wood surface, accelerating the dispersion of liquid on said surface. In line with Dai et al. (2019) [46], Niaraki and Krause [47] reported that the wood surfaces developed a larger number of interface areas after sanding. Consequently, adhesives are capable of more readily penetrating a wood's depths and dispersing more uniformly across its surface.

3.2. Contact Angle and Dynamic Wettability

The capability of a liquid to generate a contact interface with a solid surface is referred to as wetting [48]. The wetting characteristics of wood are classified based on the contact angle (θ) formed between liquid droplets and the surface of the wood [11,15,48,49]. The roughness of a wood surface is directly linked to its wettability. A higher surface roughness corresponds to a greater surface hydrophilicity, resulting in more effective wetting and a lower contact angle [3,11,24,50].

The wettability of wood has a role in the construction of an adhesive system. The wettability of the aquades and MF-based adhesives applied to various Jabon wood surfaces in our study could be quantitatively investigated with the S/G model [10,11]. Table 3 shows the initial contact angles (θ_i), equilibrium contact angles (θ_e), contact angle reduction ratio, *K* value, and R squared values for all the wood surfaces and liquid treatments examined. The change in contact angle over time during the liquid absorption process is a reducing function [10]. Liquid penetration and spreading occur in conjunction with the formation of a contact angle at the wood surface when a liquid drop is applied on it. In the early phases, until the third minute of the wetting process, there is a rapid decrease in the contact angle of a liquid drop (Figure 5). The aquades treatment exhibits the highest average contact angle reduction ratio (64.35–72.16%) compared to MF-0 and MF-1. The contact angle reduction ratio of MF-0 (42.83–51.18%) is slightly higher than that of MF-1 (41.21–44.55%). The effect angle gradually diminishes over time, ultimately achieving a state of relative equilibrium [10]. As shown in Table 3 and Figure 5, generally, adhesives based on MF exhibit higher initial contact angles and equilibrium contact angles than those of aquades. In the aquades treatment, the equilibrium contact angle is reached after 8.46–11.13 s. Meanwhile, the treatments MF-0 and MF-1 take 5.25–5.95 and 7.80–11.50 s, respectively, to achieve the equilibrium contact angle. In addition, according to [51], an equilibrium contact angle is achieved on a surface when an adhesive's adhesion is counterbalanced by the solid surface's tension over time. The time required to achieve the equilibrium contact angle is impacted by various parameters, such as the initial contact angle, the contact angle reduction ratio, and the properties of the liquid employed. When addressing aquades or MF-based adhesives, the time required to reach the contact angle tends to be longer, as the initial contact angle increases and the contact angle reduction ratio decreases. By treating aquades, decreasing the time required to reach the equilibrium contact angle tends to enhance the *K* value. This indicates that the wood surface becomes easily wet [4,11,52]. During the MF-based adhesive treatments, MF-1 takes more time to reach the equilibrium contact angle compared to MF-0. The viscosity difference between the MF-0 and MF-I adhesives also exhibits an effect, alongside the initial contact angle and contact angle reduction ratio. The higher viscosity of MF-1 compared to MF-0 [1] hinders the absorption of the adhesive into the wood surface, resulting in a longer time required to reach the equilibrium contact angle.

Table 3. Non-linear fitting results for aquades and MF-based adhesives on Jabon wood surfaces.

Liquid	Surface Type	Time to Reach the Equilibrium Contact Angle	θ_i	θ_e	Contact Angle Reduction Ratio	Wettability	
		(s)	(°)	(°)	(%)	*K* Value	R^2
Aquades	T	11.13	75.5	26.914	64.35	0.269	0.981
	R	8.46	84.77	26.790	68.40	0.385	0.998
	T/R	8.93	86.4	24.056	72.16	0.422	0.991
MF-0	T	5.25	92.10	49.902	45.82	0.331	0.875
	R	5.74	102.65	58.682	42.83	0.262	0.904
	T/R	5.95	108.06	52.752	51.18	0.320	0.930
MF-1	T	7.95	120.13	66.611	44.55	0.212	0.962
	R	7.80	112.99	65.087	42.40	0.203	0.956
	T/R	11.50	126.08	74.124	41.21	0.136	0.977

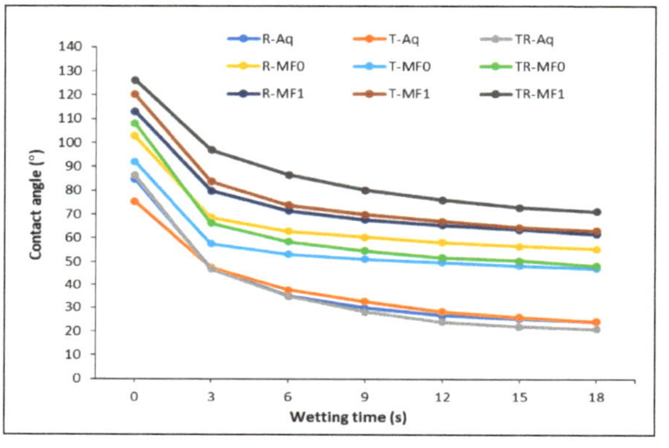

Figure 5. Contact angle changes as a function of time for aquades (Aq), MF-0, and MF-1 on different Jabon wood surfaces.

The contact angle of the modified MF (MF-1) is higher than that of MF-0, which is commercially available. A high wettability can be determined by contact angles below 90 degrees, which signifies that the liquid efficiently wets the wood surface. A contact angle exceeding 90 degrees signifies an insufficient wettability, wherein the liquid fails to sufficiently lubricate the wood surface [4,11,53]. Previous studies [40,54] reported that an increase in surface roughness results in a decrease in the contact angle, which subsequently leads to enhanced wettability and bonding performance. According to the findings of the current study and the above theory [40,54], the MF-0 adhesive has a higher bonding performance than MF-1 because its contact angle is lower. A higher *K* value indicates a more wettable surface [11]. Overall, Table 3 shows that the MF-0 adhesive exhibits a higher *K* value (0.262–0.331) than the MF-1 adhesive (0.136–0.212), indicating that the MF-0 adhesive wets the surface of Jabon wood more easily than the MF-1 adhesive. This result also shows that the wetting model had R squared values greater than 0.904 across all the wood samples examined in our study. Consequently, the S/G wetting model could be used to precisely characterize the hydration process of MF-based adhesives on Jabon wood surfaces. Nevertheless, prior research [1] shows that the MF-1 adhesive exhibits a superior bonding performance in terms of block shear strength and delamination compared to MF-0. The observed phenomenon can be attributed to the higher viscosity of MF-1 in comparison to MF-0 (Table 1). This viscosity factor restricts the absorption of the MF-1 adhesive into the wood surface, resulting in a greater contact angle of MF-1 compared to MF-0. There are other factors, in addition to the contact angle and the wettability of the wood surface, that impact the performance of wood bonding when a particular type of adhesive is utilized.

Figure 6a,b show how changes in the surface roughness of Jabon wood affect the equilibrium contact angle and wettability of MF-based adhesives. In general, the surface roughness variances between Jabon wood's three surface areas did not have much of an impact on the equilibrium contact angle. The equilibrium contact angle decreased slightly after the aquades treatment as roughness increased on the radial and semi-radial surfaces. Meanwhile, in the MF-0 and MF-1 treatments, the equilibrium contact angle fluctuated due to the increased surface roughness of Jabon wood. As the wood's surface roughness increased, the K value consequently increased in the aquades treatment. Prior studies [3,4] reported that rougher surfaces tended to result in lower θ_e and higher K values. Liquid permeated and diffused more extensively across the wood samples as surface roughness increased [11]. Furthermore, according to de Moura and Hernández (2006) [42], surfaces that have been sanded provide optimal circumstances for the spreading of liquids due to the presence of scratches caused by the abrasive grains. These scratches enhance the conduction of liquids through capillarity.

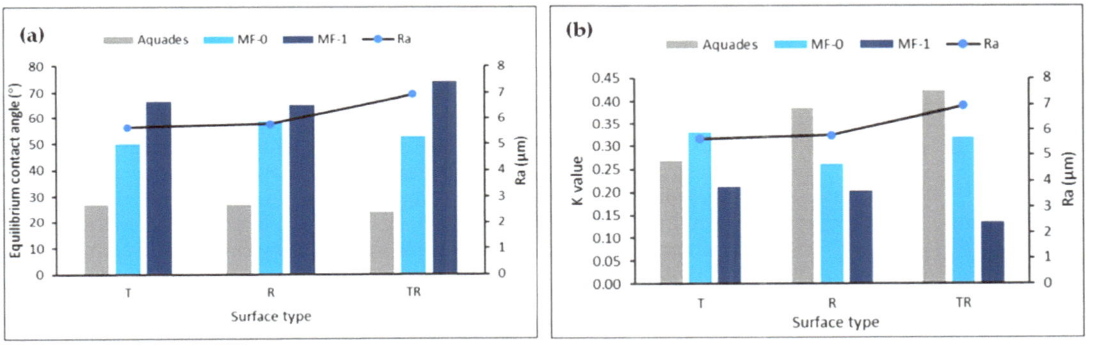

Figure 6. (**a**) Equilibrium contact angle for aquades, MF-0, MF-1, and Ra value on different Jabon wood surfaces, and (**b**) the K value for aquades, MF-0, MF-1, and Ra value on different Jabon wood surfaces.

3.3. Wood–Adhesive Interphase Analysis

The analysis of a wood–adhesive interphase entails the examination of the boundary between the wood surface and the adhesive substance. The wetting process of an adhesive on a solid surface has three main steps [55]: the establishment of interfacial adhesion at the surface; the spreading of an adhesive as a liquid flow over a solid surface; and the infiltration of a liquid into the inner regions of a porous solid. The interphase is essential in determining the overall performance and longevity of the adhesive bond. Several analytical techniques can be utilized for this objective, and DMA can be employed to investigate the viscoelastic characteristics of the interface between the adhesive and wood in question [1,38]. The interphase region exhibits an irregular or diverse layer [56]. Figure 7 displays the illustration of the wood–adhesive interphase of MF resins on Jabon wood. The interphase, within the context of wood bonding, is important in determining the strength and durability of the adhesive bond. In addition, according to [56], an adhesive's bonding performance with wood elements is substantially impacted by the degree of the adhesive's penetration into the porous network of interconnected layers.

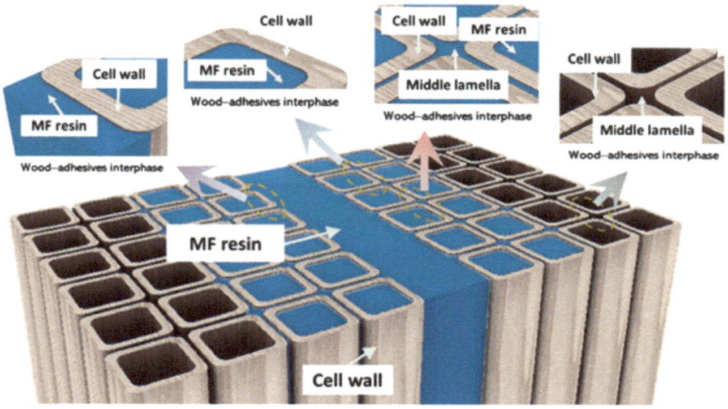

Figure 7. Illustration of woodadhesives interphase of melamine formaldehyde (MF) resins on Jabon wood.

DMA involves the application of varying levels of stress to a material and the subsequent measurement of the resulting strain. Stress refers to the amount of force exerted on a certain region, whereas strain represents the extent of deformation or change in length compared to a material's initial length. Figure 8a shows that MF-1 had a greater stress–strain curve compared to MF-0. A greater stress–strain curve suggests more stiffness in a material, indicating that the material is more resistant to deformation when subjected to dynamic loading. This result is in line with the stiffness of wood–MF resin as a function of temperature (Figure 8b). Wood–MF-0 reached a maximum stiffness of 957 N/m at 123.0 °C, while the wood–MF-1 had a maximum stiffness of 2734 N/m at 110.5 °C. The stiffness of wood adhesives is crucial in determining the overall structural integrity and performance of bonded wood products [57]. This study showed that MF-1 had a greater stiffness and could produce a better performance in its bonded wood products compared to MF-0.

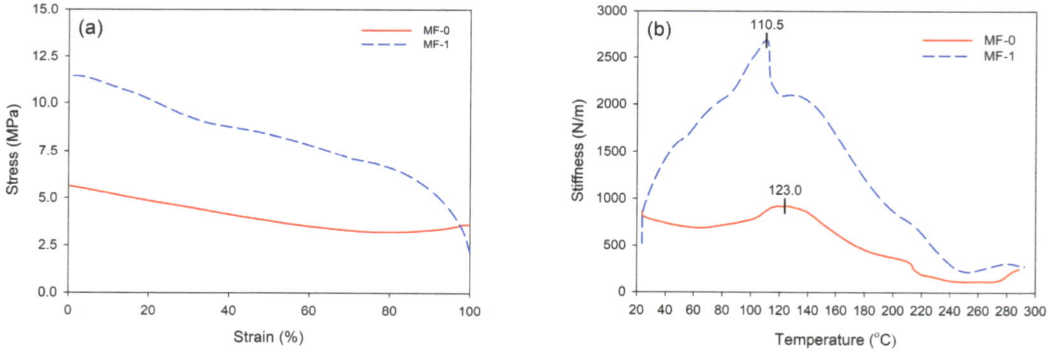

Figure 8. DMA results of wood–MF resin adhesive: (**a**) stress–strain curve and (**b**) stiffness as a function of temperature.

In line with the above results, the stiffness of wood adhesives is commonly quantified by the storage modulus (E′), which assesses an adhesive's capacity to retain elastic energy. The storage modulus quantifies the inherent stiffness or rigidity of a material. Figure 9 displays the DMA thermograms of the wood–MF resin adhesives. Wood–MF-0 had a maximum storage modulus (E'_{max}) of 12,650 MPa at a temperature of 128.9 °C, while wood–MF-1 had a maximum storage modulus (E'_{max}) of 22,950 MPa at a temperature of 113.5 °C. The results showed that MF-1 had a greater storage modulus at a lower temperature compared to MF-0. In contrast, the loss modulus of the wood–MF-0 adhesive was higher than that of wood–MF-1. The loss modulus of a wood–adhesive interphase pertains to the capacity of the interface between the wood and the adhesive to release energy when subjected to dynamic loading conditions [57]. Various factors, such as adhesive formulation, curing conditions, wood surface preparation, and unique wood features, affect the loss modulus of the wood–adhesive interphase. The interface between wood and an adhesive is pivotal for the overall efficacy of wood–adhesive bonding as it influences parameters such as bond strength, endurance, and resilience to external conditions. Like the loss modulus, the tangent delta (tanδ) is a metric that quantifies the extent of damping or energy dissipation in a substance. It is commonly employed to describe the viscoelastic properties of a system. Tan δ in the wood–adhesive interphase context indicates the proportion of a material's viscous (dissipative) reaction to its elastic (storage) response when subjected to dynamic loading circumstances.

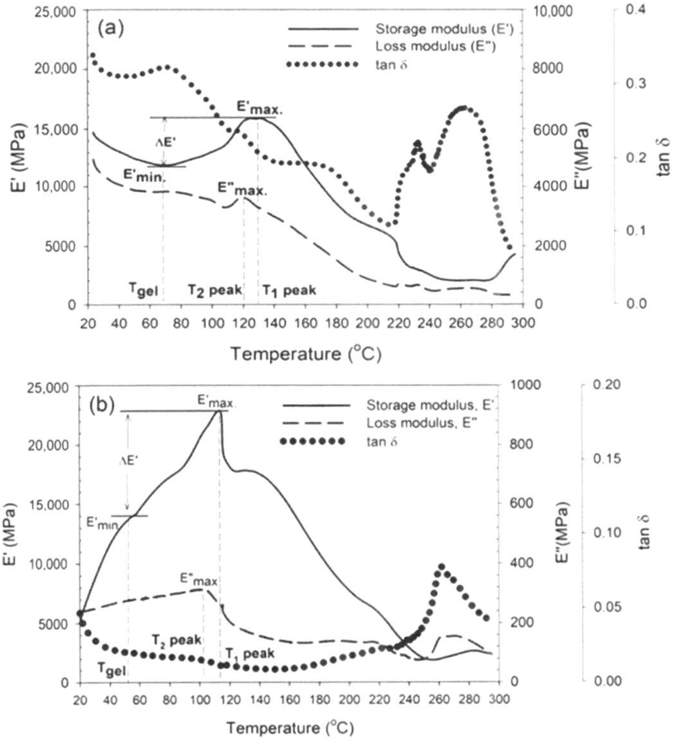

Figure 9. DMA thermograms of wood–MF resin adhesive: (**a**) MF-0 and (**b**) MF-1.

The block shear strength results revealed that the adhesive strength of MF-0 with glue spreads of 250 g/m^2, 280 g/m^2, and 300 g/m^2 was determined to be 1.03 MPa, 1.94 MPa, and 2.13 MPa, respectively (Figure 10). The shear strength values of the MF-0 samples were 30% lower than those of the MF-1 samples. The maximum block shear strength of MF-1 was 3.14 MPa, which was obtained with a glue application rate of 300 g/m^2. This adhesive achieved bonding strengths of 1.56 MPa and 2.57 MPa when applied at glue spreads of 250 g/m^2 and 280 g/m^2, respectively. These results were in accordance with the results of the DMA (Figure 9), revealing that MF-1 has a greater storage modulus compared to MF-0.

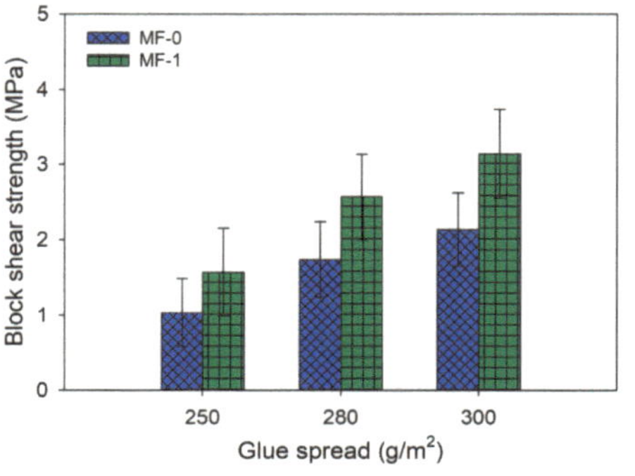

Figure 10. Bonding strength of MF resin adhesive at different amounts of glue spread.

4. Conclusions

The surface roughness and dynamic wettability of modified melamine formaldehyde-based adhesive on Jabon wood was investigated. Jabon wood treated with a P-100-grit sander had Ra values of 5.62 μm on the tangential surface (T), 5.77 μm on the radial surface (R), and 6.94 μm on the semi-radial surface (T/R). The tangential surface had a lower Ra value than the radial or semi-radial surfaces, indicating that the roughness level of Jabon wood on the tangential surface was preferable to those on the radial and semi-radial surfaces. The MF-based adhesives exhibited higher initial contact angles and equilibrium contact angles than the aquades. However, the speed of acquiring the equilibrium contact angle for the MF-based adhesives tended to be faster than that of aquades. The contact angle of the modified MF (MF-1) exceeded that of MF-0. On the other hand, the MF-0 adhesive exhibited a higher K value compared to the MF-1 adhesive, indicating that the MF-0 adhesive wets the surface of Jabon wood more easily than the MF-1 adhesive. The viscosity factor restricted the absorption of the MF-1 adhesive into the wood surface, resulting in a higher contact angle of MF-1 compared to MF-0. There are other factors, in addition to the contact angle and the wettability of the wood surface, that impact the performance of wood bonding when a particular type of adhesive is utilized. MF-1 exhibited a higher stress–strain curve and stiffness than MF-0, suggesting that it offers superior performance for bonded wood products and greater resistance to deformation under dynamic loading circumstances compared to MF-0.

Author Contributions: Conceptualization and methodology, Y.A., L.K., N.N. and M.A.R.L.; writing—original draft preparation, Y.A., L.K. and M.A.R.L.; writing—review and editing, L.K., N.N. and M.A.R.L.; field and laboratory work, data curation, Y.A., M.A.R.L. and U.A.; supervision, L.K., N.N., M.A.R.L., E.T.B. and W.D.; project administration, U.A.; formal analysis, Y.A., L.K. and M.A.R.L. All authors have read and agreed to the published version of the manuscript.

Funding: The financial support for this research was provided by The Indonesian Ministry of Education, Culture, Research, and Technology (KEMENDIKBUDRISTEK) through the Doctoral Dissertation Research Grant of IPB University for the fiscal year 2023. This grant's contract number is 15857/IT3.D10/PT.01.02/P/T/2023.

Institutional Review Board Statement: Not applicable.

Data Availability Statement: The authors can verify that the data substantiating the conclusions of this study are accessible within the paper. The primary data that substantiate the conclusions of this investigation can be obtained from the relevant author upon reasonable inquiry.

Acknowledgments: The authors declare their appreciation of the National Research and Innovation Agency (BRIN) for their support during this study. They also express their sincere thanks to Sudarmanto, Narto, and Dhiya Khairunisa for their invaluable assistance with the laboratory tasks and data processing.

Conflicts of Interest: The authors declare no conflicts of interest.

References

1. Amin, Y.; Adji, R.P.; Lubis, M.A.R.; Nugroho, N.; Bahtiar, E.T.; Dwianto, W.; Karlinasari, L. Effect of Glue Spread on Bonding Strength, Delamination, and Wood Failure of Jabon Wood-Based Cross-Laminated Timber Using Cold-Setting Melamine-Based Adhesive. *Polymers* **2023**, *15*, 2349. [CrossRef] [PubMed]
2. Bekhta, P.; Krystofiak, T.; Proszyk, S.; Lis, B. Evaluation of Dynamic Contact Angle of Loose and Tight Sides of Thermally Compressed Birch Veneer. *Drv. Ind.* **2018**, *69*, 387–394. [CrossRef]
3. Martha, R.; Dirna, F.C.; Hasanusi, A.; Rahayu, I.S.; Darmawan, W. Surface Free Energy of 10 Tropical Woods Species and Their Acrylic Paint Wettability. *J. Adhes. Sci. Technol.* **2020**, *34*, 167–177. [CrossRef]
4. Yu, Q.; Pan, X.; Yang, Z.; Zhang, L.; Cao, J. Effects of the Surface Roughness of Six Wood Species for Furniture Production on the Wettability and Bonding Quality of Coating. *Forests* **2023**, *14*, 996. [CrossRef]
5. Buyuksari, U.; Akbulut, T.; Guler, C.; As, N. Wettability and Surface Roughness of Natural and Plantation-Grown Narrow-Leaved ASH (*Fraxinus Angustifolia* Vahl.) Wood. *BioResources* **2011**, *6*, 4721–4730. [CrossRef]
6. Sinn, G.; Gindl, M.; Stanzl-Tschegg, S. Changes in the Surface Properties of Wood Due to Sanding. *Holzforschung* **2004**, *58*, 246–251. [CrossRef]
7. Hendarto, B.; Shayan, E.; Ozarska, B. Analysis of Roughness of a Sanded Wood Surface. *Int. J. Adv. Manuf. Technol.* **2006**, *28*, 775–780. [CrossRef]
8. Cahyono, T.D.; Wahyudi, I.; Priadi, T.; Febrianto, F.; Ohorella, S. Sudut Kontak Dan Keterbasahan Dinamis Kayu Samama Pada Berbagai Pengerjaan Kayu. *J. Tek. Sipil ITB* **2017**, *24*, 209–216. [CrossRef]
9. Walinder, M. Wettability Phenomena on Wood: Factors Influencing Measurements Of Wood Wettability. Ph.D. Thesis, KTH-Royal Institute of Technology, Department of Manufacturing Systems Wood Technology and Processing, Stockholm, Sweden, 2000; p. 70.
10. Shi, S.Q.; Gardner, D.J. Dynamic Adhesive Wettability of Wood. *Wood Fiber Sci.* **2001**, *33*, 58–68.
11. Darmawan, W.; Nandika, D.; Noviyanti, E.; Alipraja, I.; Lumongga, D.; Gardner, D.; Gérardin, P. Wettability and Bonding Quality of Exterior Coatings on Jabon and Sengon Wood Surfaces. *J. Coat. Technol. Res.* **2018**, *15*, 95–104. [CrossRef]
12. Mu, Q.; Zhang, Q.; Yu, W.; Su, M.; Cai, Z.; Cui, K.; Ye, Y.; Liu, X.; Deng, L.; Chen, B.; et al. Robust Multiscale-Oriented Thermoresponsive Fibrous Hydrogels with Rapid Self-Recovery and Ultrafast Response Underwater. *ACS Appl. Mater. Interfaces* **2020**, *12*, 33152–33162. [CrossRef] [PubMed]
13. Mu, Q.; Cui, K.; Wang, Z.J.; Matsuda, T.; Cui, W.; Kato, H.; Namiki, S.; Yamazaki, T.; Frauenlob, M.; Nonoyama, T.; et al. Force-Triggered Rapid Microstructure Growth on Hydrogel Surface for on-Demand Functions. *Nat. Commun.* **2022**, *13*, 6213. [CrossRef] [PubMed]
14. Qin, Z.; Gao, Q.; Zhang, S.; Li, J. Surface Free Energy and Dynamic Wettability of Differently Machined Poplar Woods. *Bio. Res.* **2014**, *9*, 3088–3103. [CrossRef]
15. Karlinasari, L.; Lestari, A.T.; Priadi, T. Evaluation of Surface Roughness and Wettability of Heat-Treated, Fast-Growing Tropical Wood Species Sengon (*Paraserianthes Falcataria* (L.) I.C.Nielsen), Jabon (*Anthocephalus Cadamba* (Roxb.) Miq), and Acacia (*Acacia Mangium* Willd.). *Int. Wood Prod. J.* **2018**, *9*, 142–148. [CrossRef]
16. Lestari, A.T.; Wahyuningsih, E.; Syaputra, M.; Anwar, H.; Suparyana, P.K.; Ritonga, F.N. Wettability and Treatability of Sengon (*Paraserianthes Falcataria* (L.) I.C. Nielsen) Wood from NTB. *IOP Conf. Ser. Earth Environ. Sci.* **2021**, *891*, 012015. [CrossRef]
17. Vaziri, M.; Karlsson, O.; Abrahamsson, L.; Lin, C.F.; Sandberg, D. Wettability of Welded Wood-Joints Investigated by the Wilhelmy Method: Part 1. Determination of Apparent Contact Angles, Swelling, and Water Sorption. *Holzforschung* **2021**, *75*, 65–74. [CrossRef]
18. Huang, Y.; Qi, Y.; Zhang, Y.; Zhu, R.; Zhang, Y.; Yu, W. Surface Properties of Novel Wood-Based Reinforced Composites Manufactured from Crushed Veneers and Phenolic Resins. *Maderas Cienc. Y Tecnol.* **2019**, *21*, 185–196. [CrossRef]
19. Ashori, A. Effects of Extractives Removal on the Performance of Clear Varnish Coatings on Boards. *J. Compos. Mater.* **2015**, *50*, 3019–3024. [CrossRef]
20. Kúdela, J. Wetting of Wood Surface by a Liquids of a Different Polarity. *Wood Res.* **2014**, *59*, 11–24.
21. Santoni, I.; Pizzo, B. International Journal of Adhesion & Adhesives Effect of Surface Conditions Related to Machining and Air Exposure on Wettability of Different Mediterranean Wood Species. *Int. J. Adhes. Adhes.* **2011**, *31*, 743–753. [CrossRef]
22. Gardner, D.J.; Generalla, N.C.; Gunnells, D.W.; Wolcott, M.P. Dynamic Wettability of Wood. *Langmuir* **1991**, *7*, 2498–2502. [CrossRef]
23. Wang, S.; Zhang, Y.; Xing, C.; Trocknungsverfahrens, E. Effect of Drying Method on the Surface Wettability of Wood Strands. *Holz. Roh. Werkst* **2007**, *65*, 437–442. [CrossRef]
24. Yuan, Y.; Lee, T.R. *Contact Angle and Wetting Properties*; Springer: Berlin/Heidelberg, Germany, 2013; ISBN 9783642342431.
25. Peng, Z.; Jiang, X.; Si, C.; Joao Cárdenas-Oscanoa, A.; Huang, C. Advances of Modified Lignin as Substitute to Develop Lignin-Based Phenol-Formaldehyde Resin Adhesives. *ChemSusChem* **2023**, *16*, e202300174. [CrossRef] [PubMed]
26. Wiesner, F.; Klippel, M.; Dagenais, C.; Dunn, A.; Östman, B.; Janssens, M.L.; Kagiya, K. Requirements for Engineered Wood Products and Their Influence on the Structural Fire Performance. In Proceedings of the WCTE 2018 World Conference on Timber Engineering, Seoul, Republic of Korea, 20–23 August 2018.
27. Flatscher, R.B.G.; Schickhofer, A.R.G. Cross Laminated Timber (CLT): Overview and Development. *Eur. J. Wood Wood Prod.* **2016**, *74*, 331–351. [CrossRef]

28. Li, M.; Zhang, S.; Gong, Y.; Tian, Z.; Ren, H. Gluing Techniques on Bond Performance and Mechanical Properties of Cross-Laminated Timber (Clt) Made from Larix Kaempferi. *Polymers* **2021**, *13*, 733. [CrossRef] [PubMed]
29. Lan, P.; Yang, R.; Mao, H.Y.; Cui, J.Q.; Brosse, N. Production of Melamine Formaldehyde Resins Used in Impregnation by Incorporation of Ethylene Glycol and Caprolactam with High Flexibility, Storage Stability, and Low Formaldehyde Content. *BioResources* **2019**, *14*, 9916–9927. [CrossRef]
30. Frihart, C.R. Introduction to Special Issue Wood Adhesives: Past, Present, and Future. *For. Prod. J.* **2015**, *65*, 4–8. [CrossRef]
31. Santos, J.; Pereira, J.; Paiva, N.; Ferra, J.; Magalhães, F.D.; Martins, J.M.; de Carvalho, L.H. Impact of Condensation Degree of Melamine-Formaldehyde Resins on Their Curing Behavior and on the Final Properties of High-Pressure Laminates. *Proc. Inst. Mech. Eng. Part C J. Mech. Eng. Sci.* **2021**, *235*, 484–496. [CrossRef]
32. Yuningsih, I.; Rahayu, I.S.; Lumongga, D. Wettability and Adherence of Acrylic Paints on Long and Short Rotation Teaks. *Wood Mater. Sci. Eng.* **2019**, *15*, 229–236. [CrossRef]
33. Hartono, R.; Iswanto, A.H.; Herawati, E.; Suramana, R.E.; Sutiawan, J.; Amin, Y.; Sumardi, I. The Improvement of Sumatran Elephant (Elephas Maximus Sumatranus) Dung Particleboard Characteristics Using Bamboo Layering. *Polymers* **2022**, *14*, 3330. [CrossRef]
34. Stalder, A.F.; Melchior, T.; Müller, M.; Sage, D.; Blu, T.; Unser, M. Low-Bond Axisymmetric Drop Shape Analysis for Surface Tension and Contact Angle Measurements of Sessile Drops. *Colloids Surfaces A Physicochem. Eng. Asp.* **2010**, *364*, 72–81. [CrossRef]
35. Mu, Q.; Zhang, Q.; Gao, L.; Chu, Z.; Cai, Z.; Zhang, X.; Wang, K.; Wei, Y. Structural Evolution and Formation Mechanism of the Soft Colloidal Arrays in the Core of PAAm Nanofibers by Electrospun Packing. *Langmuir* **2017**, *33*, 10291–10301. [CrossRef] [PubMed]
36. Mu, Q.; Hu, J. Polymer Mechanochemistry: From Single Molecule to Bulk Material. *Phys. Chem. Chem. Phys.* **2023**, *26*, 679–694. [CrossRef] [PubMed]
37. Chen, B.; Zhang, S.; Zhang, Q.; Mu, Q.; Deng, L.; Chen, L.; Wei, Y.; Tao, L.; Zhang, X.; Wang, K. Microorganism Inspired Hydrogels: Fermentation Capacity, Gelation Process and Pore-Forming Mechanism under Temperature Stimulus. *RSC Adv.* **2015**, *5*, 91937–91945. [CrossRef]
38. Lubis, M.A.R.; Labib, A.; Sudarmanto; Akbar, F.; Nuryawan, A.; Antov, P.; Kristak, L.; Papadopoulos, A.N. Influence of Lignin Content and Pressing Time on Plywood Properties Bonded with Cold-Setting Adhesive Based on Poly (Vinyl Alcohol), Lignin, and Hexamine. *Polymers* **2022**, *14*, 2111. [CrossRef] [PubMed]
39. Hamzah, N.A.; Razak, N.A.A.; Karim, M.S.A.; Salleh, S.Z. Validation of a Roughness Parameters for Defining Surface Roughness of Prosthetic Polyethylene Pe-Lite Liner. *Sci. Rep.* **2022**, *12*, 2636. [CrossRef]
40. Candan, Z.; Büyüksarı, U.; Korkut, S.; Unsal, O.; Nevzat, C. Wettability and Surface Roughness of Thermally Modified Plywood Panels. *Ind. Crops Prod.* **2012**, *36*, 434–436. [CrossRef]
41. Sulaiman, O.; Hashim, R.; Subari, K.; Liang, C.K. Effect of Sanding on Surface Roughness of Rubberwood. *J. Mater. Process. Technol.* **2009**, *209*, 3949–3955. [CrossRef]
42. De Moura, L.F.; Hernández, R.E. Evaluation of Varnish Coating Performance for Three Surfacing Methods on Sugar Maple Wood. *For. Prod. J.* **2006**, *56*, 130–136.
43. Zhong, Z.W.; Hiziroglu, S.; Chan, C.T.M. Measurement of the Surface Roughness of Wood Based Materials Used in Furniture Manufacture. *Meas. J. Int. Meas. Confed.* **2013**, *46*, 1482–1487. [CrossRef]
44. Ugulino, B.; Hernández, R.E. Analysis of Sanding Parameters on Surface Properties and Coating Performance of Red Oak Wood. *Wood Mater. Sci. Eng.* **2018**, *13*, 64–72. [CrossRef]
45. Cool, J.; Hernández, R.E. Improving the Sanding Process of Black Spruce Wood for Surface Quality and Water-Based Coating Adhesion. *For. Prod. J.* **2011**, *61*, 372–380. [CrossRef]
46. Dai, Q.; Li, M.; Khonsari, M.M.; Huang, W.; Wang, X. The Thermocapillary Migration on Rough Surfaces. *Lubr. Sci.* **2019**, *31*, 163–170. [CrossRef]
47. Rezaee Niaraki, P.; Krause, A. Correlation between Physical Bonding and Mechanical Properties of Wood Plastic Composites: Part 1: Interaction of Chemical and Mechanical Treatments on Physical Properties. *J. Adhes. Sci. Technol.* **2020**, *34*, 744–755. [CrossRef]
48. Von Fraunhofer, J.A. Adhesion and Cohesion. *Int. J. Dent.* **2012**, *2012*, 951324. [CrossRef] [PubMed]
49. Karlinasari, L.; Adzkia, U.; Sudarsono, A.S.; Larasatie, P.; Amin, Y.; Nugroho, N. Surface Characteristics and Acoustical Properties of Bamboo Particle Board Coated with Polyurethane Varnish. *Forests* **2021**, *12*, 1285. [CrossRef]
50. Piao, C.; Winandy, J.E.; Shupe, T.F. From Hydrophilicity To Hydrophobicity: A Critical Review: Part I. Wettability and Surface Behavior. *Wood Fiber Sci.* **2010**, *42*, 490–510.
51. Whitesides, G.M.; Biebuyck, H.A.; Folkers, J.P.; Prime, K.L. Acid-Base Interactions in Wetting. *J. Adhes. Sci. Technol.* **1991**, *5*, 57–69. [CrossRef]
52. Basri, E.; Martha, R.; Damayanti, R.; Rahayu, I.; Darmawan, W.; Gérardin, P. Durability and Wettability of Varnishes on the Modified and Aged Surfaces of Short Rotation Teak Wood. *Pigment Resin Technol.* **2022**, *in press*. [CrossRef]
53. Gavrilovic-Grmusa, I.; Dunky, M.; Miljkovic, J.; Djiporovic-Momcilovic, M. Influence of the Viscosity of UF Resins on the Radial and Tangential Penetration into Poplar Wood and on the Shear Strength of Adhesive Joints. *Holzforschung* **2012**, *66*, 849–856. [CrossRef]
54. Darmawan, W.; Ginting, M.B. Influence of Surface Roughness of Ten Tropical Woods Species on Their Surface Free Energy, Varnishes Wettability and Bonding Quality. *Pigment. Resin Technol.* **2020**, *49*, 441–447. [CrossRef]

55. Kim, M.; Park, B.D. Effects of Molecular Weight of Urea–Formaldehyde Resins on Wettability and Adhesion at Wood Surface, Interphase, and Plywood. *Wood Sci. Technol.* **2022**, *56*, 1675–1703. [CrossRef]
56. Ülker, O. Wood Adhesives and Bonding Theory. In *Adhesives—Applications and Properties*; InTechOpen: London, UK, 2016. [CrossRef]
57. Karliati, T.; Lubis, M.A.R.; Dungani, R.; Maulani, R.R.; Hadiyane, A.; Rumidatul, A.; Antov, P.; Savov, V.; Lee, S.H. Performance of Particleboard Made of Agroforestry Residues Bonded with Thermosetting Adhesive Derived from Waste Styrofoam. *Polymers* **2024**, *16*, 543. [CrossRef] [PubMed]

polymers

Article

An Investigation of the Mechanical Properties of Ti Films Reinforced with Wood Composites by Growing Ti Particles on a Wood Substrate

Wenhui Bao [1,2,3], Yini Tan [1], Ziyi Ying [1], Rui Xue [1], Xiaojiang Xu [1], Shuangping Duan [1,2], Haizhuan Lin [1,2] and Hui Chen [1,2,*]

1 College of Architecture and Energy Engineering, Wenzhou University of Technology, Wenzhou 325035, China; nefuwenhui@163.com (W.B.); linhaizhuan@wzut.edu.cn (H.L.)
2 Wenzhou Key Laboratory of Intelligent Lifeline Protection and Emergency Technology for Resilient City, Wenzhou 325035, China
3 Key Laboratory of Bio-based Material Science & Technology, Northeast Forestry University, Ministry of Education, Hexing Road 26, Harbin 150040, China
* Correspondence: chenhui0306@wzu.edu.cn

Abstract: Table tennis racquet blades (TTRBs) are specialized wood materials known for their excellent mechanical properties. As one of the widely used physical vapor deposition technologies, magnetron sputtering has become the most effective method for preparing various thin film materials. In this study, the surface of the TTRB is coated with a Ti film with different thicknesses by magnetron sputtering to improve the performance of the TTRB. The surface roughness, crystal structure, viscoelasticity of the TTRB were analyzed by means of non-contact surface profilometry, X-ray diffraction (XRD), and dynamic mechanical analysis (DMA). In order to effectively test TTRB properties, three types of testing devices were designed, including free-fall rebound, laser vibration measurement, and the dynamic rebound test. The results reveal that the deposition of a Ti film on the surface of the TTRB improves the rigidity and rebound efficiency of the TTRB. Under optimized conditions, the initial amplitude, vertical rebound distance, and rebound rate can reach 2.1 μm, 23.7 cm, 13.7%, respectively, when the deposition thickness is 5 μm. It is anticipated that the modification and the corresponding detection methods developed in this study can foster innovative product development, standardize the TTRB industry, and contribute to the advancement of table tennis.

Keywords: wood; Ti film; magnetron sputtering; laser vibration measurement; dynamic rebound test; dynamic mechanical analysis

Academic Editor: Petar Antov

Received: 27 January 2025
Revised: 17 February 2025
Accepted: 20 February 2025
Published: 22 February 2025

Citation: Bao, W.; Tan, Y.; Ying, Z.; Xue, R.; Xu, X.; Duan, S.; Lin, H.; Chen, H. An Investigation of the Mechanical Properties of Ti Films Reinforced with Wood Composites by Growing Ti Particles on a Wood Substrate. *Polymers* **2025**, *17*, 583. https://doi.org/10.3390/polym17050583

1. Introduction

As a renewable biomass material, wood is widely used in daily life and primarily serves engineering purposes [1,2]. Although various new buildings and decorative materials have been developed [3], wood remains irreplaceable in some specific areas of application, such as in music and sport fields, due to its unique mechanical [4,5], physical [6,7], and esthetic properties [8,9]. As one of the well-known Olympic games [10], table tennis has been a globally recognized sport for over 100 years, appearing not only in everyday life but also in international competitions. A table tennis racket is composed of table tennis racket blades (TTRBs), sponge, and rubber [11]. A TTRB commonly consists of a single layer of wood veneer, formed through gluing and compression molding [12,13]. More commonly, it is made from five-layer or seven-layer blades. The purpose of the singular plywood is to create a core layer at the center of the TTRB, referred to as the core board. The thickness of the core board typically ranges from 2.5 to 5 mm. On both sides of the core board, there are the splint board and the panel board, as illustrated in Figure 1a. The splint board, also known as the force material, has a thickness of 0.5 to 1 mm and is used to balance the force conduction between the panel board and the core board. The combination of the harder surface material and the softer force material enhances control performance during play, allowing for the better execution of various batting techniques. A thicker force material results in a faster ball rebound from the blade but reduces the ability to absorb the ball, leading to less control. The panel board is the thinnest layer, with a thickness ranging from 0.2 to 0.5 mm, and the selection and thickness of the materials for each layer play a crucial role in determining the overall softness and hardness of the TTRB. A TTRB is generally classified into two types according to the way of holding, i.e., the horizontal plate and the straight plate. The horizontal plate can be further divided into a straight handle and a waist handle. The straight racket handle is particularly conducive for fast attack techniques such as smashing, dunking, and slicing. In competitions, the waist-shaped handle is commonly used by athletes who specialize in fast-paced attacking play. This is because the front section of the racket's bottom plate is thinner, while the tail is wider and thicker. This design provides athletes with a unique grip sensation, offering a balance of looseness and tightness, which enhances control and maneuverability during fast attacks. Since the end of the 19th century, the original TTRB was made of a paper material (Figure 1b) and has gradually evolved to incorporate canvas, glass sandpaper, and multi-layer natural wood [14]. Nowadays, with the emergence of new materials, as well as changes in game rules, athletes increasingly prefer lightweight and flexible TTRBs [15]. Manufacturer have improved the properties of TTRBs by adding materials such as a carbon layer (Figure 1c), man-made fibers (Figure 1e), and photosensitive resin to the blades [16]. Overall, the manufacturer's innovation mainly focuses on three aspects: shape design, material matching, and adhesive development, but the breakthrough progress in these areas is very slow [17].

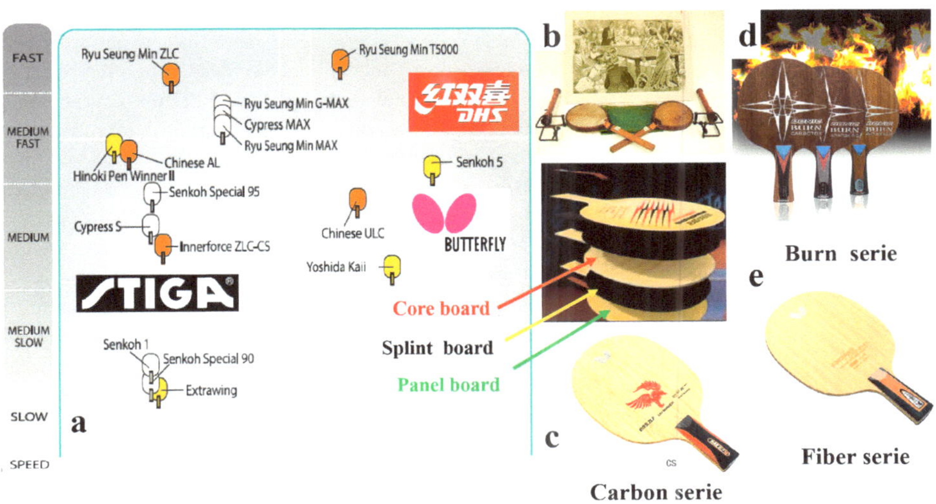

Figure 1. (**a**) The TTRBs are classified according to the subjective feel of the athlete. (**b**) The equipment of the initial origin of table tennis. (**c**) The schematic illustrations of the TTRB by adding a carbon layer. (**d**) The schematic illustrations of the TTRB hardening at a high temperature. (**e**) The schematic illustrations of the TTRB using man-made fibers.

From the perspective of wood modification [6,18], modified wood is typically defined as any wood that has undergone chemical [19,20], physical [21,22], or thermal processes [23,24] to enhance its properties [25]. The modification process can improve the dimensional stability of wood [26], reduce its hydrophilicity, and affect various mechanical properties (e.g., hardness) [27]. Today, several wood modification techniques have been commercialized, including acetylation [4], furfurylation, charring, and thermal modification [8,28]. The Burn series is a mainstream product produced by DONIC Co., Ltd. (Verklingen, Germany). (Figure 1d). The stiffness and dimensional stability of the TTRB increased after treatment at 200 °C for 24 h [29]. The carbon fiber series is the mainstream product of STIGA Co., Ltd. (Stockholm, Sweden). The mechanical properties of the TTRB are improved by adding artificial fiber cloth to the wood. Butterfly is also one of the manufacturers of high-end table tennis supplies, with its ZLC series featuring high-density woven carbon fiber in the TTRB. In addition, manufacturers also focus on the production process, glue types [30], handle design, material collocation, and other aspects of TTRB innovation. Last year, the DHS Company approached our lab to chemically modify TTRBs, aiming to reducing the cost of production by replacing precious wood with low-quality alternatives. Therefore, the goal is to enhance the performance of low-quality wood TTRBs to meet the demands of high-level athletes. The approach we adopted was to create rigid structures by immersing the wood in a low-molecular-weight resin solution, leveraging the crosslinking of the resin with the hydroxyl groups in the cell wall of the wood. The results showed that the weight of the chemically modified TTRB ranged from 150 to 160 g, while the weight of a commonly used TTRB was between 75 and 90 g. A 150 g table tennis racket base is too heavy for a professional athlete. Although the elastic modulus of the chemically modified TTRB was significantly improved in the experiments, the increased weight is not conducive for practical application. In other words, the idea of simply increasing the modulus of elasticity is undesirable for the application of TTRBs. Since the previously mentioned method has disadvantages such as a cumbersome production process, high cost, and a lack of innovation, there is a clear need to explore a new approach to improve the performance of TTRBs.

The International Table Tennis Federation requires that the natural wood content of the TTRBs should be at least 85%. The National Standardization Administration issued the standard GB/T 23115 [31] to regulate the quality of TTRBs. This standard evaluates TTRBs from three aspects: weight, thickness, and appearance, and it categorizes the TTRBs into two types, namely superior quality and first-class quality. However, this standard is somewhat broad given the wide variety of products on the market. In order to further enhance their application, manufacturers have developed a series of indicators for their respective TTRBs. These include the number of layers, racket face size, weight, thickness, sweet spot area, initial amplitude, vibration attenuation time, handle type, and other conventional evaluations. The most widely used evaluation indicators are categorized according to the player's style, typically classified into offensive, defensive, and all-round types [32]. Typically, the performance of the TTRB is mainly based on the subjective evaluation of the athletes [33]. So far, well-known brand companies, such as STIGA and Butterfly, characterize TTRBs based on their performance in areas such as controllability, softness, stiffness, flexibility, power, precision, and ease of play (Figure 1a). However, the TTRB species, TTRB and rubber sponge coordination, and the level of the player also affect the table tennis ball track and the player's feeling [34]. Due to numerous uncertainties, information on the testing methods and analytical studies of TTRBs remain very limited. Lionel et al. [26] conducted a study comparing vibration modes and frequencies obtained through simulations and experimentation, allowing for the validation of a finite element model for the racket blade. Yoichi et al. [35] investigated the effect of racket mass and the stroke rate on the kinematics and kinetics of the trunk and racket arm during a table tennis topspin backhand. Their study found that, regardless of the racket mass, the racket speed at impact was significantly lower at high ball frequencies compared to low frequencies. Peter et al. [28] introduced an approach for estimating ball speed and spin in table tennis using a single racket-mounted inertial sensor. This method is beneficial for both beginners and professional players, aiding in the analysis of techniques and tactics during training and competition. Nicolas et al. [36] investigated the impact of a table tennis ball on the polymeric racket layers under various incident angles and spin conditions. Zhou et al. [37] developed an intelligent table tennis racket with adjustable stiffness based on anisotropic electrorheological elastomers. This smart racket is designed to adapt to different playing styles and the physical conditions of users, standardize the training of unconventional table tennis techniques, and expand the application of electrorheological elastomers (EREs) in smart wearable devices and soft robotics. So far, most research on TTRBs has focused on the biomechanics of table tennis maneuvers and dynamic mechanical analysis using finite element models [38–40]. There is still a lack of knowledge regarding the evaluation of TTRBs and the development of effective testing methods.

As a physical vapor deposition (PVD) method [41], magnetron sputtering offers numerous advantages, such as high purity; strong adhesion; high deposition rate; excellent uniformity on large-area substrates; ease of automation; the ability to sputter a wide range of metals, alloys, or compounds; and extensive applicability [42]. Its deposition is achieved by rapidly colliding ionized inert gas atoms with the surface of the negatively biased target under a high electric field, thus inducing the ejection of atoms which then condense on a substrate and eventually generate a membrane [43]. As we know, the structure of biomass materials is typically porous and anisotropic, making it difficult to create a high vacuum environment. Due to advancements in sputtering power supplies and vacuum technology, PVD technology can now be widely applied to biomass substrates [44]. The radio frequency power supply allows the sputtering film to occur in a lower vacuum environment, while molecular pumps can quickly achieve the desired vacuum state. The most commonly used and easily prepared sputtered materials are pure metals, which are produced using single-target magnetron sputtering. This method allows for the tailoring of chemical and physical properties through parameter optimization by using only one target. In recent years, magnetron sputtering metal films applied to the surface of biomass substrates have endowed them with properties such as conductivity, wettability, electromagnetic shielding, etc., expanding their potential application fields. For instance, Jiao et al. [45] reported the development of a core–shell-structured composite consisting of cotton-derived carbon fibers and nano-copper. Li et al. [46] studied the deposition of Cu films with proper thickness on wood surfaces, finding it to be an effective method for achieving a rougher surface texture. Liang et al. [47] fabricated the superamphiphobic-functionalized CuO microflower/$Cu(OH)_2$ nanorod array hierarchical structure on a Cu–bamboo surface as a rough coating via a facile alkali-assisted surface oxidation technique. Cai et al. [48] fabricated a self-bonded natural fiber product with high hydrophobic and electromagnetic interference shielding performance, developed using the chemical etching treatment and magnetron sputtering of a Cu film. Wan et al. [49] reported that a Cu film was deposited on the surface of a paper via magnetron sputtering, and the Cu/paper was also electro-oxidized using the cyclic voltammetry method to convert the superficial metallic Cu into the highly electrochemically active Cu_2O. Inspired by these impactful studies, it is of great importance to explore the use of magnetron sputtering to integrate wood with a metal film for the development of novel TTRBs.

In this paper, we proposed an effective method to deposit Ti layers with varying thicknesses ranging from 0.5 to 5 μm on a seven-ply plywood TTRB panel using magnetron sputtering. In addition, the evaluation criteria for the TTRB are discussed, and the rebound distance detection equipment is designed to evaluate the performance of the TTRB. Dynamic mechanical analysis was used to measure the mechanical properties of samples, while non-contact profilometry and XRD were applied to examine the surface roughness and crystalline structures of the prepared samples. A laser vibrometer was used to analyze and evaluate the vibration behavior and rigidity of the TTRB after being struck by a ping-pong ball.

2. Materials and Methods

2.1. Materials

The TTRB production process is generally divided into the following steps: material selection, glue plate, drying and aging, and sand milling. Most of the wood is selected with a low output rate of diameter cutting, and the selected plate should be textured straight, arranged densely and evenly, and without color differences, knots, mineral lines, and other defects. The TTRB samples were fabricated using the commonly used materials of Ayous (*Triplochiton scleroxylon*) [50] and Limba (*Terminalia superba*) [51]. The core board and splint board samples of Ayous were prepared with a size of 300 × 200 mm (L × T) and the thickness ranging from 2.5 mm to 1.0 mm. The panel board samples of Limba had a size of 300 × 200 × 0.8 mm (L × T × R). These samples were oven-dried (24 h, 103 ± 2 °C) to a constant weight. All wood materials were purchased from DHS Co., Ltd. (Shanghai, China). Bisphenol epoxy resins with a viscosity of 200–300 mPas were purchased from Dow Co., Ltd. (Midland, Kansas City, MO, USA). The mix ratio of the epoxy and the cure agent was 3:1. The Ti target (99.999% purity) was supplied by Beijing New Material Company Limited. (Beijing, China).

2.2. Preparation of Ti Coating on the Wood Surface

All Ti films were deposited with a radio frequency magnetron sputtering system. The target used was a Ti (99.999% purity) plate with a diameter of 5 cm. Sputtering was carried out in argon (Ar-99.995% purity), and the sputtering system was equipped with a diffusion pump backed by a rotary pump to pump down the sputtering system to a base pressure of 3×10^{-3} Pa. The thicknesses of the Ti films were checked in situ with a quartz crystal monitor located near the wood substrate during the sputtering process. The wood substrate was pre-conditioned at room temperature, and the sputtering power was 100 W with a fixed target substrate distance of 6 cm.

2.3. Preparation of TTRB Samples

The TTRB is processed according to the production procedures of forming, gluing, aging, cold pressing, health preservation, milling, and pasting. Firstly, the Ayous core board is sanded to 2.5 mm thick, and symmetrical cross-cross billets are used, that is, the middle core layer is made of Ayous, the adjacent upper and lower layers are made of Ayous splints, and the upper and lower layers are made of Limba veneers [52]. Secondly, seven-ply plywood samples were made by the DHS Company under the following conditions: 150 g/m^2 adhesive coverage for each surface; the open assembly time was 15–20 min at room temperature, 1 min/mm cold pressing time at 30 °C, and a pressure of 1.2 MPa [53]. After conditioning at room temperature for 1 day, the plywood was then cut into specimens. The deviation of sample quality and thickness, respectively, does not exceed ±3 g, 0.5 mm, in line with standard GB/T 23115. The results are summarized in Table 1.

Table 1. List of samples characterized.

Sample	0 nm-Ti TTRB	100 nm-Ti TTRB	500 nm-Ti TTRB	1 μm-Ti TTRB	2 μm-Ti TTRB	5 μm-Ti TTRB
Weight (g)	85 g ± 1	84 g ± 2	84 g ± 2	85 g ± 2	86 g ± 3	89 g ± 2
Thickness (mm)	6.3 ± 0.2	6.4 ± 0.1	6.3 ± 0.3	6.5 ± 0.3	6.3 ± 0.2	6.4 ± 0.1

2.4. Characterization

The surface structure of the treated samples was determined using X-ray powder diffraction (XRD, Philiphs, PW 1840 diffractometer, Kyoto, Japan), operating with Cu-K radiation at a scan rate of 4°/min, an accelerating voltage of 40 kV, and an applied current of 30 mA ranging from 5° to 80°. Non-contact surface profilometry (Wyko, Greenbackville, VA, USA) was used to calculate the surface roughness [54,55]. Viscoelasticity was measured using the DMA Q800 (TA, Newcastle, DE, USA) analyzer at room temperature. The single cantilever mode was selected, and the measurement was carried out on a rectangular cross-sectional bar with dimensions of 35 × 10 × 0.8 mm^3 (L × T × R) [56]. Vertical rebound distance (Figure 2a) and dynamic rebound distance (Figure 2b) testing require the TTRB to be pneumatically fixed to the handle. The steps are as follows: Select 4 TTRBs as repeat samples; the weight of each TTRB must not differ by 5 g, and the total weight is within the range of 85~90 g. The weight of each ping-pong ball is controlled at 2.6~2.7 g. The serving speed of the table tennis machine is 4 m/s, 6 m/s, 10 m/s, and 12 m/s, respectively, and the test area is kept closed to reduce the influence of air flow on the test results. Each measurement was performed five times independently of one another.

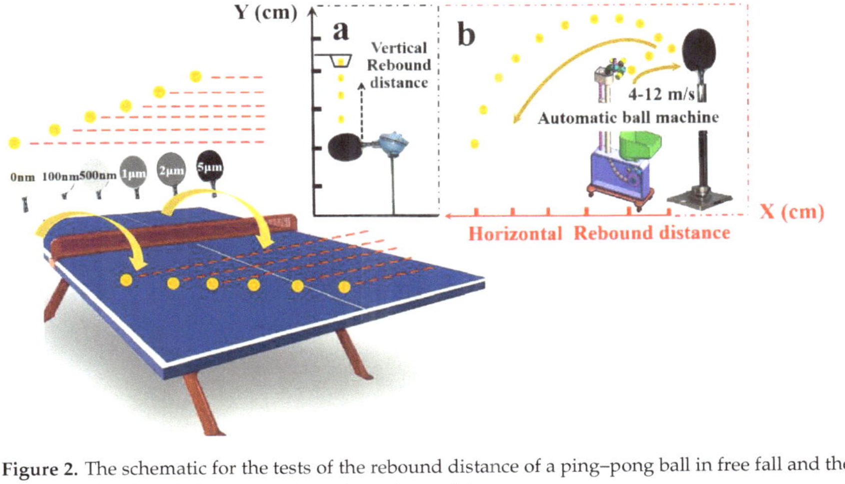

Figure 2. The schematic for the tests of the rebound distance of a ping–pong ball in free fall and the vibration of TTRB (**a**). Dynamic rebound test device (**b**).

3. Results

3.1. XRD Spectrum and Roughness of the Ti Coating on the Wood Surface

The XRD spectrum of the film is affected by various factors such as sputtering power, temperature, and the substrate material. The effect of film thickness of the Ti film on the phase structure was discussed in Reference. Figure 3 presents the X-ray diffraction patterns of pristine wood and Ti-treated wood samples of varying thicknesses. As a fiber material, the diffraction peaks at 15° and 22° correspond to the (101) and (002) crystal planes of cellulose in the wood [57]. The films of varying thicknesses exhibited a strong Ti (002) orientation. The (100) and (101) diffraction peaks along with the (103) peak at diffraction angles of 34.5°, 40.2°, and 70.9°, respectively, intensified with the increase in film thickness, while other diffractions did not. Thickness affects the degree of crystallization but does not alter the hexagonal structure of the Ti film. Thin film deposition is a continuous and complex process, which includes three stages, i.e., the film-forming process through the island-shaped membrane, mesh membrane, and continuous membrane. The diffusion and migration of sputtered atoms on the surface of the substrate play a crucial role in determining the microstructure of the film.

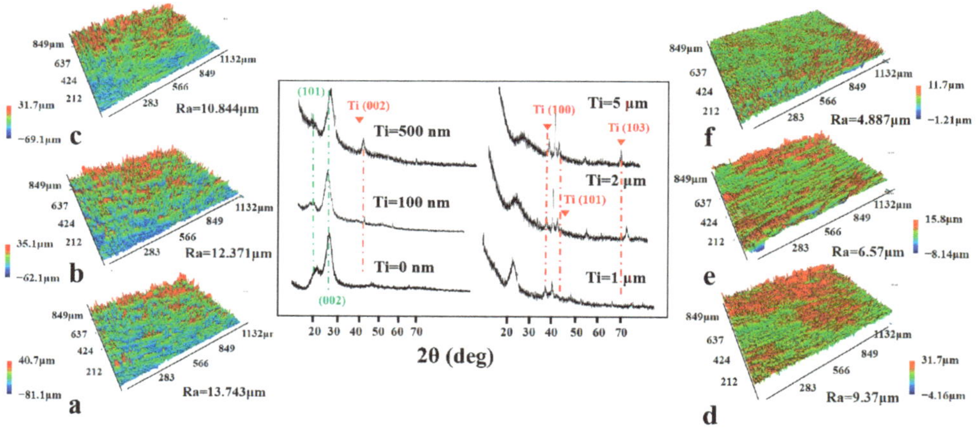

Figure 3. XRD spectra and 3D scan image of the surfaces of the pristine wood (**a**), (**b**–**f**) the Ti−treated wood with different deposition thicknesses of 100 nm, 500 nm, 1 μm, 2 μm, and 5 μm, respectively.

In order to evaluate the different thicknesses of the film on the surface roughness of the sample, a non-contact surface profiler test method was used. In Figure 3, the three-dimensional image illustrates that the roughness of the original wood surface was 13.74 μm; the blue concave area was more widely distributed, and the concave surface was due to the texture of the wood cell wall and cell cavity [58]. As the thickness of the sputtered film increased, the roughness gradually decreased. The roughness of the 5 μm thick Ti wood was 4.887 μm. There were a few red raised areas on the predominantly green smooth surfaces, indicating that the sputtered nanoparticles were mainly deposited in the cell cavity of the wood, thereby reducing the surface roughness [59]. The changes in roughness revealed that the Ti atoms preferentially deposited on the lower regions of the wood surface. Although the lateral diffusion of energy gradually spread on the surface, the vacancy on the surface was occupied so that the cell cavities of the wood were filled, and the wood surface was flat [60].

3.2. *Dynamic Mechanical Analysis of a Ti/Wood Composite*

DMA test results, including the stress–strain curves and the storage and loss data, are illustrated in Figure 4 [61]. Figure 4a shows the stress–strain curves of the wood with Ti films of different thicknesses, where the elastic modulus of samples increased significantly due to the increased slope [62]. The elastic modulus of 0 nm, 100 nm, 500 nm, 1 μm, 2 μm, and 5 μm thick film samples were measured as 33 MPa, 36.5 MPa, 34.78 MPa, 40.5 MPa, 41 MPa, and 41.5 MPa, respectively. These changes were due to the dislocations within the grain structure [63]. The interaction of dislocations formed a dislocation network, which inhibited further grain movement and consequently reinforced the films. However, when the Ti film thickness reached 1 μm or more, changes in the Ti–wood elastic modulus became less obvious, and dislocation activity most likely occurred simultaneously with grain boundary slip and diffusion [64]. Furthermore, the dislocations propagated through the grains and were eventually absorbed by the grain boundaries [65]. The storage modulus of the pristine wood was 3200 MPa, increasing by 18% to 3800 MPa for the 5 μm Ti–wood at a strain of 0.015% (Figure 4b). The loss modulus also increased linearly [66]. Thus, the rigidity of the metal film can improve the elastic modulus of wood samples. Based on this result, we aim to make use of these changes in the TTRB. Then, the three-point bending dynamic mechanical tests were performed on the Ti-TTRB samples. In Figure 4a, the elastic moduli of TTRB samples with different thicknesses of Ti films were measured as 22 MPa, 26.5 MPa, 28.2 MPa, 25.5 MPa, 23 MPa, and 22.5 MPa, respectively. It was clearly demonstrated that the stress–strain curve changes, and the Ti-TTRB samples with different thicknesses were irregular. Due to the very high modulus of the TTRB sample, the stiffness of the Ti film had a negligible effect compared to the much larger modulus of the substrate. From another point of view, the storage modulus can be released under alternating strains. However, the storage modulus of TTRB samples in the experiment did not exhibit the consumption state when the strain increased (Figure 4b). This result suggested that the TTRB sample does not yield under the driving force. Therefore, this testing method does not work for TTRBs.

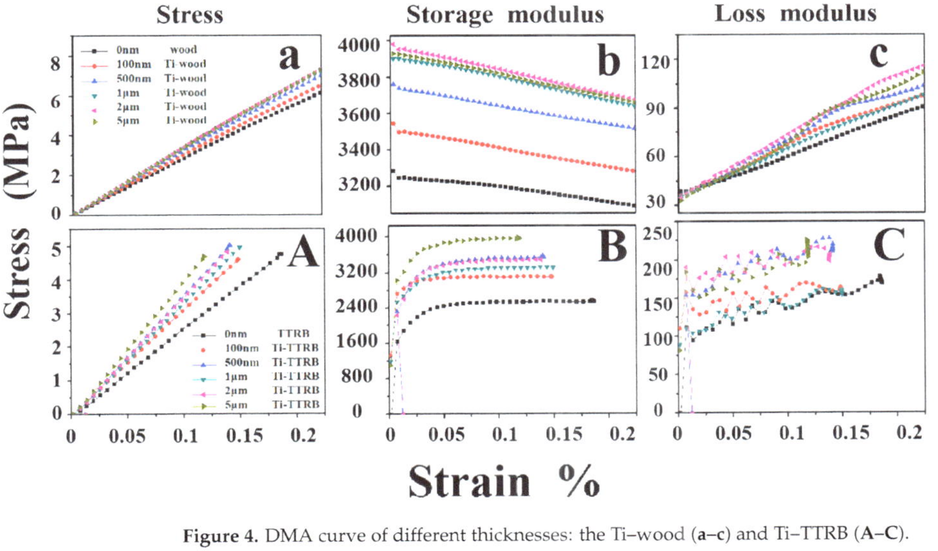

Figure 4. DMA curve of different thicknesses: the Ti–wood (**a**–**c**) and Ti–TTRB (**A**–**C**).

3.3. Rebound Distance and TTRB Vibrations

As shown in Figure 5c, a TTRB testing method was developed, in which the TTRB was fixed on a leveled countertop under a constant force. The table tennis ball was dropped freely from a height of 30 cm to hit the TTRB. The vertical rebound height of the table tennis ball after its first rebound, the vibration image of the TTRB after first hitting, the fall point of the ball, and the vibration point were examined and recorded. As shown in Figure 5a, the TTRB can oscillate around the equilibrium position, reciprocating its vibration after the ball hits it. The vibration attenuation curve was not completely regular, primarily because, when the TTRB was subjected to external forces, the different layers of the TTRB experienced varying stress force superposition effects due to the inhomogeneity of the TTRB. The external force first hits the panel layer, and the hardness of the layer decides how deep to swallow the ball. The term "swallowing" the ball means that the TTRB hits the ball and gives feedback to the athlete's hand feel [32,67]. The description of the hand feel was also three levels, namely, hard, soft, and control. The hand feel was defined as the initial amplitude of TTRB vibrations, while the amplitude representing the characteristic range and intensity of the vibrations (Figures S1 and S2). In the red circle in Figure 5a, the different thicknesses of the Ti films of the TTRB exhibited varying initial amplitudes, and the amplitude reduced from 2.6 μm (0 nm Ti-TTRB) to 2.1 μm (5 μm Ti-TTRB) as the film thickness increased, indicating that the rigid metal film can effectively improve the hardness of TTRBs (Figure S3). Figure 5b shows the vertical rebound height (VRH) of the ball, showing that as the film thickness increased, the VRH increased from 20.8 cm (pristine TTRB) to 21.9 cm (500 nm Ti film TTRB) and further increased to 23.7 cm (5 μm Ti film TTRB). These results suggest that the compatibility between the Ti film and the TTRB significantly enhances the mechanical performance of the composite.

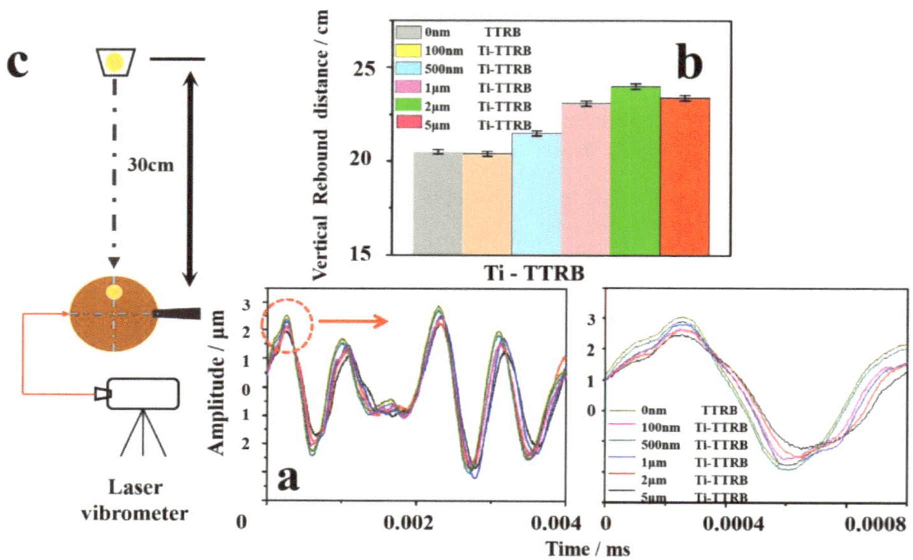

Figure 5. (**a**,**b**) Vibration curve and vertical rebound distance images of TTRBs with different thicknesses of Ti films. (**c**) The schematic diagram of ball and laser vibration measurement.

3.4. Dynamic Rebound Test

This test method simulated the actual use environment of the TTRB. The dynamic rebound device is shown in Figure 6b. The TTRB was hit by the table tennis ball at different speeds from an automatic ball machine. The same quality of table tennis balls was selected for the impact tests. Each test distance datum represents the average rebound distance of 50 balls. The height of the TTRB from the ground was 150 cm. The TTRB was fixed on a vertical countertop under a constant force. The test area remained closed to minimize the impact of air movement on the trajectory of the table tennis ball. Figure 6a shows the horizontal rebound distance (HRD), where the TTRB was hit by the table tennis ball at different speeds. When the ball hits the TTRB at a speed of 4 m/s, the HRDs of TTRBs with different Ti film thicknesses of 0 nm, 100 nm, 500 nm, 1 μm, 2 μm, and 5 μm were 182 cm, 198 cm, 208 cm, 211 cm, 213 cm, and 215 cm, respectively. The rebound distance increased due to the enhanced hardness of the TTRB provided by the metal Ti film. Meanwhile, when the ball speed was 8 m/s, the Ti-TTRB displayed the same advantage in rebound distance compared to the pristine TTRB. The rebound rates of the TTRB with five different Ti film thicknesses increased by 7%, 9%, 13%, 14%, and 13.7%, respectively. However, when the ball speeds were 10 m/s and 12 m/s, the effect of Ti film hardness on the improvement in the rebound distance of the TTRB was negligible. The main reason probably was due to the special structure of the TTRB. When the TTRB suffered a greater impact, the impact from the external delivery concentrated in the middle of the core board, causing all the energy of the TTRB to be activated (Figure S4). Although the metal film had a certain hardness, in a high-intensity application environment, the surface hardness could not enhance the overall hardness of the TTRB.

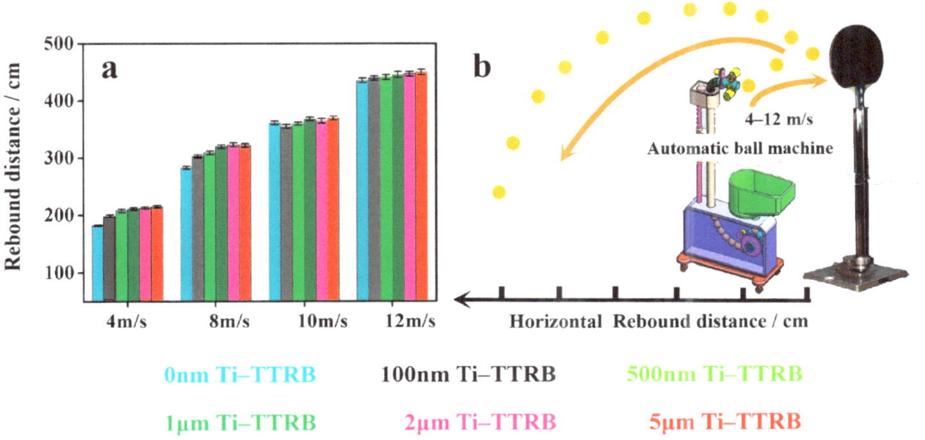

Figure 6. (**a**) Dynamic horizontal rebound distance of the TTRB with different thicknesses of Ti films. (**b**) The schematic diagram of the test rebound distance detection equipment.

4. Conclusions

In this study, an easy and efficient method for TTRB surface hardening was developed through the magnetron sputtering process. The effects of the thickness of the Ti film on the wood surface were explored. Compared to the pristine wood, the modulus of the wood with different Ti coating thicknesses increased by 5.1% to 24%. Furthermore, three types of TTRB mechanical property testing devices were developed, including free fall rebound, laser vibration, and dynamic rebound tests. The results indicate that the VRH of the different thicknesses of Ti-TTRB increased by 5% to 13%, and the amplitude reduced by 12% to 19% due to the rigidity of the Ti film. The 5 μm thick Ti films deposited on the TTRB surface exhibited the best mechanical properties, with the initial amplitude and VRH reaching 2.1 μm and 23.7 cm, respectively. Moreover, in the dynamic rebound test, the rebound rate increased by 13.7% when the ball speed was 8 m/s. However, as the hitting speed increased, the rigidity effect of the Ti film gradually decreased. When the ball speed was above 10 m/s, there was almost no difference in the rebound rate of TTRB with different thicknesses of Ti films. In conclusion, the panel board is a key factor for determining the hardness of the TTRB, which is mainly reflected by two indicators: initial amplitude and rebound distance. The increased hardness of the TTRB helps improve the speed of the ball and the player's hitting feeling, which is suitable for offensive players. The findings in this work pave the way for innovative developments in TTRBs and the unification of TTRB evaluation standards.

Supplementary Materials: The following supporting information can be downloaded at https://www.mdpi.com/article/10.3390/polym17050583/s1, Table S1: List of samples characterized; Figure S1: Vertical rebound distance images of TTRB of different brand types; Figure S2: Vibration curve images of TTRB of different brand types; Figure S3: Vibration curve images of TTRB of different brand types; Figure S4: Dynamic horizontal rebound distance of TTRB with different brand types.

Author Contributions: Conceptualization, W.B. and H.C.; Data Curation, R.X. and Z.Y.; Formal Analysis, W.B. and X.X.; Investigation, S.D. and H.L.; Methodology, W.B. and H.C.; Project Administration, H.C.; Supervision, H.C. and S.D.; Validation, W.B., H.C.; Visualization, H.C.; Writing—Original Draft, W.B. and Y.T.; Writing—Review and Editing, W.B. and S.D. All authors have read and agreed to the published version of the manuscript.

Funding: This research was funded by the 2024 Higher Education Research Project (No. KT2024023).

Institutional Review Board Statement: Not applicable.

Data Availability Statement: The original contributions presented in this study are included in the article and Supplementary Material. Further inquiries can be directed to the corresponding author.

Conflicts of Interest: The authors declare no conflicts of interest.

References

1. Zhu, H.; Luo, W.; Ciesielski, P.N.; Fang, Z.; Zhu, J.Y.; Henriksson, G.; Himmel, M.E.; Hu, L. Wood-Derived Materials for Green Electronics, Biological Devices, and Energy Applications. *Chem. Rev.* **2016**, *116*, 9305–9374. [CrossRef]
2. Qin, Q.; Zeng, S.; Duan, G.; Liu, Y.; Han, X.; Yu, R.; Huang, Y.; Zhang, C.; Han, J.; Jiang, S. "Bottom-up" and "top-down" strategies toward strong cellulose-based materials. *Chem. Soc. Rev.* **2024**, *53*, 9306–9343. [CrossRef] [PubMed]
3. Kopac, J.; Sali, S. Wood: An important material in manufacturing technology. *J. Mater. Process. Technol.* **2003**, *133*, 134–142. [CrossRef]
4. Li, T.; Zhu, M.; Yang, Z.; Song, J.; Dai, J.; Yao, Y.; Luo, W.; Pastel, G.; Yang, B.; Hu, L. Wood Composite as an Energy Efficient Building Material: Guided Sunlight Transmittance and Effective Thermal Insulation. *Adv. Energy Mater.* **2016**, *6*, 1601122. [CrossRef]
5. Li, Z.; Chen, C.; Xie, H.; Yao, Y.; Zhang, X.; Brozena, A.; Li, J.; Ding, Y.; Zhao, X.; Hong, M.; et al. Sustainable high-strength macrofibres extracted from natural bamboo. *Nat. Sustain.* **2021**, *5*, 235–244. [CrossRef]
6. Garemark, J.; Ram, F.; Liu, L.; Sapouna, I.; Cortes Ruiz, M.F.; Larsson, P.T.; Li, Y. Advancing Hydrovoltaic Energy Harvesting from Wood through Cell Wall Nanoengineering. *Adv. Funct. Mater.* **2022**, *33*, 2208933. [CrossRef]
7. Gao, X.; Zhu, D.; Fan, S.; Rahman, M.Z.; Guo, S.; Chen, F. Structural and mechanical properties of bamboo fiber bundle and fiber/bundle reinforced composites: A review. *J. Mater. Res. Technol.* **2022**, *19*, 1162–1190. [CrossRef]
8. Zhao, X.; Liu, Y.; Zhao, L.; Yazdkhasti, A.; Mao, Y.; Siciliano, A.P.; Dai, J.; Jing, S.; Xie, H.; Li, Z.; et al. A scalable high-porosity wood for sound absorption and thermal insulation. *Nat. Sustain.* **2023**, *6*, 306–315. [CrossRef]
9. Li, Z.; Chen, C.; Mi, R.; Gan, W.; Dai, J.; Jiao, M.; Xie, H.; Yao, Y.; Xiao, S.; Hu, L. A Strong, Tough, and Scalable Structural Material from Fast-Growing Bamboo. *Adv. Mater.* **2020**, *32*, 1906308. [CrossRef] [PubMed]
10. Wegst, U.G.K. Bamboo and Wood in Musical Instruments. *Annu. Rev. Mater. Res.* **2008**, *38*, 323–349. [CrossRef]
11. Liu, J.Q.; Wang, B.; Zhao, X.; Dou, Y. The Application of Rubber Materials on Table Tennis Racket. *Appl. Mech. Mater.* **2013**, *473*, 116–120. [CrossRef]
12. Castanié, B.; Peignon, A.; Marc, C.; Eyma, F.; Cantarel, A.; Serra, J.; Curti, R.; Hadiji, H.; Denaud, L.; Girardon, S.; et al. Wood and plywood as eco-materials for sustainable mobility: A review. *Compos. Struct.* **2024**, *329*, 117790. [CrossRef]
13. Lirya Silva, L.C.; Lima, F.O.; Monteiro, S.N.; Azevedo, A.R.G.; Christoforo, A.L.; Ferreira, B.S.; Goveia, D.; de Campos, C.I. The use of nano-structured cellulose to improve plywood: A review. *J. Mater. Res. Technol.* **2025**, *35*, 392–401. [CrossRef]
14. Sugiyama, S.; Kiuchi, M.; Yanagimoto, J. Application of semisolid joining—Part 4 glass/metal, plastic/metal, or wood/metal joining. *J. Mater. Process. Technol.* **2008**, *201*, 623–628. [CrossRef]
15. Arifin, A.M.T.; Hassan, M.F.; Ismail, A.E.; Rahim, M.Z.; Ibrahim, M.R.; Haq, R.H.A.; Rahman, M.N.A.; Yunos, M.Z.; Amin, M.H.M. Investigation on Suitability of Natural Fibre as Replacement Material for Table Tennis Blade. *IOP Conf. Ser. Mater. Sci. Eng.* **2017**, *226*, 012037. [CrossRef]
16. Al-Oqla, F.M.; Hayajneh, M.T.; Nawafleh, N. Advanced synthetic and biobased composite materials in sustainable applications: A comprehensive review. *Emergent Mater.* **2023**, *6*, 809–826. [CrossRef]
17. Ding, Y.; Pang, Z.; Lan, K.; Yao, Y.; Panzarasa, G.; Xu, L.; Lo Ricco, M.; Rammer, D.R.; Zhu, J.Y.; Hu, M.; et al. Emerging Engineered Wood for Building Applications. *Chem. Rev.* **2022**, *123*, 1843–1888. [CrossRef] [PubMed]
18. Montanari, C.; Li, Y.; Chen, H.; Yan, M.; Berglund, L.A. Transparent Wood for Thermal Energy Storage and Reversible Optical Transmittance. *ACS Appl. Mater. Interfaces* **2019**, *11*, 20465–20472. [CrossRef]

19. Li, J.; Chen, C.; Chen, Q.; Li, Z.; Xiao, S.; Gao, J.; He, S.; Lin, Z.; Tang, H.; Li, T.; et al. Kilogram-scale production of strong and smart cellulosic fibers featuring unidirectional fibril alignment. *Natl. Sci. Rev.* **2024**, *11*, nwae270. [CrossRef] [PubMed]
20. Zhang, K.; Zhai, W.; Cao, Z.; Wang, Y.; Li, L.; Li, J.; Liu, J.; Xie, Y.; Gan, W. A sustainable pore wall strengthening for strong and fire-retardant nanopolymerised wood. *Chem. Eng. J.* **2024**, *480*, 147971. [CrossRef]
21. Fu, Z.; Lu, Y.; Wu, G.; Bai, L.; Barker-Rothschild, D.; Lyu, J.; Liu, S.; Rojas, O.J. Wood elasticity and compressible wood-based materials: Functional design and applications. *Prog. Mater. Sci.* **2025**, *147*, 101354. [CrossRef]
22. Gan, W.; Wang, Y.; Xiao, S.; Gao, R.; Shang, Y.; Xie, Y.; Liu, J.; Li, J. Magnetically Driven 3D Cellulose Film for Improved Energy Efficiency in Solar Evaporation. *ACS Appl. Mater. Interfaces* **2021**, *13*, 7756–7765. [CrossRef]
23. Liu, Y.; Lv, Z.; Zhou, J.; Cui, Z.; Li, W.; Yu, J.; Chen, L.; Wang, X.; Wang, M.; Liu, K.; et al. Muscle-Inspired Formable Wood-Based Phase Change Materials. *Adv. Mater.* **2024**, *36*, 2406915. [CrossRef]
24. Zhang, T.; Qu, J.; Wu, J.; Jiao, F.Z.; Li, C.; Gao, F.L.; Liu, J.; Yu, Z.Z.; Li, X. All-In-One Self-Floating Wood-Based Solar-Thermal Evaporators for Simultaneous Solar Steam Generation and Catalytic Degradation. *Adv. Funct. Mater.* **2024**, *34*, 2403505. [CrossRef]
25. Zelinka, S.L.; Altgen, M.; Emmerich, L.; Guigo, N.; Keplinger, T.; Kymäläinen, M.; Thybring, E.E.; Thygesen, L.G. Review of Wood Modification and Wood Functionalization Technologies. *Forests* **2022**, *13*, 1004. [CrossRef]
26. Li, T.; Song, J.; Zhao, X.; Yang, Z.; Pastel, G.; Xu, S.; Jia, C.; Dai, J.; Chen, C.; Gong, A.; et al. Anisotropic, lightweight, strong, and super thermally insulating nanowood with naturally aligned nanocellulose. *Sci. Adv.* **2018**, *4*, eaar3724. [CrossRef] [PubMed]
27. Zhang, K.; Cai, L.; Nilghaz, A.; Chen, G.; Wan, X.; Tian, J. Enhancing output performance of surface-modified wood sponge-carbon black ink hygroelectric generator via moisture-triggered galvanic cell. *Nano Energy* **2022**, *98*, 107288. [CrossRef]
28. Biswas, S.K.; Sano, H.; Shams, M.I.; Yano, H. Three-Dimensional-Moldable Nanofiber-Reinforced Transparent Composites with a Hierarchically Self-Assembled "Reverse" Nacre-like Architecture. *ACS Appl. Mater. Interfaces* **2017**, *9*, 30177–30184. [CrossRef] [PubMed]
29. Hill, C.A.S. *Wood Modification: Chemical, Thermal and Other Processes*; Stevens, C.V., Ed.; John Wiley & Sons, Inc.: Hoboken, NJ, USA, 2006; p. 260.
30. Li, C.; Lei, H.; Wu, Z.; Xi, X.; Du, G.; Pizzi, A. Fully Biobased Adhesive from Glucose and Citric Acid for Plywood with High Performance. *ACS Appl. Mater. Interfaces* **2022**, *14*, 23859–23867. [CrossRef]
31. Deng, J.; Wei, X.; Zhou, H.; Wang, G.; Zhang, S. Inspiration from table tennis racket: Preparation of rubber-wood-bamboo laminated composite (RWBLC) and its response characteristics to cyclic perpendicular compressive load. *Compos. Struct.* **2020**, *241*, 112135. [CrossRef]
32. Brich, Q.; Casals, M.; Crespo, M.; Reid, M.; Baiget, E. Quantifying Hitting Load in Racket Sports: A Scoping Review of Key Technologies. *Int. J. Sports Physiol. Perform.* **2024**, *19*, 519–532. [CrossRef] [PubMed]
33. Raab, M.; Masters, R.S.; Maxwell, J.P. Improving the 'how' and 'what' decisions of elite table tennis players. *Hum. Mov. Sci.* **2005**, *24*, 326–344. [CrossRef]
34. Manin, L.; Poggi, M.; Havard, N. Vibrations of table tennis racket composite wood blades: Modeling and experiments. *Procedia Eng.* **2012**, *34*, 694–699. [CrossRef]
35. Iino, Y.; Kojima, T. Effect of the racket mass and the rate of strokes on kinematics and kinetics in the table tennis topspin backhand. *J. Sports Sci.* **2015**, *34*, 721–729. [CrossRef] [PubMed]
36. Rinaldi, R.G.; Manin, L.; Moineau, S.; Havard, N. Table Tennis Ball Impacting Racket Polymeric Coatings: Experiments and Modeling of Key Performance Metrics. *Appl. Sci.* **2019**, *9*, 158. [CrossRef]
37. Zhou, X.; Wang, L.; Huang, D.; Liang, Y.; Shi, Q.; Yaying, H.; Zhang, M.; Pu, H.; Wen, W.; Wu, J. Smart Table Tennis Racket with Tunable Stiffness for Diverse Play Styles and Unconventional Technique Training. *Adv. Mater. Technol.* **2021**, *6*, 2100535. [CrossRef]
38. Carius, D.; Kaminski, E.; Clauß, M.; Schewe, Y.; Ryk, L.; Ragert, P. Quantifying motor adaptation in a sport-specific table tennis setting. *Sci. Rep.* **2024**, *14*, 601. [CrossRef] [PubMed]
39. Mao, C.; Liu, T.; Li, X.; Lu, Z.; Li, Z.; Xing, K.; Chen, L.; Sun, Y. A Comparative Biomechanical Analysis of Topspin Forehand against Topspin and Backspin in Table Tennis. *Appl. Sci.* **2023**, *13*, 8119. [CrossRef]
40. Iino, Y.; Kojima, T. Kinematics of table tennis topspin forehands: Effects of performance level and ball spin. *J. Sports Sci.* **2009**, *27*, 1311–1321. [CrossRef] [PubMed]
41. Kelly, P.J.; Arnell, R.D. Magnetron sputtering: A review of recent developments and applications. *Vacuum* **1999**, *56*, 159–172. [CrossRef]
42. Zhang, Y.; Ma, H.; Yi, M.; Shen, Z.; Yu, X.; Zhang, X. Magnetron-sputtering fabrication of noble metal nanodots coated TiO_2 nanoparticles with enhanced photocatalytic performance. *Mater. Des.* **2017**, *125*, 94–99. [CrossRef]
43. Liang, J.; Liu, Q.; Li, T.; Luo, Y.; Lu, S.; Shi, X.; Zhang, F.; Asiri, A.M.; Sun, X. Magnetron sputtering enabled sustainable synthesis of nanomaterials for energy electrocatalysis. *Green Chem.* **2021**, *23*, 2834–2867. [CrossRef]
44. Li, Y.; Cui, J.; Shen, H.; Liu, C.; Wu, P.; Qian, Z.; Duan, Y.; Liu, D. Useful spontaneous hygroelectricity from ambient air by ionic wood. *Nano Energy* **2022**, *96*, 107065. [CrossRef]

45. Jiao, Y.; Wan, C.; Zhang, W.; Bao, W.; Li, J. Carbon Fibers Encapsulated with Nano-Copper: A Core–Shell Structured Composite for Antibacterial and Electromagnetic Interference Shielding Applications. *Nanomaterials* **2019**, *9*, 460. [CrossRef]
46. Bao, W.; Zhang, M.; Jia, Z.; Jiao, Y.; Cai, L.; Liang, D.; Li, J. Cu thin films on wood surface for robust superhydrophobicity by magnetron sputtering treatment with perfluorocarboxylic acid. *Eur. J. Wood Wood Prod.* **2018**, *77*, 115–123. [CrossRef]
47. Bao, W.; Jia, Z.; Cai, L.; Liang, D.; Li, J. Fabrication of a superamphiphobic surface on the bamboo substrate. *Eur. J. Wood Wood Prod.* **2018**, *76*, 1595–1603. [CrossRef]
48. Wang, Q.; Xiao, S.; Shi, S.Q.; Xu, S.; Cai, L. Self-bonded natural fiber product with high hydrophobic and EMI shielding performance via magnetron sputtering Cu film. *Appl. Surf. Sci.* **2019**, *475*, 947–952. [CrossRef]
49. Wan, C.; Jiao, Y.; Bao, W.; Gao, H.; Wu, Y.; Li, J. Self-stacked multilayer FeOCl supported on a cellulose-derived carbon aerogel: A new and high-performance anode material for supercapacitors. *J. Mater. Chem. A* **2019**, *7*, 9556–9564. [CrossRef]
50. Simo-Tagne, M.; Rémond, R.; Rogaume, Y.; Zoulalian, A.; Perré, P. Characterization of sorption behavior and mass transfer properties of four central africa tropical woods: Ayous, Sapele, Frake, Lotofa. *Maderas. Cienc. Tecnol.* **2016**, online. [CrossRef]
51. Mahnert, K.-C.; Adamopoulos, S.; Koch, G.; Militz, H. Topochemistry of heat-treated and N-methylol melamine-modified wood of koto (*Pterygota macrocarpa* K. Schum.) and limba (*Terminalia superba* Engl. et. Diels). *Holzforschung* **2013**, *67*, 137–146. [CrossRef]
52. Chawla, V.; Jayaganthan, R.; Chawla, A.K.; Chandra, R. Microstructural characterizations of magnetron sputtered Ti films on glass substrate. *J. Mater. Process. Technol.* **2009**, *209*, 3444–3451. [CrossRef]
53. Chen, C.; Kuang, Y.; Zhu, S.; Burgert, I.; Keplinger, T.; Gong, A.; Li, T.; Berglund, L.; Eichhorn, S.J.; Hu, L. Structure–property–function relationships of natural and engineered wood. *Nat. Rev. Mater.* **2020**, *5*, 642–666. [CrossRef]
54. Zhu, S.; Kumar Biswas, S.; Qiu, Z.; Yue, Y.; Fu, Q.; Jiang, F.; Han, J. Transparent wood-based functional materials via a top-down approach. *Prog. Mater. Sci.* **2023**, *132*, 101025. [CrossRef]
55. Jia, C.; Chen, C.; Mi, R.; Li, T.; Dai, J.; Yang, Z.; Pei, Y.; He, S.; Bian, H.; Jang, S.H.; et al. Clear Wood toward High-Performance Building Materials. *ACS Nano* **2019**, *13*, 9993–10001. [CrossRef]
56. Ogasawara, T.; Tsuda, T.; Takeda, N. Stress–strain behavior of multi-walled carbon nanotube/PEEK composites. *Compos. Sci. Technol.* **2011**, *71*, 73–78. [CrossRef]
57. Zhu, M.; Li, Y.; Chen, F.; Zhu, X.; Dai, J.; Li, Y.; Yang, Z.; Yan, X.; Song, J.; Wang, Y.; et al. Plasmonic Wood for High-Efficiency Solar Steam Generation. *Adv. Energy Mater.* **2017**, *8*, 1701028. [CrossRef]
58. Gao, H.; Li, Y.; Xie, Y.; Liang, D.; Li, J.; Wang, Y.; Xiao, Z.; Wang, H.; Gan, W.; Pattelli, L.; et al. Optical wood with switchable solar transmittance for all-round thermal management. *Compos. Part B Eng.* **2024**, *275*, 111287. [CrossRef]
59. Wan, J.; Song, J.; Yang, Z.; Kirsch, D.; Jia, C.; Xu, R.; Dai, J.; Zhu, M.; Xu, L.; Chen, C.; et al. Highly Anisotropic Conductors. *Adv. Mater.* **2017**, *29*, 1703331. [CrossRef] [PubMed]
60. Jiang, Q.; Lu, H.M. Size dependent interface energy and its applications. *Surf. Sci. Rep.* **2008**, *63*, 427–464. [CrossRef]
61. Ye, X.; Wang, H.; Zheng, K.; Wu, Z.; Zhou, H.; Tian, K.; Su, Z.; Tian, X. The interface designing and reinforced features of wood fiber/polypropylene composites: Wood fiber adopting nano-zinc-oxide-coating via ion assembly. *Compos. Sci. Technol.* **2016**, *124*, 1–9. [CrossRef]
62. Pitt, K.; Lopez-Botello, O.; Lafferty, A.D.; Todd, I.; Mumtaz, K. Investigation into the material properties of wooden composite structures with in-situ fibre reinforcement using additive manufacturing. *Compos. Sci. Technol.* **2017**, *138*, 32–39. [CrossRef]
63. Meng, Z.-Y.; Chen, L.; Zhong, H.-Y.; Yang, R.; Liu, X.-F.; Wang, Y.-Z. Effect of different dimensional carbon nanoparticles on the shape memory behavior of thermotropic liquid crystalline polymer. *Compos. Sci. Technol.* **2017**, *138*, 8–14. [CrossRef]
64. Fuentes, C.A.; Brughmans, G.; Tran, L.Q.N.; Dupont-Gillain, C.; Verpoest, I.; Van Vuure, A.W. Mechanical behaviour and practical adhesion at a bamboo composite interface: Physical adhesion and mechanical interlocking. *Compos. Sci. Technol.* **2015**, *109*, 40–47. [CrossRef]
65. Lee, J.-Y.; Kumar, V.; Tang, X.-W.; Lee, D.-J. Mechanical and electrical behavior of rubber nanocomposites under static and cyclic strain. *Compos. Sci. Technol.* **2017**, *142*, 1–9. [CrossRef]
66. Dong, J.; Jia, C.; Wang, M.; Fang, X.; Wei, H.; Xie, H.; Zhang, T.; He, J.; Jiang, Z.; Huang, Y. Improved mechanical properties of carbon fiber-reinforced epoxy composites by growing carbon black on carbon fiber surface. *Compos. Sci. Technol.* **2017**, *149*, 75–80. [CrossRef]
67. Zhang, S.; Chen, G.; Wu, Q.; Li, X. The Interplay Between Table Tennis Skill Development and Sports Performance: A Comprehensive Review. *Pac. Int. J.* **2023**, *6*, 150–156. [CrossRef]

polymers

Article

Preparation of Microcrystalline Cellulose/N-(2-aminoethyl)-3-Aminopropyl Methyl Dimethoxysilane Composite Aerogel and Adsorption Properties for Formaldehyde

Yaning Li [1,2,†], Zhongzheng Liu [1,2,†], Chuanxi Chi [1,2], Bin Yuan [1,2], Yang Zhang [1,2], Guiquan Jiang [1,2,*] and Jianxi Song [1,2,*]

1 Key Laboratory of Wooden Materials Science and Engineering of Jilin Province, Beihua University, Jilin City 132013, China; liyaning@beihua.edu.cn (Y.L.); liuzhongzheng@beihua.edu.cn (Z.L.); chichuanxi@beihua.edu.cn (C.C.); yuanbin@beihua.edu.cn (B.Y.); zhangyang@beihua.edu.cn (Y.Z.)
2 Key Laboratory for the Structure and Function of Polysaccharides in Traditional Chinese Medicine, Beihua University, Jilin City 132013, China
* Correspondence: jgq@beihua.edu.cn (G.J.); songjianxi@beihua.edu.cn (J.S.)
† These authors contributed equally to this work.

Abstract: Air pollution is related to the development of the national economy and people's livelihoods. Formaldehyde, as one of the main pollutants in the air, affects people's physical and mental health. In order to remove formaldehyde and better protect the health of residents, it is necessary to develop efficient adsorption materials. In this study, APMDS-modified cellulose composite aerogel microcrystalline was investigated. The adsorption of formaldehyde by the MCC/APMDS (Microcrystalline Cellulose/N-(2-aminoethyl)-3- Aminopropyl Methyl Dimethoxysilane) composite aerogel mainly relied upon the reaction of the protonated $-NH_3{}^+$ group in APMDS with formaldehyde to form a Schiff base to achieve the effect of deformaldehyde. Meanwhile, the modification of the aerogel reduced the pore volume and specific surface area, and the average pore size increased to 14.56 nm, which enhanced the adsorption capacity of formaldehyde, and the adsorption amount reached 9.52 mg/g. This study provides valuable information for the preparation of adsorbent materials with high formaldehyde adsorption capacity for air purification.

Keywords: cellulose; aerogel; adsorption; formaldehyde

Citation: Li, Y.; Liu, Z.; Chi, C.; Yuan, B.; Zhang, Y.; Jiang, G.; Song, J. Preparation of Microcrystalline Cellulose/N-(2-aminoethyl)-3-Aminopropyl Methyl Dimethoxysilane Composite Aerogel and Adsorption Properties for Formaldehyde. *Polymers* **2023**, *15*, 3155. https://doi.org/10.3390/polym15153155

Academic Editor: Iolanda De Marco

Received: 6 July 2023
Revised: 19 July 2023
Accepted: 21 July 2023
Published: 25 July 2023

1. Introduction

According to research, per capita indoor stay time usually exceeds 90% [1–4]. Formaldehyde is the poster child for indoor pollutants, and more than 65% of formaldehyde is used in the production of synthetic resins in building materials [5]. There is significant evidence that formaldehyde has a positive and important effect on the development of nasopharyngeal cancer [6–8]. Indoor formaldehyde pollution sources are widespread. Urea-formaldehyde resins, melamine resins, phenolic resins used as adhesives for furniture boards [9–11], paint, lampblack and smoking in home kitchens [12] are the most common sources [13]. In addition, formaldehyde will make indoor residents experience headaches, nausea, and mucosal inflammation; nasopharynx, eye and throat irritation; asthma and allergic rhinitis; eczema, fatigue, etc. [13,14]. Long-term inhalation of formaldehyde can also lead to lower learning and work efficiency and have an unfavorable effect on the physical and mental health of residents [15–17].

So far, several indoor formaldehyde removal strategies have been developed, such as adsorption, plasma-catalyzed decomposition, photo-catalytic degradation and plant purification methods. Adsorption is one of the most convenient methods for indoor formaldehyde removal. Its advantages mainly lie in its fast removal rate, high efficiency, simple operation, energy savings and low cost. Usually, adsorption methods are divided into physical adsorption and chemical adsorption methods. Physical adsorption relies on weak intermolecular forces between the adsorbent and formaldehyde, and therefore, its stability is poor. In contrast, chemical adsorption relies on the surfactant functional group of the adsorbent forming a strong chemical bond with HCHO. As a result, it is highly stable and selective. Therefore, many adsorbents, such as SiO_2 [18,19], Al_2O_3 [20], MOFs [21] and microporous carbonaceous materials [22,23], have been widely studied for indoor formaldehyde adsorption and removal. In recent years, various novel materials, such as active carbon fibers (ACFs), carbon nanotubes (CNTS), graphene, metal–organic skeletons (MOF) and porous organic polymers (pops) [7,9,22,24,25], have been introduced as effective adsorbents for removing HCHO. The adsorption performance of nitrogen-containing porous materials for HCHO is better than that of common adsorbents, and the adsorption performance is closely related to the type and content of surface nitrogen [26]. Therefore, surface modification of adsorbents is an effective way to improve adsorption capacity by increasing the adsorption's active site and the selectivity of HCHO.

Amino-modified cellulose aerogel adsorbents show a high adsorption capacity of HCHO. APMDS is often used for the functionalization of cellulose nanofibers to enhance the adsorption capacity of cellulose aerogel. APMDS-modified cellulose aerogels are usually achieved through two different methods. One is the synthesis of amino-silanized cellulose by freeze-drying cellulose and APMDS suspension, resulting in the the preparation of amino-silanized cellulose aerogel. Li et al. [27] reacted CNF suspension with APMDS suspension to prepare APMDS-modified CNF and finally prepared modified CNF aerogel through ultrasonic treatment and freeze-drying. Another approach is to functionalize cellulose hydrogels or cellulose membranes in APMDS solution and then proceed with supercritical carbon dioxide drying or freeze-drying to prepare. Zhang et al. [28] reacted prepared cellulose hydrogel spheres in 12 wt% APMDS solution for 12 h (100 °C) and finally obtained amino-modified A-NCC aerogel by freeze-drying.

Compared with traditional adsorption materials, nano-cellulose-based aerogels have better adsorption capacity, and their adsorption performance for formaldehyde can be further improved by means of amino functionalization. As an adsorbent, nano-cellulose-based aerogels are a potential field for current and future research.

2. Preparation of MCC/APMDS Composite Aerogel

Dissolving 0.1 g microcrystalline cellulose in 5 g 60 wt% LiBr solution (1:100), we magnetically stirred the solution until it was evenly dispersed. We then heated it in a 140 °C oil bath on high for 20 min until a clear homogeneous solution was formed. We poured the hot cellulose suspension into the mold and then used an ultrasonic cleaner to remove air bubbles from the cellulose suspension. After that, we cooled to room temperature to form a gel. We soaked the formed gel in deionized water and rinsed it with excessive deionized water until lithium bromide could not be detected by silver nitrate solution. The gel was then solvent displaced using ethanol and tert-butanol in an attempt to replace the water in the hydrogel with tert-butanol. Finally, it was frozen in an ultra-low-temperature freezer at −80 °C for 12 h. After removing the frozen sample, we placed it in a freeze dryer to obtain cellulose aerogel.

The cellulose hydrogels and tert-butanol (mass ratio 1:4) prepared above were placed in a 250 mL round-bottomed flask, and the pH of the mixture was adjusted to 4–5 by adding acetic acid to facilitate the hydrolysis of APMDS. Then, a certain amount of APMDS (0, 2, 4, 6 and 8 wt%, based on the total mass of the cellulose hydrogel and tert-butanol mixture) was added to this mixture, and the mixture was reacted at 90 °C for 4 h. After the reaction was completed, the APMDS and its self-condensation were removed by repeated washing three to five times with the tert-butanol solution. Meanwhile, solvent replacement was carried out, and the gel after solvent replacement was freeze-dried for 24 h to obtain MCC/APMDS composite aerogel. The above composite aerogels modified by APMDS with different mass fractions (2, 4, 6 and 8 wt%) were denoted as ACC-1, ACC-2, ACC-3 and ACC-4, respectively. The aerogel prepared with a mass fraction of 0 wt% was denoting as CAB aerogel.

According to the above method, 6 wt% APMDS was added to the mixture solution of cellulose hydrogel and tert-butanol (mass ratio 1:4). The reaction temperature was controlled at 90 °C, and the reaction time was set as 1, 2, 3, 4 and 5 h. After the reaction was completed, the MCC/APMDS composite aerogel with different reaction times (1, 2, 3, 4 and 5 h) was washed, solvent replacement was carried out, and the gel after solvent replacement was freeze-dried in accordance with the above methods to obtain ACC-1 h, ACC-2 h, ACC-3 h, ACC-4 h and ACC-5 h, respectively.

According to the above method, 6 wt% APMDS was also added to the mixture solution of cellulose hydrogel and tert-butanol (mass ratio 1:4). The reaction time was controlled as 4 h, and the reaction temperature was set as 60, 70, 80, 90 and 100 °C. After the reaction was complete, the MCC/APMDS composite aerogel with different reaction temperatures (60, 70, 80, 90 and 100 °C) was washed, solvent replacement was carried out, and the gel after solvent replacement was freeze-dried according to the above methods. Finally, the composite aerogel samples were obtained as ACC-60, ACC-70, ACC-80, ACC-90 and ACC-100, respectively.

3. Characterization Details

The morphology of the sample was observed by scanning electron microscope (JSM 7600F, Ltd., Tokyo, Japan), where the current was 10 μA, and the acceleration voltage was 5 kV. The structure of the sample was analyzed by an infrared spectrometer, where the scanning range was 4000–400 cm^{-1}, and the resolution was 4 cm^{-1}. After drying the aerogel sample at 60 °C for 12 h, the pore structure of the sample was determined by a specific surface area and porosity meter (ASAP 2020, Micromeritics Instrument Corporation, Norcross, Norcross, GA, USA). An X-ray diffractometer (D/max-2500VL/PC, Rigaku Corporation, Tokyo, Japan) was used to characterize the crystalline state of the corresponding aerogel samples and cellulose raw materials, where the acceleration voltage was 40 KV, the current was 50 mA, the scanning speed was 4°/min, and the scanning range was 5°–40°. The thermal stability of the corresponding aerogel samples and cellulose samples was analyzed by a thermos gravimetric analyzer (NETZSCH STA409PC, Netzsch Gerätebau GmbH, Selb, Germany), where the heating rate was 10 K/min, and the measuring range was 40–600 °C. X-ray photoelectron spectroscopy (THERMO, Thermo Fisher Scientific, Wilmington, NC, USA) was used to analyze the surface of the cellulose aerogel samples before and after modification.

4. Analysis of Adsorption of Gaseous Formaldehyde

A static adsorption experiment of gaseous formaldehyde was carried out on the prepared sample at room temperature. The 0.05 g aerogel sample was weighed and then dried in a vacuum drying oven at 80 °C for 12 h. We then put it into the transition chamber of a glove box (170 L). Weighing out 20 mg of paraformaldehyde, we put this into a round-bottomed flask, heat it at 70 °C in a water bath for 1 h and passed the formaldehyde gas generated by the reaction into the reaction chamber of the box. After the concentration of gaseous formaldehyde in the reactor became stable for 1 h, we opened the transition chamber and placed the weighed aerogel sample into the reaction chamber for adsorption. We placed the portable formaldehyde detector (PPM HTV in the UK) into the reaction chamber of the glove box in advance. After the aerogel sample began to adsorb, the sampling interval was 30 min in the early stage and 60 min in the late stage of adsorption. We then determined the concentration of formaldehyde and calculated the adsorption amount. The adsorption amount of formaldehyde was calculated using the following formula:

$$q_t = \frac{(C_0 - C_t)V}{m} \tag{1}$$

In the formula, q_t is the adsorption amount of formaldehyde at t time, mg/g; C_0 and C_t are the initial concentration and t time concentration of formaldehyde, mg/m^3; m is the mass of the aerogel, mg; V is the volume of the glove box, m^3.

5. Results and Discussion

5.1. SEM Analysis

Figure 1 shows a comparison of the microscopic morphologies observed by SEM of the CAB aerogel (Figure 1a) and the ACC-3 (Figure 1b) composite aerogel. It can be seen that the addition of APMDS did not significantly change the shape of the internal structure of the cellulose aerogel. The aerogel samples all showed a three-dimensional network structure with irregularly shaped nano-scale pores. The aerogel observed in Figure 1 is highly porous, indicating that the pore morphology did not collapse during processing. The CAB aerogel had a low density (21.8 mg/cm^3) and high porosity (98.2%) compared with the uniform porosity of the CAB aerogel, as with the introduction of APMDS, a large number of pores in the aerogel are blocked and cross linked, forming a flaky structure. This phenomenon may be due to the introduction of APMDS, which promotes the aggregation of cellulose during freezing and requires a certain space for the cross linking and aggregation of amine-loaded groups, thereby causing the pores to be blocked. From the comparison, it was found that Si and N elements appeared on ACC-3, which indicated that ADMPS had been successfully grafted onto the aerogel.

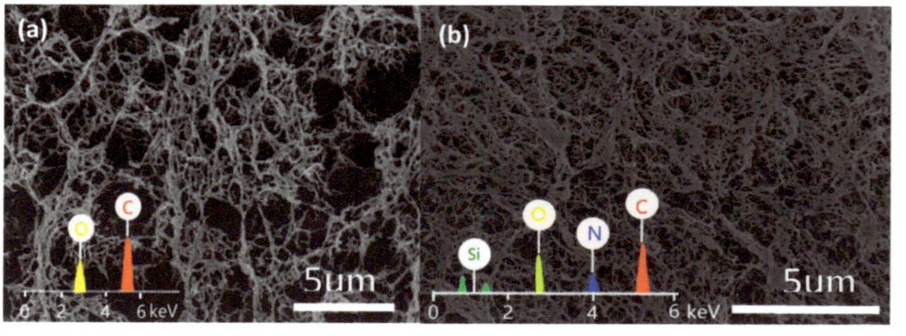

Figure 1. SEM-EDS diagram of CAB (**a**) and MCC/APMDS (**b**) composite aerogel.

5.2. N_2 Sorption Isotherms and BET Analysis, FT-IR Analysis, XRD Analysis

As shown in Figure 2a, similar isotherm curves were obtained, and the aerogel before and after modification had H1-type retention rings of type IV isotherms, indicating the typical mesoporous structure of the aerogel. P/P0 formed a large retention ring structure in the range of 0.8 to 1.0, which revealed the existence of the mesoporous structure in the aerogels, with abundant mesoporous and microporous. In addition, the pore size distribution curve shown in Figure 2b shows that the pore size of the aerogel before and after modification was concentrated in the range of 2–20 nm, and the peak value of the MCC/APMDS composite aerogel moved to the left compared to the CAB aerogel. This is possibly due to the change in pore structure distribution resulting from the grafting of APMDS to the cellulose chain. After APMDS was grafted onto the cellulose, the specific surface area and pore distribution of the aerogel were changed. After APMDS modification, the pore volume and specific surface area of the aerogel were reduced. By using BJH and BET methods to calculate the characteristics of pore parameters, it was concluded that the specific surface area and pore volume of the CAB aerogel and the ACC-3 aerogel were 127.1 m^2/g and 0.272 cm^3/g and 102.8 m^2/g and 0.212 cm^3/g, respectively. The average pore size of the modified aerogel increased from 12 nm to 14.56 nm, meaning that some may be covered. The results are consistent with those obtained from the examination of SEM images.

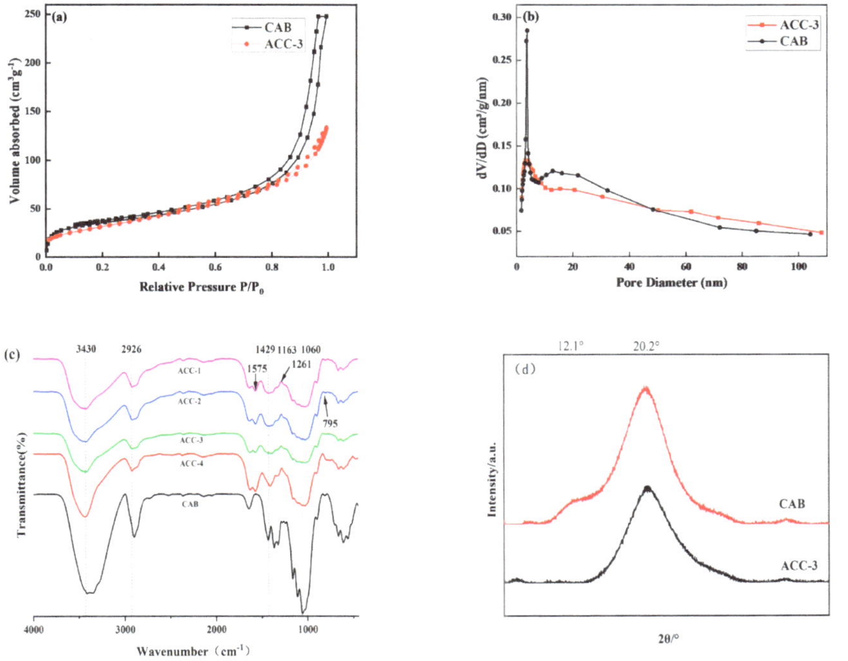

Figure 2. N_2 adsorption−desorption isotherm (**a**) and pore size distribution (**b**) of aerogels. FT-IR diagram of CAB and MCC/APMDS composite aerogel (**c**); XRD diagram of CAB and MCC/APMDS composite aerogel (**d**).

As shown in Table 1, the aerogel before and after modification is a lightweight material with high porosity. However, the density of the aerogel after amination modification with APMDS increased, and its porosity decreased. This may be because the introduction of APMDS leads to an increase in the mass of the cellulose aerogel, the blockage of pores inside the aerogel and a decrease in spatial structure, which leads to an increase in the density and a decrease in the porosity of the modified aerogel.

Table 1. Physical properties of CAB and MCC/APMDS composite aerogel.

Sample	BET Surface Area /(m^2 g^{-1})	Average Aperture/nm	Total Pore Volume/($cm^3 \cdot g^{-1}$)	Density (mg/cm^3)
CAB	127.1	12	0.272	21.8
ACC-3	102.8	14.56	0.212	26.7

Figure 2c shows the FTIR spectra of the CAB aerogel and the aerogel modified with APMDS. As shown, the infrared absorption spectra of both the modified and unmodified aerogels show bands typical of cellulose. For example, the broad absorption peak at 3430 cm^{-1} is the stretching vibration peak of hydroxyl groups between cellulose molecules. The absorption peak near 2926 cm^{-1} is the stretching vibration peak of C–H, and the bending vibration peak of –CH_2 is the absorption peak near 1429 cm^{-1}. The absorption peak near 1163 cm^{-1} is the asymmetric stretching vibration peak of C–O–C ether in the pyran ring. The absorption peak around 1060 cm^{-1} can be attributed to the absorption vibration peak of C–O. These results show that the aerogel modified by APMDS still has the matrix materials of cellulose.

In addition, some important changes in the infrared spectrum were observed after modification with APMDS compared to the IR spectra of the CAB aerogel. There is a new peak at 1261 cm^{-1}, which is usually related to C–O–Si tensile vibration. Due to the introduction of the N–H bond after modification, with the increase in APMDS concentration, the band intensity around 2926 cm^{-1} increases. In addition, two new bands were observed at 795 and 1575 cm^{-1}, which were attributed to N–H bending vibration and N-H stretching vibration of $-NH_2$ after amination. All these results showed that APMDS was successfully grafted onto CNF, and the primary amine group of APMDS remained intact.

Figure 2d shows the XRD curves of the CAB and ACC-3 aerogel samples. In Figure 2d, diffraction peaks at $2\theta = 12.1°$ and $20.2°$ can be seen, and these diffraction peaks correspond to the of type II cellulose, respectively. The characteristic curves of the ACC-3 aerogel show the characteristic peaks of cellulose. This indicates that the crystal structure of cellulose was not changed by the introduction of APMDS.

5.3. XPS Analysis, TG Analysis

Figure 3 shows the XPS spectra using the CAB and MCC/APMDS composite aerogels, respectively. It can be seen from the broad sweep spectra that an amino silane grafting reaction occurred in the MCC/APMDS composite aerogel after modification with APMDS, and the Si element appeared in the broad sweep spectra, which is consistent with the results of infrared analysis. The low-resolution spectra of the unmodified CAB aerogel showed that C and O atoms were the main components in the aerogel, whereas N and Si silicon elements were detected in the modified aerogel, indicating successful grafting of APMDS with cellulose. As can be seen from the high-resolution maps of C1s and N1s of ACC-3 (illustration in Figure 3b), there are four types of carbon bonds: C–C (C1, 284.8 eV), C–O (C2, 286.4 eV), O–C–O (C3, 287.8 eV) and C–N (C4, 288.9 eV), respectively. The presence of the N element is divided into two forms, namely, $-NH_2$ and $-NH_3^+$. This indicates that the primary amino group was protonated during chemical grafting. The energy peak position of O in Figure 3c is 531 eV, the peak shape of O is a slightly asymmetric peak, and a low-energy shoulder slit can be observed, suggesting that oxygen is present in the form of hydroxide ions (OH–), which is speculated to be the result of Schiff base formation.

As shown in Figure 3d, the MCC/APMDS composite aerogel has two thermal degradation processes. The first thermal degradation process is between 50 °C and 180 °C. In this temperature range, cellulose aerogel has slight thermal degradation, and the absorbed water inside the aerogel overflows, resulting in slight mass loss. Thermal degradation was observed between 200 °C and 400 °C, which may be related to the breaking of cellulose glycosides bonds; the inside of the aerogel collapsed and degraded, which produced a lot of water and gas, and then the cellulose aerogel lost a lot of weight. The residual weight loss amount of the aerogel modified with APMDS is larger than that of the CAB aerogel. At the same time, with the increase in the use of APMDS, the greater the content of nitrogen, and the greater the residual weight loss amount. This may be due to the possible existence of the Si element in APMDS; the modified CNC aerogel generates refractory substances, such as SiO_2, during thermal degradation, thus increasing the thermal degradation residues of the aerogel. Because the higher the content of nitrogen, the more refractory substances produces, so the weight loss residue of the MCC/APMDS composite aerogel increases, and the maximum residue reaches 25.6%. The results also indirectly indicated that AEAPMDS was successfully grafted onto cellulose. In addition, the thermos gravimetric curve of the modified aerogel is much smoother than that of the unmodified aerogel, probably because the introduction of APMDS causes the chemical bond energy of the CAB aerogel to increase, which requires higher thermal degradation energy, and the degradation process is more moderate.

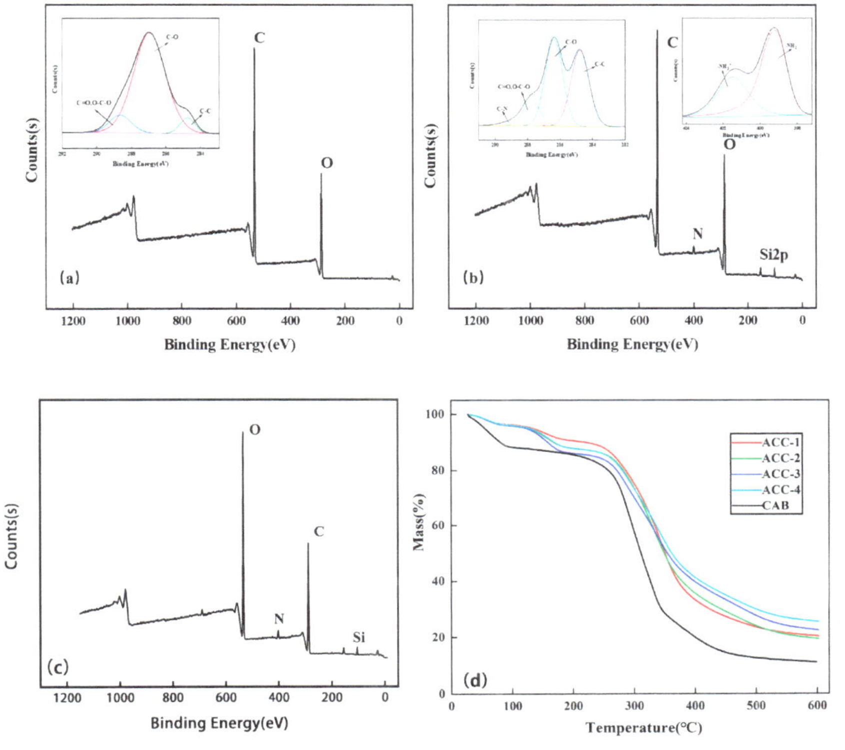

Figure 3. XPS spectra: (**a**) CAB pre-adsorption; (**b**) ACC-3 pre-adsorption; (**c**) ACC-3 post-adsorption; TG spectra of CAB and MCC/APMDS composite aerogel (**d**).

5.4. Effect of Different Reaction Conditions on Nitrogen Content of MCC/APMDS Composite Aerogel Samples

A single-factor method was used to investigate the effects of the APMDS dosage (based on the total mass of the mixture solution), the reaction time and the reaction temperature on the nitrogen content of the MCC/APMDS composite aerogel. At the same time, the nitrogen content of MCC/APMDS composite aerogel samples was detected by X-ray photoelectron spectroscopy, and the final results were obtained as shown in Figure 4a.

Figure 4a shows the effect of APMDS usage on the nitrogen content of the MCC/APMDS composite aerogel when the reaction time is 4 h and the reaction temperature is 90 °C. It can be seen from Figure 4a that with the increase in the amount of APMDS added, the nitrogen content of the sample increases rapidly. However, when the amount of APMDS was greater than 6%, there was almost no increase in nitrogen content. This may be because the higher amount of APMDS may lead to the rapid formation of APMDS self-condensation polymer, which is not conducive to the reaction of APMDS with cellulose. Therefore, nitrogen content increased slowly after the application of 6 wt% APMDS. The results showed that the appropriate amount of APMDS was 6 wt%, and the nitrogen content of the MCC/APMDS composite aerogel reached 8.98 wt% when 6 wt% APMDS was used.

Figure 4b shows the effect of reaction time on the nitrogen content of the MCC/APMDS composite aerogel when the amount of APMDS is 6 wt% and the reaction temperature is 90 °C. As shown in Figure 4b, the nitrogen content of the MCC/APMDS composite aerogel increased rapidly with the increase in reaction time before 3 h. But at 4 h, almost all reached their own plateau. This may be because before the reaction time of 3 h, due to the reaction of a large number of reactive hydroxyl groups on cellulose and the high content of free amino silanols formed by the hydrolysis of APMDS in the reaction medium, the content of nitrogen elements increases rapidly. After the reaction time reaches 4 h, the reactive hydroxyl groups or APMDS reactions on cellulose are exhausted. As a result, the nitrogen content of the MCC/APMDS composite aerogel remained constant after 4 h of reaction. The results showed that the nitrogen content of the MCC/APMDS composite aerogel reached the maximum when the reaction time was 4 h.

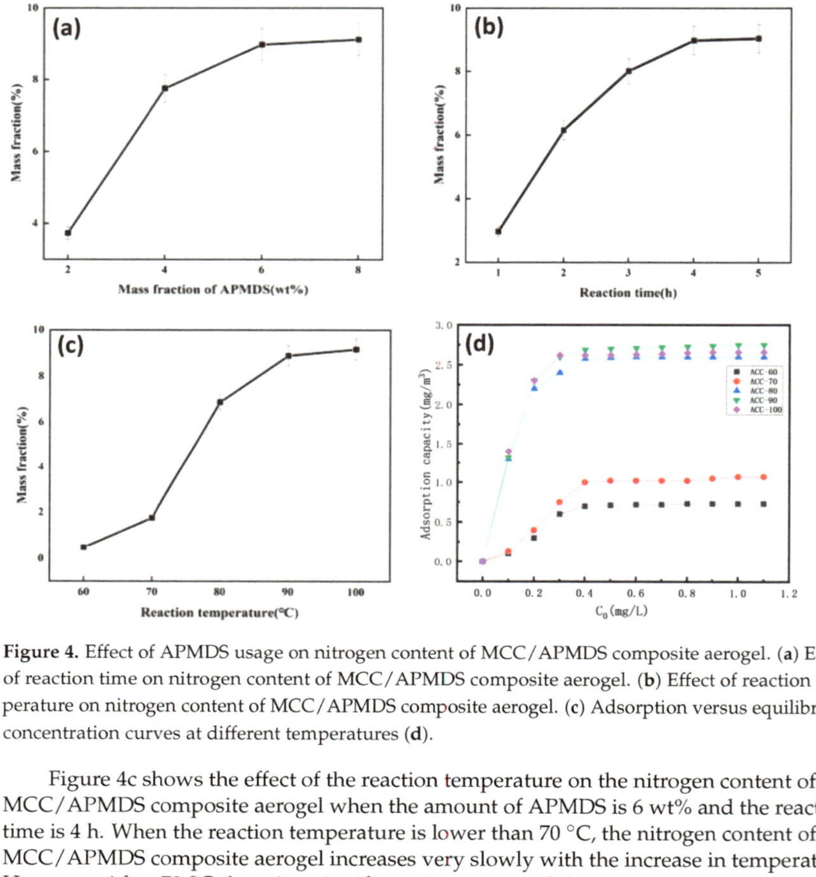

Figure 4. Effect of APMDS usage on nitrogen content of MCC/APMDS composite aerogel. (**a**) Effect of reaction time on nitrogen content of MCC/APMDS composite aerogel. (**b**) Effect of reaction temperature on nitrogen content of MCC/APMDS composite aerogel. (**c**) Adsorption versus equilibrium concentration curves at different temperatures (**d**).

Figure 4c shows the effect of the reaction temperature on the nitrogen content of the MCC/APMDS composite aerogel when the amount of APMDS is 6 wt% and the reaction time is 4 h. When the reaction temperature is lower than 70 °C, the nitrogen content of the MCC/APMDS composite aerogel increases very slowly with the increase in temperature. However, After 70 °C there is a significant increase until the reaction temperature rises to 90 °C. After 90 °C, little effect of the reaction temperature on the amine load was observed. This is because APMDS is prone to hydrolysis to form amino silanols even at room temperature, so the results suggest that temperatures higher than 70 °C are needed to facilitate the reaction of amino silanols with hydroxyl groups on cellulose. The rapid increase in the nitrogen content of the MCC/APMDS composite aerogel after the reaction temperature was 70 °C may be due to the increase in temperature, which increases the rate constant of the reaction of APMDS with the hydroxyl group on the cellulose. At temperatures greater than 90 °C, the nitrogen content of the MCC/APMDS composite aerogel has little effect. From Figure 4d, it can be seen that as the equilibrium concentration increases, the amount of adsorption is greater and greater. It remains stable after reaching a certain concentration, and the highest amount of adsorption is found when the temperature is 90 °C. These results indicate that the optimum temperature is 90 °C.

From the above results, it can be seen that the nitrogen content of the MCC/APMDS composite aerogel reaches its maximum when the amount of APMDS is 6 wt%, the reaction time is 4 h, and the reaction temperature is 90 °C.

5.5. Formaldehyde Adsorption Performance Analysis

The adsorption results of formaldehyde gas by the MCC/APMDS composite aerogel are shown in Figure 5. The experiment was conducted at 25 °C, and the initial concentration of formaldehyde was 5.0 mg/m^3. Figure 5a–c show the results of the adsorption of formaldehyde gas by the MCC/APMDS composite aerogel by the amount of APMDS, reaction time and reaction temperature, respectively. It can be seen from the Figure 4 that after formaldehyde adsorption by the MCC/APMDS composite aerogel for 2 h, the sample tends to be saturated, and its maximum adsorption capacity is 9.52 mg/g. As can be seen from Figure 5a, the formaldehyde adsorption capacity of the modified aerogel increased with the increase in the concentration of the modifier. When the amount of APMDS reached 8 wt%, the adsorption capacity of the aerogel for formaldehyde gas hardly increased compared with that of the 6 wt%. As can be seen from Figure 5b, with the increase in reaction time, the adsorption amount of formaldehyde by the aerogel increases, and when the reaction time reaches 5 h, the formaldehyde adsorption amount hardly increases. It can be seen from Figure 5c that the formaldehyde adsorption capacity increases with the increase in temperature. When the reaction temperature is lower than 80 °C, the formaldehyde adsorption capacity of the aerogel increases slowly, and when the temperature reaches 100 °C, the formaldehyde adsorption capacity hardly increases. This phenomenon is related to the change in nitrogen content in the MCC/APMDS composite aerogel.

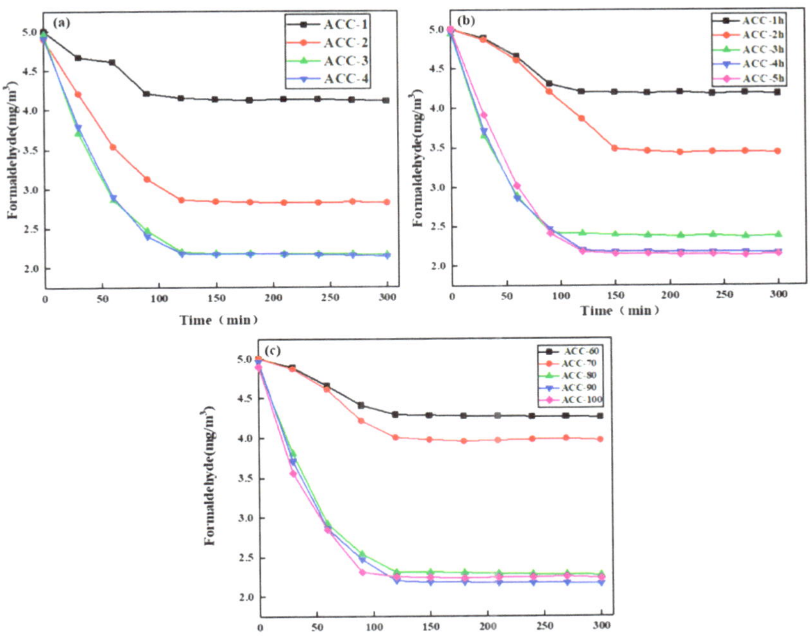

Figure 5. Change in formaldehyde concentration.

Table 2 shows that the adsorption amount is related to the nitrogen content, and the higher the nitrogen content, the stronger the adsorption in a certain range. As can be seen from Tables 3 and 4, the sample prepared under the conditions of an APMDS dosage of 6 wt%, a reaction time of 4 h and a reaction temperature of 90 °C had the largest average pore size (14.56 nm) and the largest formaldehyde adsorption capacity (9.52 mg/g). After many repeated tests, this sample was found to be able to keep the adsorption amount and the average pore size error very small and to remain stable.

Table 2. Analysis of experimental errors for different variables at 95 per cent confidence level.

Variant	Tolerance Range (Nitrogen Content%)				
APMDS usage (2, 4, 6, 8%)	3.73 ± 0.1865	7.75 ± 0.3875	8.98 ± 0.449	9.12 ± 0.456	
Response time (1 h, 2 h, 3 h, 4 h, 5 h)	2.97 ± 0.1485	6.15 ± 0.3075	8.01 ± 0.4005	8.98 ± 0.449	9.04 ± 0.452
Reaction temperature (60 °C, 70 °C, 80 °C, 90 °C, 100 °C)	0.48 ± 0.024	1.76 ± 0.088	6.87 ± 0.3435	8.89 ± 0.4445	8.89 ± 0.458

Table 3. Mean pore size and adsorption properties of aerogels of different variables tested.

MCC/APMDS Aerogel	Average Pore Size (nm)	Adsorption Capacity (mg/g)
2%APMDS, 4 h, 90 °C	14.21	9.33
4%APMDS, 4 h, 90 °C	14.43	9.36
6%APMDS, 4 h, 90 °C	14.56	9.52
8%APMDS, 4 h, 90 °C	14.53	9.49
6%APMDS, 4 h, 60 °C	13.91	9.41
6%APMDS, 4 h, 70 °C	14.19	9.45
6%APMDS, 4 h, 80 °C	14.23	9.52
6%APMDS, 4 h, 100 °C	14.51	9.33
6%APMDS, 1 h, 90 °C	13.88	8.16
6%APMDS, 2 h, 90 °C	14.13	8.98
6%APMDS, 3 h, 90 °C	14.37	9.24
6%APMDS, 5 h, 90 °C	14.52	9.33

Table 4. Error analysis of repeated tests at 95 per cent confidence level.

6%APMDS, 4 h, 90 °C MCC/APMDS Aerogel	Average Pore Size (nm)	Adsorption Capacity (mg/g)
Test 1	14.56	9.49
Test 2	14.59	9.52
Test 3	14.54	9.54
Test 4	14.58	9.50
Test 5	14.53	9.55
Average value	14.56	9.52
Standard deviation	0.0228	0.0228
Tolerance range	14.56 ± 0.0102	9.52 ± 0.0102

As can be seen from Figure 6a,b, after the adsorption of formaldehyde by the MCC/APMDS composite aerogel, part of the microporous pores become smaller and covered up, which proves that the aerogel modified by APMDS has good adsorption performance on formaldehyde. The black line in Figure 7a and the purple line in Figure 7b both represent element C. The different N and H distributions lead to the appearance of two different morphologies, $-NH_3^+$ and $-NH_2$. According to Figure 7a,b, when the MCC/APMDS composite aerogel adsorbed formaldehyde, the content of $-NH_3^+$ is obviously reduced, which indicates that the positively charged nitrogen groups are closer to formaldehyde molecules. This indicates that the adsorption effect of the MCC/APMDS composite aerogel on formaldehyde mainly depends on the protonated $-NH_3^+$ group of APMDS to react with formaldehyde to produce Schiff bases to achieve the effect of formaldehyde removal.

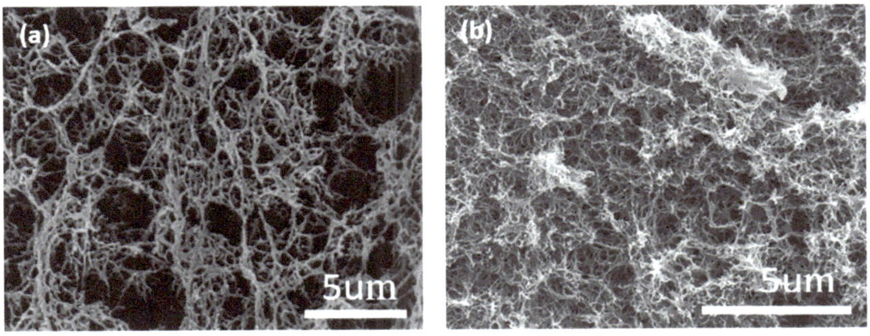

Figure 6. SEM diagram of sample ACC-3 before (**a**) and after (**b**) adsorption.

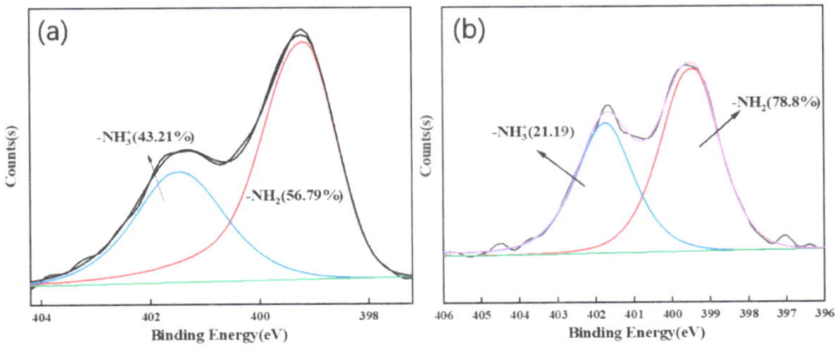

Figure 7. N1s high-resolution pattern of samples before and after ACC-3 adsorption: (**a**) before adsorption; (**b**) post-adsorption.

6. Conclusions and Outlook

In this paper, MCC/APMDS composite aerogels were successfully prepared by heating, solvent replacement and freeze-drying using 2 wt% cellulose hydrogel as a matrix material and modified with aminosilane (APMDS). The modified MCC/APMDS composite aerogels underwent an aminosilane grafting reaction, and the adsorption effect of formaldehyde was mainly dependent upon the protonated $-NH_3^+$ group of APMDS reacting with formaldehyde to produce Schiff bases to achieve the effect of formaldehyde removal. Meanwhile, the modification of the aerogel resulted in a reduction in pore volume and specific surface area. An increase in the average pore size to 14.56 nm also led to a stronger adsorption capacity for formaldehyde, and the adsorption amount could reach 9.52 mg/g. This study provides valuable information for the preparation of adsorbent materials with a high efficiency of formaldehyde adsorption capacity for air purification.

Author Contributions: Y.L.: Methodology, Investigation, Writing—original draft; Z.L.: Conceptualization, Visualization, Writing—review and editing; C.C.: Software, Data curation; B.Y.: Formal analysis, Validation; Y.Z.: Data curation; G.J.: Formal analysis, Resources; J.S.: Funding acquisition, Project administration. All authors have read and agreed to the published version of the manuscript.

Funding: This research was funded by Jilin Province fund project (Grant No. YDZJ202201ZYTS672), China; the Jilin Provincial Development and Reform Commission Project (Grant No. 2021C036-9), China; the Jilin City Science and Technology Innovation Development Plan Project (Grant No. 20200104081), China; and the project of Jilin Provincial Department of Education (Grant No. JJKH20220079KJ), China. The Article Processing Charge was covered by the Beihua University.

Institutional Review Board Statement: This manuscript does not contain any studies involving human or animal subjects.

Data Availability Statement: No new data were created or analyzed in this study.

Conflicts of Interest: There are no conflict to declare.

References

1. Dai, X.; Liu, J.; Yin, Y.; Ren, H.; Chen, X. Modeling and controlling indoor formaldehyde concentrations in apartments: On-site investigation in all climate zones of China. *Build. Environ.* **2018**, *127*, 98–106. [CrossRef]
2. Huang, Y.; Su, T.; Wang, L.; Wang, Z.; Liu, W. Evaluation and characterization of volatile air toxics indoors in a heavy polluted city of northwestern China in wintertime. *Sci. Total Environ.* **2019**, *662*, 470–480. [CrossRef] [PubMed]
3. Zhang, B.; Hu, X.; Zhang, Y.; Feng, G.; Yu, Y. Research progress on indoor formaldehyde pollution and its influencing factors in China, a Review. *IOP Conf. Ser. Earth Environ. Sci.* **2021**, *692*, 3. [CrossRef]
4. Wang, J.T.; Zhong, Z.M. Study on adsorption of formaldehyde by aminated Chitosan/silica aerogels. *Chem. Enterp. Manag.* **2018**, *16*, 40–44.
5. Tang, X.; Bai, Y.; Duo, N.G.; Bai, C. Formaldehyde in China: Production, consumption, exposure levels, and health effects. *Environ. Int.* **2009**, *36*, 1210–1224. [CrossRef] [PubMed]
6. Huangfu, Y.; Lima, N.M.; O'Keefe, P.T.; Zhang, J. Whole house emission rates and loss coefficients of formaldehyde and other VOCs as a function of air change rate. *Indoor Air* **2020**, *30*, 831–844.
7. Zhang, Z.F.; Zhang, X.; Zhang, X.M.; Wu, Y.Y.; Cui, L. Indoor occurrence and health risk of formaldehyde, toluene, xylene and total volatile organic compounds derived from an extensive monitoring campaign in Harbin, a megacity of China. *Chemosphere* **2020**, *250*, 126324. [CrossRef]
8. Wu, L.R.; Zhang, L.X.; Yu, F.; Chen, J.H.; Ma, J. Facile composite of amino carbon nanotubes / graphene: Preparation and adsorption for gaseous formaldehyde. *China Environ. Sci.* **2015**, *35*, 3251–3256.
9. Liu, Z.; Wu, P. Preparation of Chitosan/Cellulose Aerogel Beads and Its Formaldehyde Gas Adsorption Performance. *Chem. Ind. For. Prod.* **2017**, *37*, 27–35.
10. Piekarski, C.M.; Francisco, A.; Luz, L.; da Silva, P.H.M. Life cycle assessment of medium-density fiberboard (MDF) manufacturing process in Brazil. *J. Clean. Prod.* **2017**, *575*, 103–111. [CrossRef]
11. Lopes-Silva, D.A.; Rocc Lahr, F.A.; Varandal, D.; Ferreira-Leitão, V.S.; Ormond da Silva, J. Environmental performance assessment of the melamine-urea-formaldehyde (MUF) resin manufacture: A case study in Brazil. *J. Clean. Prod.* **2015**, *96*, 299–307. [CrossRef]
12. Canha, N.; Lage, J.; Coutinho, J.T.; Freitas, M.C.; Almeida-Silva, M. Comparison of indoor air quality during sleep in smokers and non-smokers' bedrooms: A preliminary study. *Environ. Int.* **2015**, *85*, 236–241. [CrossRef] [PubMed]
13. Salthammer, T.; Menteşe, S.; Marutzky, R. Formaldehyde in the Indoor Environment. *Chem. Rev.* **2010**, *110*, 2536–2572. [CrossRef]
14. Becerra, J.A.; Lizana, J.; Gil, M.; Tapia, A.; González, M.; Vera, T. Identification of potential indoor air pollutants in schools. *Environ. Sci. Pollut. Res.* **2020**, *242*, 118420–118438. [CrossRef]
15. Baloch, R.M.; Maesano, C.N.; Christoffersen, J.; Banerjee, S.; Gabriel, M.; Csobod, É.; de Oliveira Fernandes, E.; Annesi-Maesano, I.; SINPHONIE Study Group. Indoor air pollution, physical and comfort parameters related to schoolchildren's health: Data from the European SINPHONIE study. *Environ. Int.* **2015**, *78*, 79–87. [CrossRef] [PubMed]
16. Fan, G.; Xie, J.; Yoshino, H.; Zhang, H.; Li, Z.; Li, N.; Lv, Y.; Liu, J.; Zhu, S.; Yanagi, U.; et al. Concentration characteristics of gaseous carbonyl compounds in urban houses in two different climatic zones of China and health risk assessment for schoolchildren. *Sci. Total Environ.* **2014**, *472*, 1107–1114. [CrossRef]
17. Na, G.; Liu, Y. Study on formaldehyde adsorption performance of hemin modified pineapple leaf fiber. *J. Hainan Norm. Univ. (Nat. Sci.)* **2017**, *30*, 356–359.
18. Xu, Z.; Chen, Y.; Ding, S.; Song, S.; Zhang, X.; Wang, S. Microemulsion-assisted preparation of a mesoporous ferrihydrite/SiO2 composite for the efficient removal of formaldehyde from air. *J. Phys. Chem. C* **2013**, *117*, 15699–15706.
19. Yao, L.; Guo, D.; Bei, C.; Yang, X.; Yu, S. Bio-template-assisted synthesis of hierarchically hollow SiO2 microtubes and their enhanced formaldehyde adsorption performance. *J. Hazard. Mater.* **2013**, *274*, 110–116.
20. Chen, D.; Qu, Z.; Sun, Y.; Chen, C.; Zou, Z. Adsorption-desorption behavior of gaseous formaldehyde on different porous Al_2O_3 materials. *J. Colloid Interface Sci.* **2011**, *355*, 370–377. [CrossRef]
21. Kim, R.; Ryu, U.J.; Jee, S.; Kim, S.H.; Kim, H.G.; Lee, C.H. Surface coating of MOF layers on the nanocrystals of other MOFs using nanoparticle mediated nucleation for the efficient removal of formaldehyde. *J. Hazard. Mater.* **2011**, *402*, 123470. [CrossRef]
22. Liao, Y.; Pan, X. Self-indicating and high-capacity mesoporous aerogel-based biosorbent fabricated from cellulose and chitosan via co-dissolution and regeneration for removing formaldehyde from indoor air. *Environ. Sci. Nano* **2021**, *8*, 1283–1295. [CrossRef]
23. Zhang, K.; Xu, Z.; Li, W.; Liu, S. Preparation of Urea/Nicotinamide Modified Activated Carbon and Its Formaldehyde Adsorption Performance. *Chem. Ind. For. Prod.* **2020**, *40*, 75–82.
24. Yin, Y.; Wang, Z.; Wang, S.; Zhang, X.; Wu, J. Cellulose-based formaldehyde adsorbents with large capacities: Efficient use of polyethylenimine for graphene oxide stabilization in alkaline–urea system. *J. Appl. Polym. Sci.* **2019**, *136*, 47999. [CrossRef]
25. Yan, S.; Wang, B. Study on adsorption behavior of hemp charcoal/PET functional fiber on formaldehyde. *Synth. Technol. Appl.* **2018**, *33*, 2023070445.
26. Shen, W.; Wei, B.; Deng, Y.; Sun, Y.; Wang, J. Nitrogen-containing porous carbons: Synthesis and application. *J. Mater. Chem. A* **2015**, *3*, 13108–13128. [CrossRef]

27. Li, Y.; Jia, P.; Xu, J.; Li, Y.; Liu, C. The Aminosilane Functionalization of Cellulose Nanofibrils and the Mechanical CO_2 Adsorption Characteristics of Their Aerogel. *Ind. Eng. Chem. Res.* **2020**, *59*, 2874–2882. [CrossRef]
28. Zhang, T.; Zhang, Y.; Jiang, H.; Yan, B.; Chen, Y. Aminosilane-grafted spherical cellulose nanocrystal aerogel with high CO_2 adsorption capacity. *Environ. Sci. Pollut. Res. Int.* **2019**, *26*, 16716–16726. [CrossRef]

polymers

MDPI

Article

A Study on the Evaluation of Thermal Insulation Performance of Cellulose-Based Silica Aerogel Composite Building Materials

Jeo Hwang [1], Yoonmi Kim [2], Jooyoung Park [2] and Dongho Rie [3,*]

1 Graduate School of Safety Engineering, Incheon National University, Incheon 22012, Republic of Korea; hjojjang@inu.ac.kr
2 Department of Safety Engineering, Incheon National University, Incheon 22012, Republic of Korea; 1231yunmi@naver.com (Y.K.); wndud3539@naver.com (J.P.)
3 Fire Disaster Prevention Research Center, Incheon National University, Incheon 22012, Republic of Korea
* Correspondence: riedh@inu.ac.kr

Abstract: Buildings utilize both inorganic and organic insulation materials to conserve energy and prevent heat loss. However, while exhibiting excellent thermal insulation performance, organic insulation materials increase the risk of fire due to the emission of intense heat and toxic smoke in the event of a fire. Conversely, inorganic insulation materials are characterized by a lower thermal insulation performance, leading to an increase in the weight of the building with extensive use. Therefore, the necessity for research into new insulation materials that address the drawbacks of existing ones, including reducing weight, enhancing fire resistance, and improving thermal insulation performance, has been recognized. This study focuses on evaluating the enhancement of the thermal insulation performance using novel building materials compared to conventional ones. The research methodology involved the incorporation of porous aerogel powders into paper-based cellulose insulation to improve its insulating properties. Samples were prepared in standard 100×100 mm^2 panel forms. Two control groups were utilized: a pure control group, where specimens were fabricated using 100% recycled cardboard for packaging, and a mixed control group, where specimens were produced using a mixture ratio of 30 wt% ceramic binder and 40 wt% expandable graphite. Experimental group specimens were prepared by increasing the aerogel content from 200 to 1000 mL under each condition of the control groups (pure and mixed) after mixing. The thermal insulation performance of the specimens was evaluated in terms of thermal conductivity and thermal diffusivity according to ISO 22007-2 (for solids, paste, and powders). Through this study, it was found that the thermal insulation performances of the pure control and experimental groups improved by 16.66%, while the mixed control and experimental groups demonstrated a 17.06% enhancement in thermal insulation performance with the addition of aerogel.

Keywords: cellulose; thermal conductivity; thermal diffusivity; building materials

Citation: Hwang, J.; Kim, Y.; Park, J.; Rie, D. A Study on the Evaluation of Thermal Insulation Performance of Cellulose-Based Silica Aerogel Composite Building Materials. *Polymers* **2024**, *16*, 1848. https://doi.org/10.3390/polym16131848

Academic Editor: Luis Alves

Received: 6 May 2024
Revised: 15 June 2024
Accepted: 27 June 2024
Published: 28 June 2024

1. Introduction

The recent increase in greenhouse gas emissions has led to a rapid escalation of global warming. Surface temperatures on Earth have shown a swift rise since 1970, resulting in significant climate change. In response, the international community has adopted low-carbon policies to minimize greenhouse gas emissions. Enhanced thermal insulation in buildings is anticipated to contribute to a reduction in the use of fossil fuels, thus making it a key objective of low-carbon policies.

Furthermore, the global spread of COVID-19 has significantly shifted towards non-face-to-face lifestyles, resulting in a sharp increase in parcel delivery services [1]. In 2020, the number of parcels delivered reached 3.37 billion, marking a 20.9% increase from 2.798 billion in 2019. Recycling paper, commonly disposed of after use, can save 17 trees per ton (907 kg). Research is underway to utilize cellulose, a component of paper, for producing construction materials, aiming to achieve both low-carbon emission policies and recycling objectives.

The main components of paper are cellulose, hemicellulose, lignin, and various additives. Cellulose constitutes about 40% to 60% of paper and serves as the primary structural element, providing strength and durability. Hemicellulose accounts for approximately 10% to 30% and consists of shorter fibers compared to cellulose. Lignin makes up about 15% to 30% and binds cellulose and hemicellulose together, enhancing their strength [2].

Traditional insulation materials used in buildings primarily aim to prevent energy loss. Depending on the materials used, traditional insulation can be classified as shown in Figure 1.

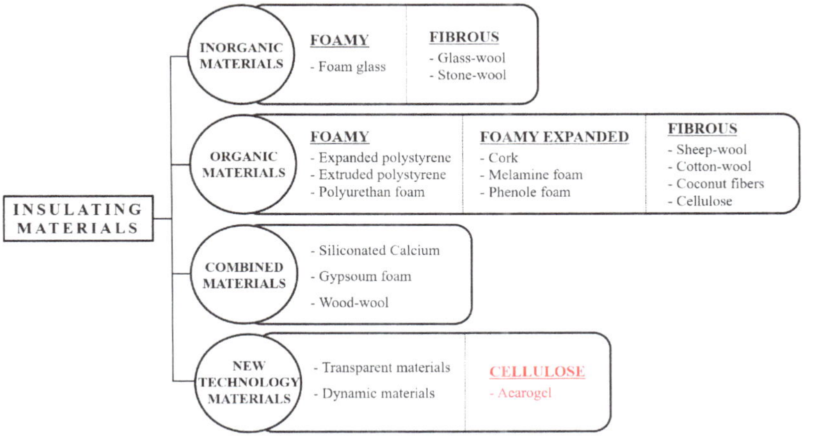

Figure 1. Classification of existing insulation materials and the material mix for this experiment.

According to a study by Han-Hsi, L., et al., the toxicity indices of organic foaming materials and polyurethane, as well as polyethylene-based building materials, evaluated using the NES-713 assessment method, were found to be higher than 10, indicating a high risk of smoke toxicity [3]. Despite their efficient energy storage capabilities, organic insulation materials are susceptible to fire hazards. Consequently, in the event of a building fire, the expansion of the fire and the emission of toxic gasses can result in significant casualties.

As awareness of the shortcomings of organic insulation materials grows, there has been an increasing use of inorganic insulation materials. However, inorganic insulation materials are characterized by a lower insulation efficiency compared to organic ones and pose challenges in installation. Hence, research on new technology materials with excellent insulation efficiency is in demand.

Aerogel is a lightweight and porous insulating material characterized by its nanoporous structure. Aerogels are composed primarily of silica, with oxygen arranged in a dual-structure, double-bonded form on both sides. Silica finds application in various fields such as nanocomposites, construction, and chemical industries. Historically, the high production cost of aerogel has hindered its widespread application as a conventional material. However, recent advancements in mass production technology have lowered production costs, leading to its utilization in both research and consumer goods.

In 1931, Steven Kistler introduced aerogel, which is lightweight and porous. Silica, a basic component of soil and rocks, including granite and basalt, is the most abundant mineral on Earth and serves as the fundamental material for aerogels [4]. Silica finds application in various fields such as nanocomposites, construction, and chemical mechanical polishing [5]. Research on silica-based aerogel manufacturing methods for commercialization is ongoing [6]. Silica aerogel exhibits outstanding insulation properties due to its ability to transmit only about 1/100th of the heat compared to ordinary glass [7]. With pore sizes ranging from 5 nm to 70 nm, silica aerogel contains a variety of cells filled with air, which block heat transfer and enhance insulation performance [8]. The thermal conductivity of silica aerogel, a porous material, is approximately 14 mW/m·K [9].

Initially, aerogel was mainly used in advanced chemical products for aerospace and chemical industries, with limited application in construction materials due to high production costs [10]. However, in the early 2000s, aerogel materials began to be used in buildings, primarily as aerogel blankets [11]. Aerogel-based construction materials, such as aerogel blankets (thermal conductivity around 15 mW/m·K) and aerogel boards (thermal conductivity around 16~19 mW/m·K), are utilized in buildings [12]. In addition to insulation performance, the sustainability of construction materials is crucial for increasing the lifespans of buildings [13].

Sambucci et al.'s study identifies the emerging use of fiber-reinforced aerogel blankets (FRABs) as an alternative insulation material for cryogenic tanks used in liquefied natural gas (LNG) transportation. They found that compared to the traditionally used back-filled perlite-based system for transporting liquefied natural gas (LNG), it allows for a thinner outer shell, stores more material, and reduces the weight of LNG transportation semitrailers [14].

Liu et al. studied the thermal conductivity of specimens as the mass fraction of expanded graphite increased from 2 wt% to 10 wt%. The measured results indicated a linear increase in thermal conductivity. Specifically, the measured value increased from 0.2055 to 0.5218 W/m·K. The thermal conductivity of the mixture with 10 wt% expanded graphite added was 3.28 times higher than before the addition of expanded graphite. The increase in thermal conductivity was found to correspond to a decrease in thermal insulation performance [15].

This study produced specimens by pulping paper to extract its main component, cellulose, and adjusting the content of porous aerogel. Specimens mixed with paper and porous aerogel were classified as the pure test group. The composite test group was also created by combining porous aerogel with 30 wt% ceramic binder and 40 wt% expanded graphite to delay fire spread [16]. The thermal insulation performance of the fabricated specimens was evaluated using the ISO 22007-2 method, through which their thermal conductivity and thermal diffusivity were measured.

2. Materials and Methods

2.1. Materials

The powdered silica aerogel used in this study was manufactured by Hongda Technologies. Table 1 shows the physical characteristics of the powdered silica aerogel provided by the manufacturer.

Table 1. Physical properties of powdered silica aerogel.

Color	White, Translucent
Density (kg/m^3)	<40
Thermal conductivity (W/mK at 25 °C)	0.011
Porosity (%)	99
Aperture (nm)	10~50
Pore volume (cm^2/g)	3.0~3.6
Particle size (μm)	1~10
Specific surface area (m^2/g)	650~1250
Speed of sound (m/s)	~100

The paper was purchased from a local recycling center in Incheon, South Korea, and was used as the primary raw material for this research. The paper used in this study was cardboard. The cardboard was recycled from paper previously used for delivery boxes. The ceramic binder used in this experiment was supplied by Seonjin Chemicals Co., Ltd., In-cheon, South Korea, under the product name 'FP-100'. The ceramic binder comprised 48% MgO, 17% SiO_2, 14.6% Al_2O_3, and other components.

Expandable graphite, obtained from Samjeong C&E in Gyeongsan, Hwa-seong, South Korea, was used in the form of a 100-mesh product. Expandable graphite is characterized by its porous structure, which expands several hundred times upon heating, leading to a layer separation phenomenon and making it suitable for use as a material for flame retardancy [17]. The addition of expandable graphite in the composite material ultimately reduces the HRR and THR values of the experimental specimens [18]. Therefore, in this study, expandable graphite and ceramic binder were used as additive materials to investigate the insulation effect of aerogel under mixed conditions.

2.2. Aerogel Cellulose Specimen Manufacturing Method

The method of specimen fabrication proceeded in four main steps. In the first step, cardboard was pulped into fine particles. In the second step, pure specimens were fabricated by mixing with aerogel. In this step, ceramic binder and expanded graphite were added and mixed for composite specimens. In the third step, specimens were molded using a molding machine and compressed with an electric compressor. In the fourth step, they were dried in a dryer maintained at 60 °C for 48 h to remove moisture. Specimens molded over four steps were then subjected to stability evaluation under conditions of 23 (±2) °C temperature and 50 (±5)% humidity for 2 days in a constant temperature and humidity chamber. Three specimens of each condition were prepared for all samples, each measuring 10 cm × 10 cm in size. Figure 2 shows the specimen fabrication process.

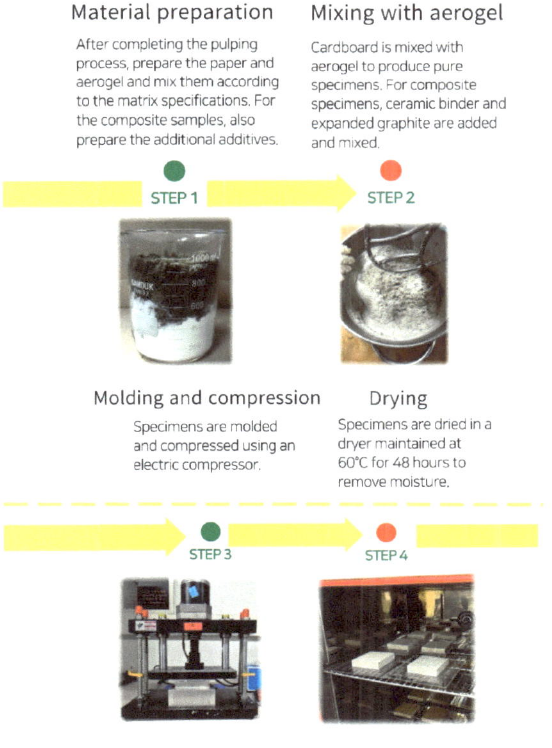

Figure 2. The overall fabrication method of the specimens.

To evaluate the insulation performance of cellulose and aerogel mixed specimens, pure control specimens devoid of any additives were prepared. As the experimental groups, specimens with increased aerogel content were fabricated from the pure control specimens. Table 2 lists the compositions of the cellulose and aerogel mixed specimens. During specimen fabrication, organic compounds were used to induce bonding between cellulose and aerogel. In this study, the paper was measured in grams (g) for the blending ratio, while the ceramic binder and expandable graphite were adjusted in proportion to the weight of the paper, maintaining the wt% unit used previously. As aerogel is inherently lightweight, volume, measured in cubic centimeters (mL), was utilized instead of weight. Similarly, organic compounds were measured in grams (g) to compose the specimen materials, consistent with the weight-based measurement of paper. In this study, a substance based on a certain amount of ethanol as an organic compound was used. Figure 3 shows the appearances of the specimens in the pure control group.

Table 2. Material-specific compositions of the pure experimental group specimens.

Test Specimens Category	Paper (g)	Aerogel (mL)	Organic Compounds (g)
A-400	150	400	600
A-600	150	600	600
A-800	150	800	600
A-1000	150	1000	600

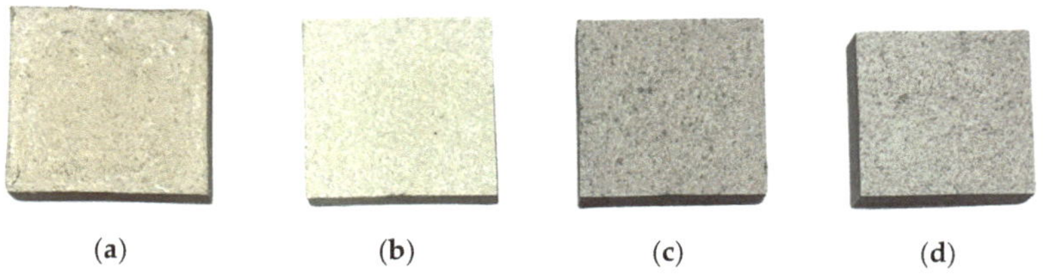

Figure 3. The shapes of mixed specimens of paper and aerogel: (**a**) A-400, (**b**) A-600, (**c**) A-800, (**d**) A-1000.

The composite control specimens were fabricated under conditions synthesized with a ceramic binder and an expandable graphite ratio of 30 wt% and 40 wt%, respectively. Composite experimental specimens were prepared by increasing the aerogel content from the composite control specimens' composition ratio. In the composite experimental specimens, the aerogel, used as a variable, was applied in quantities of 200 mL, 600 mL, and 1000 mL. The same method and unit of material mixing used for the pure control and experimental specimens were also applied to the composite control and experimental specimens. Table 3 lists the mixing ratios and components of the composite experimental specimens. Figure 4 shows the appearances of the composite experimental specimens.

Figure 4. Appearances of the fabricated specimens (**a**) with Aerogel 200 mL, (**b**) with Aerogel 600 mL, (**c**) with Aerogel 1000 mL.

Table 3. Material-specific compositions of the composite experimental group specimens.

Test Specimens Category	Paper (g)	Aerogel (mL)	Organic Compound (g)	Ceramic Binder (wt%)	Expandable Graphite (wt%)
EC-1	150	200	600	30	40
EC-2	150	600			
EC-3	150	1000			

2.3. Evaluation of Insulation Performance (Description of ISO 22007-2 Test)

The ISO 22007-2 test method typically measures thermal conductivity (λ) values within the range of 0.01 W/m·K λ < 500 W/m·K. Additionally, the value of thermal diffusivity (α) is measured within the range of 5×10^{-8} $m^2/s < \alpha < 10^{-4}$ m^2/s. The measurement method calculates the measured values according to the ISO 22007-2 standard by applying heat to the specimen and observing temperature changes [19].

The thermal performance measurement instrument used was the "MP-V" product manufactured by "Thermtest", Canada. The measurement method of the MP-V product adheres to the "ISO 22007-2 test" method. For solids, the measurement was conducted using the Transient Plane Source (TPS) sensor. The TPS sensor, designed for solid measurements, consists of a double spiral of encapsulated nickel between insulation layers. The measurement range of thermal conductivity with this sensor is wide, from 0.005 to 1800 W/m·K. The measurement procedure involves placing the specimen between the two sides of the sensor, fixing it in place, and then measuring thermal conductivity and thermal diffusivity after heating the equipment. Furthermore, for each set of three specimens prepared under the same conditions, measurements were taken at two random points per specimen. Consequently, a total of six measurements were conducted per specimen for the thermal conductivity evaluation.

3. Evaluation Results

3.1. Physical Properties of the Specimens

Table 4 presents the physical characteristics of the specimens fabricated for the experimental group. The average initial dry weight of the pure experimental group specimens was determined to be 199.8 g. After the drying process, the average weight was found to be 88.5 g. Aerogel possesses the inherent property of hydrophobicity [20], leading to induced mixing using a low concentration of organic compounds. The decrease in weight after drying confirmed the evaporation of water and organic compounds present during the manufacturing process. The moisture content evaporated during the drying process was observed to be an average of 55.7% in the pure experimental group specimens. Additionally, in the composite control group specimens, a moisture content of 58.2% was observed. Equation (1) represents the calculation method for moisture content in the experiment:

$$MC\,[\%] = [(Wm - Wd)/Wm] \times 100, \quad (1)$$

where the variables have the following meanings:

MC = Moisture content [%];
Wm = Weight of sample before dry;
Wd = Weight of sample after dry.

The area and height were measured using a specimen after drying. The molding size was constant in the molding process, and the height was measured based on the height after drying. In addition, the weight after drying was measured. Therefore, the method of calculating the density used the following formula.

$$\text{Density}\left(\frac{g}{cm^3}\right) = \frac{\text{weight after drying}(g)}{\text{Volume after drying}(cm^3)} \quad (2)$$

The density of the pure experimental group specimens averaged 0.41 g/cm^3. This value remained consistent in the composite experimental group specimens, also measuring at 0.42 g/cm^3. Panyakaew et al. conducted research utilizing coconut husks and sugarcane to produce insulation materials, achieving a density of 0.35 g/cm^3 [21], while Aliaksandr Bakatovich et al. achieved densities ranging from 0.2 to 0.25 g/cm^3 using straw-based cereals [22]. Furthermore, according to Ahn et al., specimens made from cellulose-based paper had a density of 0.25 g/cm^3 [23]. The specimens produced in this study exhibited higher densities compared to conventional insulation materials cited in previous research.

Table 4. Physical properties of the experimental specimen.

Test Specimen Category	Weight before Drying (g)	Weight after Drying (g)	Height (cm)	Area (cm^2)	Volume (cm^3)	Density (g/cm^3)	Water Content (%)
A-400	221.5	97.6	2.24	100	224	0.44	55.94
A-600	188.6	84.6	2.1	100	210	0.40	55.14
A-800	200.4	86.8	2.13	100	213	0.41	56.69
A-1000	188.6	84.8	2.05	100	205	0.41	55.04
EC-1	232.5	108.4	2.57	100	257	0.42	53.38
EC-2	231	86.1	2.08	100	208	0.41	62.73
EC-3	222.75	92.2	2.18	100	218	0.42	58.61

3.2. Structural Characteristics of the Specimens

The structural characteristics of the specimens were measured using "FE-SEM_7001F". The specimens used for the measurement were control specimens composed solely of paper. For the control specimens, both a control sample containing 1000 mg of porous aerogel and an experimental sample were used. By comparing the structural differences between the control and experimental specimens, the synthesis of aerogel was evaluated. Figure 5 shows the appearance of the control specimen.

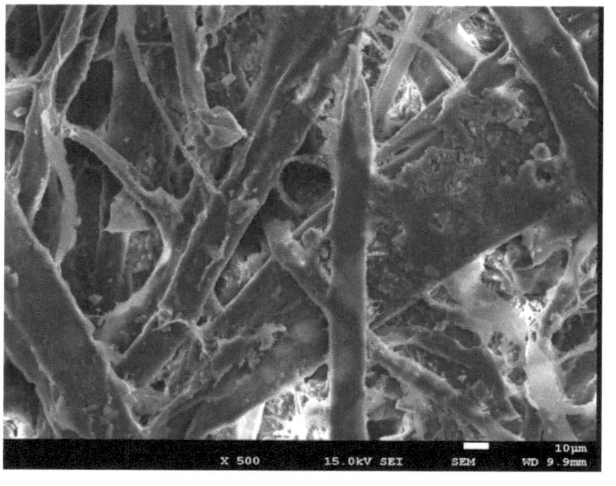

Figure 5. The appearance of the control specimen made entirely of paper.

The control specimens exhibited materials presumed to be binders used in conventional cardboard production, along with cellulose, lignin, and other substances. These specimens showed a solid structure with fibers intertwined, suggesting the absence of porous materials.

In contrast, the control specimens revealed the presence of porous aerogel. Figure 6 shows SEM images of the experimental specimens. In contrast to the control group, a spherical porous aerogel structure was observed, indicating its adsorption in the vicinity of cellulose in a spherical form.

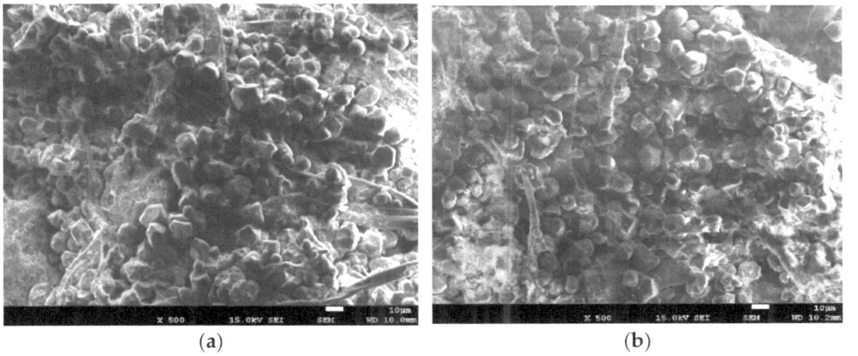

(a) (b)

Figure 6. Appearances of the control specimens composed solely of paper. (**a**) Pure test group specimen. (**b**) Composite test group specimen.

Additionally, quantitative EDS analysis was conducted on the SEM images of the experimental specimens. Table 5 presents the EDS analysis results of the experimental specimens. The analysis confirmed the presence of aerogel, based on silicon (Si) and oxygen(O).

Table 5. EDS analysis results of the test group specimens.

Element	Pure Test Group Specimen		Composite Test Group Specimen	
	Weight%	Atomic%	Weight%	Atomic%
Carbon (C)	43.86	51.56	44.04	51.93
Oxygen (O)	53.25	46.99	52.09	46.12
Silicon (Si)	2.89	1.46	3.87	1.95
Totals	100.00	-	100.00	-

3.3. *Results of Insulation Performance Evaluation*

The thermal conductivity of the pure control group specimens, without aerogel, was determined to be 0.1939 W/m·K. For the pure experimental group specimens composed of aerogel and paper, the thermal conductivity ranged from 0.1901 to 0.1616 W/m·K. Figure 7 shows the measured thermal conductivity and thermal diffusivity values for the pure control and pure experimental group specimens. When the aerogel content was 400 mL, the average thermal conductivity was 0.1901 W/m·K. For the specimens with 600 mL (A-600) aerogel content, it was 0.1766 W/m·K, for 800 mL (A-800) aerogel content, it was 0.1720 W/m·K, and for 1000 mL (A-1000) aerogel content, it was 0.1616 W/m·K. It was observed that as the aerogel content increased, the thermal conductivity decreased. During the fabrication process of the specimens made from the pure experimental group, it was noted that aerogel, suspected to have incomplete bonding due to its hydrophobicity, was expelled along with moisture. The thermal diffusivity varied between 0.239 and 0.1948 mm^2/s depending on the aerogel content.

The thermal conductivity of the composite control group specimen, mixed with 30 wt% ceramic binder and 40 wt% expanded graphite, along with aerogel, was determined to be 0.2157 W/m·K. When the aerogel content was increased to 200 mL, 600 mL, and 1000 mL in the composite control group specimens, the thermal conductivity ranged from 0.1883 to 0.1791 W/m·K. In the composite control group specimens where aerogel was mixed under the conditions of ceramic binder and expanded graphite, the thermal conductivity varied with the aerogel content: 0.1882 W/m·K for 200 mL, 0.1821 W/m·K for 600 mL, and 0.1791 W/m·K for 1000 mL. A decreasing trend in thermal diffusivity, ranging from 0.3289 to 0.2579 mm^2/s, was observed with increasing aerogel content. Figure 8 illustrates the individual thermal conductivity and thermal diffusivity of the composite control and composite experimental group specimens.

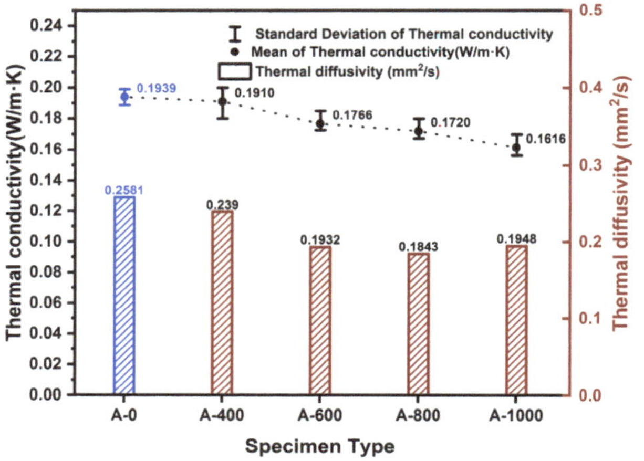

Figure 7. Changes in thermal conductivity and thermal diffusivity according to aerogel content. (Blue: composite control group, Red: composite experimental group).

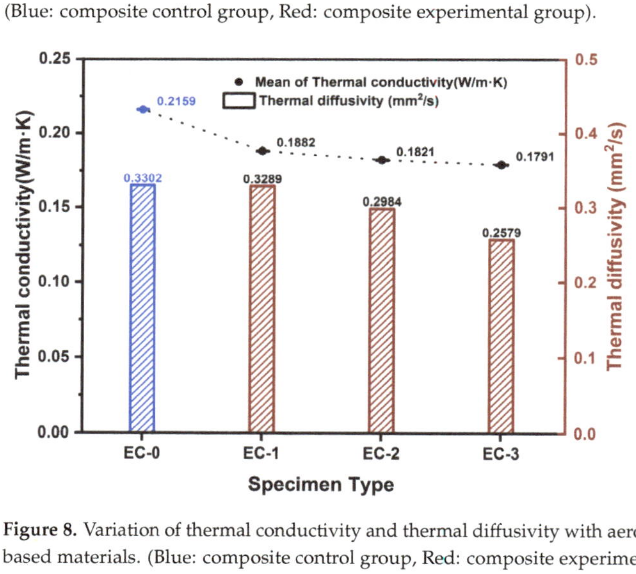

Figure 8. Variation of thermal conductivity and thermal diffusivity with aerogel content in composite-based materials. (Blue: composite control group, Red: composite experimental group).

4. Discussion

Thermal conductivity is an intrinsic property of a material that indicates its ability to transfer heat. It is typically measured in W/m·K and denoted by symbols such as k, λ, or κ. For instance, at 1 atmosphere and 293 K (20 °C), the thermal conductivity of air is approximately 0.025 W/m·K, while that of water is around 0.5918 W/m·K. Formula (2) is commonly used as the basic formula for thermal conductivity, with the most significant factor being the heat transfer quantity, denoted as Q. As this value increases, indicating better heat transfer, the thermal conductivity of the material increases accordingly. A higher K value signifies greater heat loss, as materials with higher thermal conductivity allow more heat to pass through.

$$K = \frac{Q * l}{A * \Delta T} \quad (3)$$

where K = thermal conductivity (W/(m·K)), Q = heat transfer (W), A = heat transfer area (m^2), and ΔT = temperature difference (K or °C)

According to a study by Cha et al., the thermal conductivities of conventional building materials measured using the Heat Flow Meter 436 (HFM 436) were found to be 0.1254 W/m·K for reinforced flooring, 0.2021 W/m·K for fire-resistant gypsum board, and 0.0415 W/m·K for polystyrene [24]. Through our research, the composite specimens based on recycled paper and aerogel showed the best thermal conductivity of 0.1616 W/m·K when the aerogel content was 1000 mL. This value is comparable to those suitable for flooring materials such as laminate flooring and gypsum board.

The R-value (thermal resistance) is a commonly used metric for evaluating the performance of insulation materials, expressed in units of m^2·K/W. Formula (3) expresses the method for calculating the R-value. Table 6 presents the R-value values of the experimental specimens.

$$R = \frac{L}{k} \quad (4)$$

where R = thermal resistance (m^2·K/W), L = thickness (M), and k = thermal conductivity (W/(m·K)).

Table 6. R-value values by specimens.

A Test Specimens	A-400	A-600	A-800	A-1000	EC-1	EC-2	EC-3
R-value (m^2·K/W)	0.118	0.119	0.124	0.127	0.137	0.114	0.122

According to a study by Acharya et al., the R-value of architectural materials composed of aerogel was reported to be 0.26 m^2·K/W. Figure 9 shows the R-value values of conventional architectural insulation materials and the mixed insulation materials used in this study [25]. For the pure experimental group specimens with recycled paper and 1000 mL aerogel added, the R-value was confirmed to be 0.127 m^2·K/W. Furthermore, in the case of the composite experimental group specimens under the mixing conditions of ceramic binder and expanded graphite, the specimen with 200 mL aerogel showed the highest R-value of 0.137 m^2·K/W.

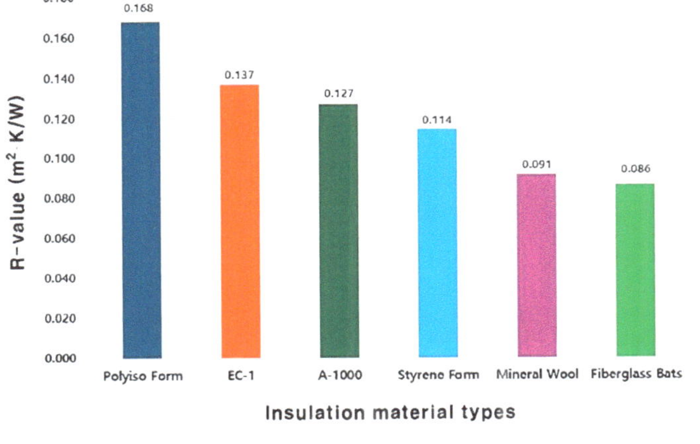

Figure 9. R-value by insulation material types.

5. Conclusions

Firstly, adding aerogel, a porous material, to recycled paper to enhance thermal insulation performance resulted in the highest value of 0.1616 W/m·K at the highest concentration of 1000 mL. This value indicates a 16.66% improvement in thermal insulation performance compared to the control group. The architectural material incorporating 1000 mL aerogel into recycled paper showed values suitable for use as flooring materials and gypsum boards in conventional architectural materials.

Secondly, the thermal conductivity value of the composite control group was confirmed to be 0.2157 W/m·K. Additionally, among the composite control group specimens, the specimen with 1000 mL of added aerogel showed a thermal conductivity value of 0.1791 W/m·K. The reduced value demonstrates a 17.06% improvement in thermal insulation performance compared to the control group.

Lastly, in terms of *R*-value, the specimen mixed with 1000 mL aerogel in the composite experimental group showed an *R*-value of 0.137 m^2·K/W. The *R*-value values of EC-1 and A-1000 were lower than those of Styrene Foam.

In this study, consistent with prior aerogel research, no significant increase in thermal conductivity was observed. This can be attributed to the chosen specimen preparation method involving compression, resulting in high specimen density. It is speculated that this high density impedes the formation of air gaps crucial for thermal insulation. Nevertheless, this study quantitatively confirms the potential for further development and commercialization of cellulose-based porous aerogel composite building materials through additional research.

Author Contributions: Conceptualization, D.R.; Methodology, D.R.; Software, J.H.; Validation, J.H., Y.K. and J.P.; Analysis, D.R. and J.H.; Resources, J.H.; Writing, J.H. and D.R.; Review and Editing, D.R. and J.H.; Visualization, J.H., Y.K. and J.P.; Supervision, D.R.; Project Administration, D.R.; Funding Acquisition, D.R. All authors have read and agreed to the published version of the manuscript.

Funding: This research received no external funding.

Institutional Review Board Statement: Not applicable.

Informed Consent Statement: Not applicable.

Data Availability Statement: Data are contained within the article.

Conflicts of Interest: The authors declare no conflicts of interest.

References

1. Unnikrishnan, A.; Figliozzi, M. A Study of the Impact of COVID-19 on Home Delivery Purchases and Expenditures. 2020. Available online: https://archives.pdx.edu/ds/psu/33410 (accessed on 29 July 2020).
2. Klemm, D.; Heublein, B.; Fink, H.; Bohn, A. Cellulose: Fascinating Biopolymer and Sustainable Raw Material. *Angew. Chem. Int. Ed.* **2005**, *44*, 3358–3393. [CrossRef] [PubMed]
3. Liang, H.-H.; Ho, M.-C. Toxicity Characteristics of Commercially Manufactured Insulation Materials for Building Applications in Taiwan. *Constr. Build. Mater.* **2007**, *21*, 1254–1261. [CrossRef]
4. Yoon, J.-K.; Koo, K.-W. A Study on Properties of Thermal Insulation Board Prepared by Porous Silica Aerogel. *Trans. Korean Inst. Electr. Eng.* **2012**, *61*, 1362–1367. [CrossRef]
5. Su, Z.; Meng, J.; Su, Y. Application of SiO2 Nanocomposite Ferroelectric Material in Preparation of Trampoline Net for Physical Exercise. *Adv. Nano Res.* **2023**, *14*, 355–362. [CrossRef]
6. Sun, L.; Liang, T.; Zhang, C.; Chen, J. The Rheological Performance of Shear-Thickening Fluids Based on Carbon Fiber and Silica Nanocomposite. *Phys. Fluids* **2023**, *35*, 032002. [CrossRef]
7. Jelle, B.P.; Baetens, R.; Gustavsen, A. Aerogel Insulation for Building Applications. *Sol-Gel Handb.* **2015**, Chapter 45, 1385–1412. [CrossRef]
8. Alvey, J.B.; Patel, J.; Stephenson, L.D. Experimental Study on the Effects of Humidity and Temperature on Aerogel Composite and Foam Insulations. *Energy Build.* **2017**, *144*, 358–371. [CrossRef]
9. Akhter, F.; Soomro, S.A.; Inglezakis, V.J. Silica Aerogels; a Review of Synthesis, Applications, and Fabrication of Hybrid Composites. *J. Porous Mater.* **2021**, *28*, 1387–1400. [CrossRef]
10. Richter, K.; Norris, P.M.; Tien, C.-L. Aerogels: Applications, Structure, and Heat Transfer Phenomena. *Annu. Rev. Heat Transf.* **1995**, *6*, 61–114. [CrossRef]

11. Ganobjak, M.; Brunner, S.; Hofmann, J.; Klar, V.; Ledermann, M.; Herzog, V.; Kämpfen, B.; Kilian, R.; Wehdorn, M.; Wernery, J. Current Trends in Aerogel Use in Heritage Buildings: Case Studies from the Aerogel Architecture Award 2021. *Gels* **2023**, *9*, 814. [CrossRef]
12. Ganobjak, M.; Brunner, S.; Wernery, J. Aerogel Materials for Heritage Buildings: Materials, Properties and Case Studies. *J. Cult. Herit.* **2020**, *42*, 81–98. [CrossRef]
13. Thapliyal, P.C.; Singh, K. Aerogels as Promising Thermal Insulating Materials: An Overview. *J. Mater.* **2014**, *2014*, 127049. [CrossRef]
14. Sambucci, M.; Savoni, F.; Valente, M. Aerogel Technology for Thermal Insulation of Cryogenic Tanks—Numerical Analysis for Comparison with Traditional Insulating Materials. *Gels* **2023**, *9*, 307. [CrossRef]
15. Liu, S.; Zhang, X.; Zhu, X.; Xin, S. A Low-Temperature Phase Change Material Based on Capric-Stearic Acid/Expanded Graphite for Thermal Energy Storage. *ACS Omega* **2021**, *6*, 17988–17998. [CrossRef] [PubMed]
16. Hwang, J.; Park, D.; Rie, D. Manufacture and Combustion Characteristics of Cellulose Flame-Retardant Plate through the Hot-Press Method. *Polymers* **2023**, *15*, 4736. [CrossRef]
17. Park, S.-J.; Kim, K.-S. A Study on Oil Adsorption of Expanded Gaphites. *Korean Chem. Eng. Res.* **2004**, *42*, 362–367. Available online: https://www.researchgate.net/publication/263645246_A_Study_on_Oil_Adsorption_of_Expanded_Gaphites (accessed on 5 September 2023).
18. Laachachi, A.; Burger, N.; Apaydin, K.; Sonnier, R.; Ferriol, M. Is Expanded Graphite Acting as Flame Retardant in Epoxy Resin? *Polym. Degrad. Stab.* **2015**, *117*, 22–29. [CrossRef]
19. *ISO 22007-2:2022*; Plastics—Determination of Thermal Conductivity and Thermal Diffusivity—Part 2: Transient Plane Heat Source (Hot Disc) Method. ISO: Geneva, Switzerland, 2022.
20. Fesmire, J.; Sass, J. Aerogel Insulation Applications for Liquid Hydrogen Launch Vehicle Tanks. *Cryogenics* **2008**, *48*, 223–231. [CrossRef]
21. Panyakaew, S.; Fotios, S. New thermal insulation boards made from coconut husk and bagasse. *Energy Build.* **2011**, *43*, 1732–1739. [CrossRef]
22. Bakatovich, A.; Davydenko, N.; Gaspar, F. Thermal insulating plates produced on the basis of vegetable agricultural waste. *Energy Build.* **2018**, *180*, 72–82. [CrossRef]
23. Ahn, C.; Park, D.; Hwang, J.; Rie, D. A Study on the Reliability of Mass, Density, and Fire Performance of Recycled Wastepaper Building Finishing Material Made with Large Wet Cellulose 3D Printers. *Sustainability* **2022**, *14*, 13090. [CrossRef]
24. Cha, J.-H.; Seo, J.-K.; Kim, S.-M. Correlation Analysis of the Thermal Conductivity Heat Flow Meter and MTPS (Modified Transient Plane Source) Method Using Wood Flooring and Wall Materials. *J. Korea Furnit. Soc.* **2011**, *22*, 118–125. Available online: https://koreascience.kr/article/JAKO201117161875734.page (accessed on 26 September 2023).
25. Acharya, A.; Joshi, D.; Gokhale, V.A. AEROGEL—A Promising Building Material for Sustainable Buildings. *Chem. Process Eng. Res.* **2013**, *9*, 1–6, ISSN 2224-7467 (Paper). ISSN 2225-0913 (Online).

polymers

MDPI

Article

Fabrication of PVA–Silica Sol Wood Composites via Delignification and Freezing Pretreatment

Rizheng Cong [1,†], Taoyang Cai [2,†], Shangjie Ge-Zhang [3,*], Hong Yang [4,*] and Chang Zhang [3]

1 School of Civil Engineering and Transportation, Northeast Forestry University, Harbin 150040, China
2 Aulin College, Northeast Forestry University, Harbin 150040, China
3 College of Science, Northeast Forestry University, Harbin 150040, China
4 Research Institute of Intelligent Control and Systems, Harbin Institute of Technology, Harbin 150040, China
* Correspondence: gzsj19@nefu.edu.cn (S.G.-Z.); yangh23@stu.hit.edu.cn (H.Y.)
† These authors contributed equally to this work.

Abstract: The efficient exploitation of planted fast-growing wood is crucial for enhancing wood resource utilization. In this study, the fast-growing poplar wood was modified by in situ impregnation through vacuum impregnation with polyvinyl alcohol and nano-silica sol as impregnation modifiers, combined with delignification–freezing pretreatment. The samples were characterized by FTIR, XRD, SEM, and the universal mechanical testing machine. The results showed that the wrinkle deformation and cracking of the wood blocks were greatly alleviated after the delignification–freezing pretreatment and the polyvinyl alcohol and nano-silica sol were successfully integrated into the wood. The resulting polyvinyl alcohol–silica sol poplar composites exhibited about 216%, 80% and 43% higher compressive strength with respect to delignified wood, natural wood and impregnated natural wood, respectively, thereby demonstrating superior mechanical properties and potential opportunities for value-added and efficient utilization of low-quality wood.

Keywords: delignification; freezing pretreatment; modified wood; compressive strength

Citation: Cong, R.; Cai, T.; Ge-Zhang, S.; Yang, H.; Zhang, C. Fabrication of PVA–Silica Sol Wood Composites via Delignification and Freezing Pretreatment. *Polymers* **2024**, *16*, 1949. https://doi.org/10.3390/polym16131949

Academic Editors: Antonio Pizzi and Antonios N. Papadopoulos

Received: 11 May 2024
Revised: 14 June 2024
Accepted: 27 June 2024
Published: 8 July 2024

1. Introduction

Forests, viewed as a natural, environmentally friendly and renewable resource, serve an invaluable role in the global carbon cycles. Apart from assisting in hydrological regulation and soil sequestration, trees continue to sequester carbon throughout their life spans, which is also a key component of climate change mitigation efforts to achieve carbon neutrality targets [1–4]. Nevertheless, the rapid expansion of society is leading to an increasing consumption of wood, and it has emerged as a pressing issue to balance harvesting and carbon sequestration for sustainable development [5]. Although fast-growing wood offers new opportunities for maintaining high wood volumes without disrupting the carbon balance in contrast to traditional wood, its short growth cycle inevitably brings about disadvantages, such as looseness of material, inferior mechanical strength and insufficient durability, which severely limits the service life of its products and restricts its large-scale application.

The delignification treatment enables the removal of lignin and hemicellulose from the wood, which on the one hand creates more channels and anchor points inside the wood, while on the other hand alters parts of the physical and mechanical properties of the wood, thus exerting an influence on the prepared composites. The alkaline delignification process selectively dissolves lignin from the wood cell walls, increasing the accessible surface area and accelerating the dissolution of lignin, among other behaviors, thereby triggering modifications in the microstructure and macroscopic properties of the wood [6,7]. However, excessive delignification may cause crinkling and splitting of the wood. The crinkling refers to the deformation of the wood resulting from the movement of moisture during drying, inducing uneven surfaces, irregular cross-sections and even shrinkage of the wood, which in turn may lead to an increase in the rate of damage to the wood product, to the detriment of further processing and subsequent application [8]. The cryogenic treatment is remarkable for its effectiveness in enhancing the structural properties and stability of the material, as well as being described as an environmentally friendly physical treatment. With the proven low-temperature pre-freezing treatment, the water in the cell cavities of the process freezes into ice, destroying microstructures such as the grain pore membrane and improving the permeability of the wood. In addition, at lower temperatures, the ice inside the wood sublimates directly into water vapor without generating liquid capillary tension, which can effectively prevent wrinkling and collapse of the wood and improve the wood properties, which can be of great assistance in subsequent comprehensive wood modification [9,10].

For further advancement of mechanical properties, researchers typically compress wood after delignification or fill it with other materials to obtain enhanced mechanical properties [11–16]. Among these, impregnation treatment is a commonly employed green and facile method for wood modification [17,18]. More recently, the introduction of inorganic nanoparticles into organic polymers has gained much attention, and organic–inorganic nanocomposites offer an effective approach to optimize decay resistance, physical and mechanical properties, UV stability, and so on [19–21]. Dong et al. prepared a polymer–SiO_2 hybrid nanocomposite solution by the sol–gel method. The wood impregnated with this composite solution has excellent compressive strength and has the potential to be used as a substitute for wood–plastic composites and high-quality solid wood materials [22]. Chang et al. used silicon dioxide–polymer hybrid material, TEOS as an inorganic precursor and HDTMS as an organic modifier to prepare wood superhydrophobic coating by sol–gel chemistry, and it was still stable in strong acid and alkali environment [23]. Hoyos-Martínez et al. successfully prepared wood fire-retardant coatings based on the bio-based phenolic resin formula of lignin, tannin and inorganic nanoparticles [24].

In this study, a PVA–silica sol-modified impregnation solution was prepared, and the poplar fast-growing wood was modified in situ by delignification–freezing pretreatment and vacuum impregnation. This method is simple and economical. The effects of the impregnation solution on the delignification–freezing treatment of samples were investigated, and the effects of the delignification–freezing treatment method and PVA–silica sol-modified impregnation solution on the chemical structure, crystallinity properties and mechanical properties of poplar fast-growing wood were discussed. This method improved the impregnation rate and obtained high-strength wood with better mechanical properties.

2. Materials and Methods

2.1. Materials

The poplar blocks (Populus ussuriensis Kom, 21 years old) were collected from a plantation forestry farm in Shijiazhuang, Hebei Province, China. The samples were extracted from wood that presented no knots, discoloration, fungi or other obvious flaws and were cut to the dimensions of 30 mm × 20 mm × 20 mm (L × T × R). All blocks were continuously sanded with 600–2000-grit metallographic sandpaper until the dust was completely removed and were subsequently stored at 20 °C and 65% relative humidity to achieve hygroscopic equilibrium. Anhydrous ethanol (99%), NaOH (>96%), Na_2SO_3 (>97%) and deionized water were supplied by Harbin Junan Medical Glass Wholesale Station; polyvinyl alcohol (PVA, Mw~27,000) was purchased from Maclean's Ltd., Shanghai, China; and silica sol (JN-30) was acquired from Shandong Youso Chemical Technology Co., Ltd., Linyi, China. All of the chemical raw materials were analytically pure and did not require further purification for use.

2.2. Delignification–Freezing Pretreatment of the Poplar Wood Blocks

An aqueous solution of 1.25 mol·L^{-1} NaOH and 0.4 mol·L^{-1} Na_2SO_3 was prepared at room temperature and magnetically stirred for 10 min to obtain a delignification solution. The poplar samples were immersed in the delignification solution and placed in a constant temperature water bath shaker and hydrothermally treated at 95 °C for 6 h under 35 rpm.

The above procedure was performed to remove the lignin and hemicellulose from the wood. The samples were picked up and washed several times with boiling deionized water until the solution was almost colorless. The cleaned delignified wood was stored at −30 °C for 12 h and then dried to a steady weight in a vacuum drying oven to obtain delignified–frozen treated wood, referred to as DW.

2.3. Preparation of PVA–Silica Sol-Modified Impregnating Solution

The PVA powders were dissolved in deionized water and stirred for 6 h at 90 °C with a collector-type thermostatic heating magnetic stirrer (Yuhua DF-101S, Gongyi City Yuhua Instrument Co., Ltd., Gongyi, China) to obtain the aqueous PVA solution of 3wt%. The aqueous PVA solution and silica sol were then mixed in a mass ratio of 3:1 and ultrasonicated for 30 min with an ultrasonic cleaner (Fuyang F-020SD, Shenzhen Fuyang Technology Group Co., Ltd., Shenzhen, China) at a power of 180 W, followed by magnetic stirring at room temperature for 6 h to yield a homogeneous and stable PVA–silica sol-modified impregnation solution.

2.4. Preparation of PVA–Silica Sol-Modified Wood Based on Delignified Lignin Wood Framework

The dried DW samples were infiltrated into the PVA–silica sol-modified impregnating solution under vacuum for 2 h and then kept at room temperature and pressure for 22 h for adequate soaking. After the completion of impregnation, the poplar samples were removed and the residual mixture solution on the surface was washed several times with deionized water, drained at room temperature and subsequently dried and cured under a vacuum at 50 °C for 24 h to obtain PVA–silica sol poplar composites (impregnated delignified wood, IDW). A detailed preparation schematic is shown in Figure 1.

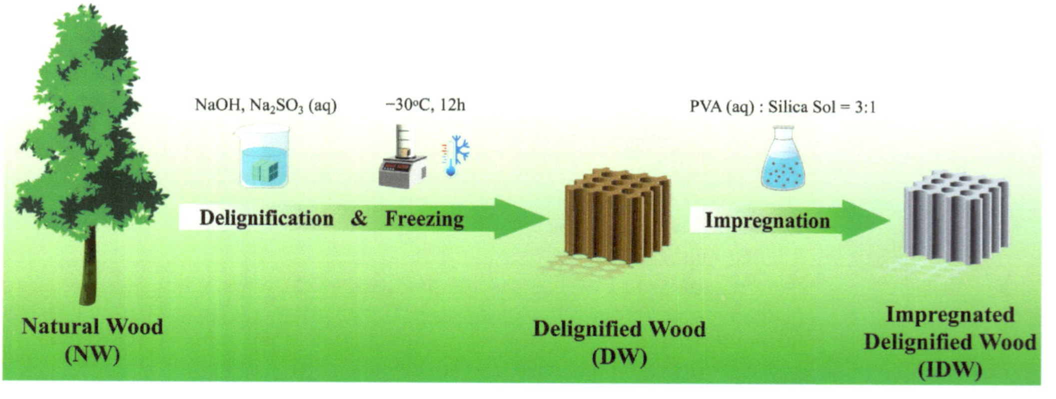

Figure 1. Schematic of the fabrication of composite wood.

2.5. Characterization

The surface morphological features of the wood samples were characterized with a cold field emission scanning electron microscope (FE-SEM, JSM-7500F, JEOL Ltd., Akishima, Japan), and the chemical elements in the samples were examined by an associated Oxford X-Max cooling energy spectrometer. The wood was cut into small slices parallel to the direction of growth, i.e., radial sections, using a pathological sectioning blade, and the cross-sectional micromorphology of NW and IDW slices at an accelerating voltage of 5.0 kV was analyzed by FE-SEM, and the homogeneous distribution of the modifier in the samples was further verified by the chemical elemental composition as tested by EDS [25–27]. The samples were ground into 200–300 mesh powders and assessed for the variations in the chemical composition of the wood using Fourier transform infrared spectroscopy (FTIR, Nicolet iS50, Thermo Fisher Scientific Inc., Waltham, MA, USA) with 32 scans in the wavelength range of 3000–400 cm^{-1} at a resolution of 4 cm^{-1}. The samples were scanned using an X-ray diffractometer (XRD, XRD-6100, Shimadzu Corporation, Kyoto, Japan), which was carried out to inspect the crystal structure by scanning samples ground to 200–300 mesh. The radiation source voltage was set at 40 kV, the radiation tube current at 30 mA, and the scanning range and measurement rate at 5–85° and 10°/min, respectively. Compression tests were conducted along the grain direction using a universal mechanical testing machine (Sanshi CMT-6305, Shenzhen, China), following the guidelines of reference standard GB/T 1935-2009 [28]. Prior to testing, samples were conditioned in a chamber set at a constant temperature of 20 °C and relative humidity of 65% to ensure uniform moisture content. Testing was carried out at room temperature (20 ± 5 °C), with a loading speed of 10 mm/min. The compression strength of five samples was calculated, and the average value was considered as the mean compression strength. The thermogravimetric analysis of the sample was carried out by using a simulated thermal analyzer (STA, Perkin Elmer STA 6000, Waltham, MA, USA), and the mass loss of the sample was measured at 30 °C to 800 °C at a heating rate of 10 °C min^{-1}. Hygroscopicity was evaluated according to GB/T 1934.1-2009 [29]; dry shrinkage also uses this standard. In addition, in order to evaluate the wettability of the composites, the static water contact angle of the samples was measured by a droplet shape analyzer (DSA100, Krüss, Hamburg, Germany), and the droplet volume of 5 μL was selected.

3. Results and Discussion

3.1. Impregnation Rate and Weight Increase Rate

The weighing was conducted using an analytical balance. The mass of the dried sample before impregnation is recorded as m_0 and the mass of the sample after impregnation as m_1. The impregnated sample is dried in a vacuum drying oven at 50 °C to a constant weight and then quickly removed and weighed to prevent the wood from absorbing moisture, and the mass of the impregnated sample after drying is recorded as m_2. The rate of impregnation and the rate of weight increase for NW and DW are calculated and compared according to the following two equations:

$$\text{Impregnation Rate} = \frac{m_1 - m_0}{m_0} \times 100\% \tag{1}$$

$$\text{Weight Increase Rate} = \frac{m_2 - m_0}{m_0} \times 100\% \tag{2}$$

Since it is known that the presence or absence of lignin significantly affects the modification effect, an excessive lignin content weakens the modification effect [30,31]. After the delignification of the samples using a mixture of NaOH and Na_2SO_3 solutions, most of the lignin and hemicellulose of the wood was removed, which will be confirmed in the FTIR and XRD patterns below [32], and it increased the porosity of the wood and provided more channels and sites for the impregnation of the modified impregnating solution, which will further improve impregnation effect [33,34]. The impregnation effect can be visualized by calculating the impregnation rate and the weight gain rate. Figure 2 shows the impregnation and weight gain rates for NW and DW, where the black bars show a 78.20% increase in impregnation rate for delignified–frozen treated wood (DW) compared to direct impregnation with NW, and the red bars show a 98.03% increase in weight gain for DW compared to NW. This indicates that the degree of impregnation of the delignin–frozen treated samples was increased and that the added modified impregnating solution was effective in cross-linking with the wood [35], resulting in an increase in weight gain.

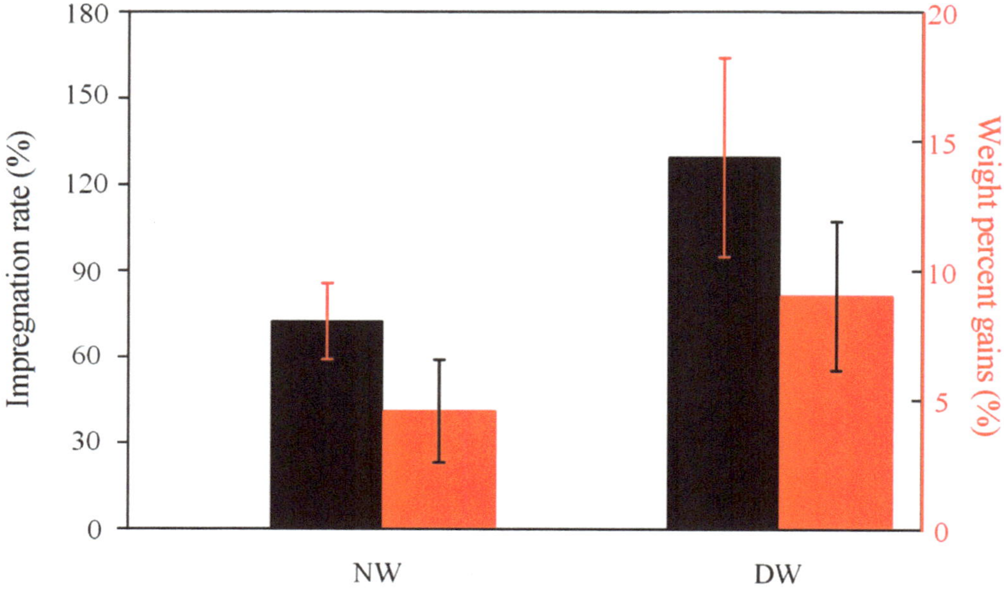

Figure 2. Impregnation and weight gain rates of NW and DW.

3.2. Micromorphology and Microstructure

The FE-SEM equipped with EDS enabled us to determine the microstructural morphology and elemental distribution of the samples more precisely. The EDS images in Figure 3b show that, despite repeated washing with boiling deionized water, a small amount of Na from the delignification solution remained uniformly on the surface of the wood, (3.27%), i.e., the wood was uniformly and profoundly delignified. By comparing the SEM images of the untreated NW radial section in Figure 3a and the DW radial section in Figure 3b, it is clear that under the action of the delignification solution, some components of wood are selectively removed, exhibiting larger and more irregular lumens and higher porosity [36,37], which provides more channels for the penetration of the PVA–silica sol-modified impregnating solution in the subsequent steps. In addition, the porous and layered structure of the poplar samples was not significantly affected by the delignification–freezing treatment, as the pre-freezing treatment resulted in "cold shrinkage" due to the loss of water from the cell walls, and the rapid sublimation of the ice prevented fluid movement and reduced macroscopic wrinkling of the wood, in addition to the fact that the expanding ice disrupted part of the microstructure of the wood, improving the permeability of the wood. In addition, the expanding ice disrupts part of the microstructure of the wood, improving the permeability of the wood and making subsequent PVA–nano-silica sols easier to infiltrate [38,39]. Figure 3c is an SEM image of the IDW, demonstrating the uniform distribution of the PVA–silica sol-modified impregnation on the wood cell walls, with no tight gaps between them. Further, as shown in Figure 3d, the filling of the grain pores with the impregnation solution can be clearly observed in the SEM image of the IDW sample at a magnification of 5000, indicating that the impregnation of the DW was adequate and complete.

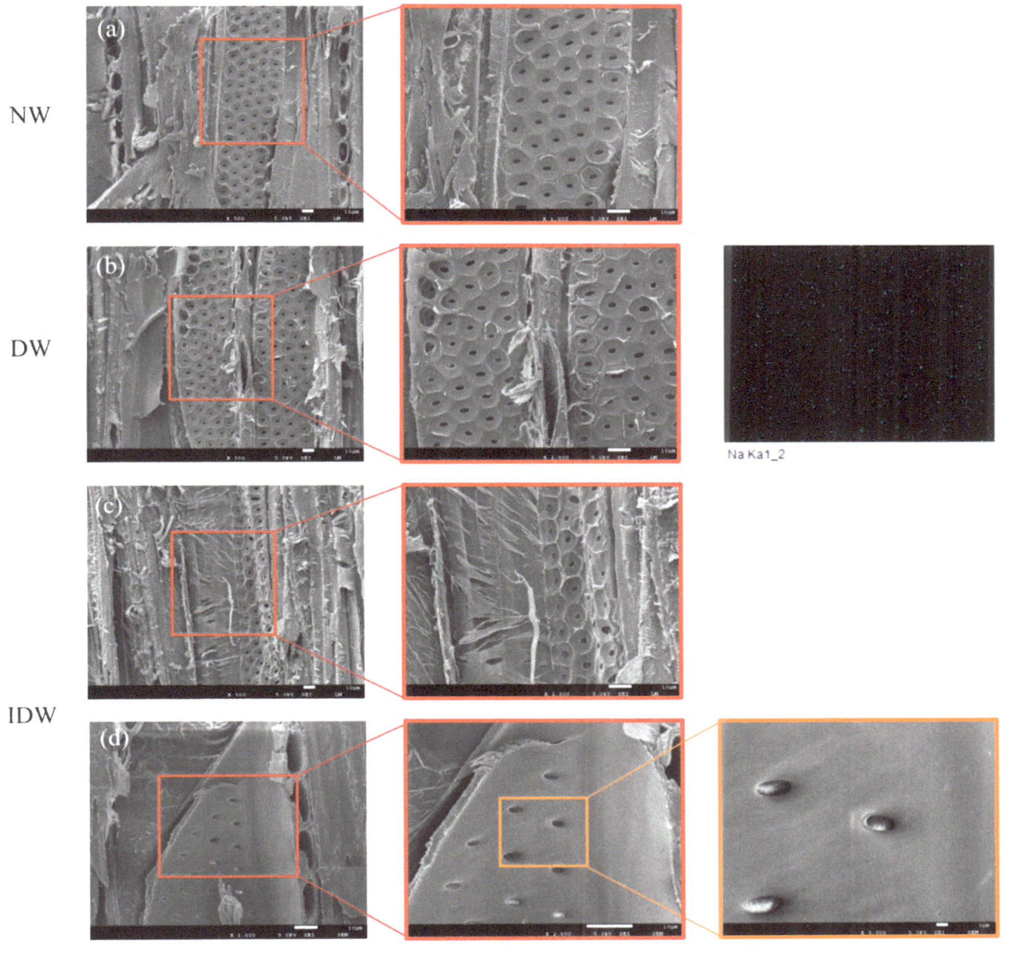

Figure 3. The SEM images and EDS images of NW (**a**), DW (**b**), IDW (**c**,**d**) at different magnifications.

3.3. Chemical Composition

The samples from the NW, DW and IDW groups were analyzed by FTIR spectroscopy to investigate the effect of delignification–freezing treatment versus modified impregnation solution on the chemical structure of poplar wood. Figure 4a illustrates the FTIR spectra of the three samples, all of which reveal an absorption vibration peak of -OH around 3333 cm^{-1}, a phenolic hydroxyl vibration peak of lignin at 1419 cm^{-1} [40] and a strong characteristic absorption peak of C-O at 1030 cm^{-1}. The samples from the DW and IDW groups show significantly enhanced -OH absorption vibrations around 3333 cm^{-1} compared to NW, which is partly due to the increased content of bound hydroxyl groups in the impregnated modified wood due to the high hydrophilic hydroxyl group of PVA itself, and partly due to the formation of more hydrogen bonds between the modified impregnate and the wood. The poplar samples from the IDW group display a weakening of the C=O stretching vibration peak at 1734 cm^{-1}. The fact that the activation energy of delignified wood is always less than that of untreated wood suggests that the C=O double bonds in the delignified wood-modified wood are more likely to break and are more likely to undergo additional reactions with the silica nanosol. As can be seen in Figure 4a, the characteristic Si-O-Si vibrational peak appears at 1589 cm^{-1} and the enhancement of the Si-O-Si vibrational peak is more pronounced for IDW. This suggests that delignin promotes the chemical reaction of the PVA–silica sol to form a cross-linked system within the wood, with increasingly larger molecules but fewer hydrophilic Si-O- groups, rendering the molecules insoluble in water and improving the hydrophobicity of the composite wood [37–40].

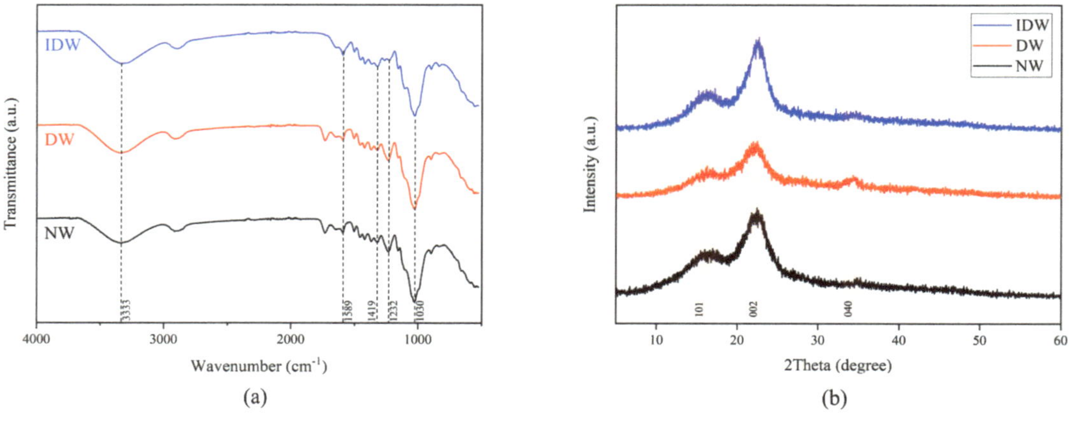

Figure 4. FTIR (**a**) and XRD (**b**) patterns of NW, DW, IDW.

The variation in the crystalline structure of cellulose in the cell walls of wood affects the size of its crystalline zones, cell parameters and crystallinity. The relative crystallinity of wood is basically the percentage of the cellulose crystalline zone in relation to the total cellulose. Variations in crystallinity are closely related to the dimensional stability, hardness, tensile strength and density of the fibers of the wood. According to the two-phase system theory of cellulose structure, in the crystalline region, the cellulose molecular chains are arranged in a directionally ordered manner, whereas in the non-crystalline region, they are irregularly arranged [41], and X-ray diffraction allows for the detection of highly crystalline cellulose molecular chains [42]. To investigate the effect of the hybridized system formed by delignification–freezing treatment and PVA–nano-silica sol on the crystal structure of the modified wood, the NW, DW and IDW groups were examined using XRD, and the results are displayed in Figure 4b. In the X-ray diffraction pattern of the wood, the highest peak is located at $2\theta = 22.42^\circ$, which is the 002 crystal plane and represents the width of the crystalline region. At $2\theta = 34.7^\circ$, which is the 040 crystal plane, represents the length of the crystal zone. The 002 crystal plane has a higher intensity than the 040 crystal plane due to the molecules within the cellulose being located mainly in planes parallel to the 002 crystal plane. A 101 wave is observed at $2\theta = 16.5^\circ$, which is the diffraction intensity of the amorphous region. A comparison of the three sets of spectra shows no significant change in the position of the diffraction peaks, i.e., the cellulose crystalline regions of NW, DW and IDW have not suffered alteration [43]. Delignin breaks the hydrogen bonds of the cellulose itself weakening the inter-molecular interactions between the cellulose molecular chains and reducing the crystallinity [44]. In general, the crystallinity of wood is positively correlated with the mechanical strength of wood [45]. Although the crystallinity of delignified DW was lower and the mechanical properties were weaker, the compressive strength of the IDW group was higher than the other groups in the mechanical property tests, which may be due to the reaction of PVA with wood cellulose after it has been immersed inside the wood to create stronger chemical bonds and improve the mechanical properties.

In summary, based on SEM-EDS images, FTIR and XRD patterns, we can infer that delignification and freezing treatments give more possibilities for PVA and silica sol to incorporate into the wood, cross-linking and filling the pores of the wood cell walls, thus ameliorating the properties of the wood.

3.4. Physical and Mechanical Properties

The compressive strength parallel to the grain means the maximum stress produced by pressure in the direction of the grain and is one of the most important indicators of the mechanical properties of wood. The mechanical properties of the NW, INW, DW and IDW groups were tested in Figure 5. The data are derived from the mean value of the compressive strength of five randomly selected samples in each group. As shown in Figure 5, the smooth grain compressive strengths of the DW, NW, INW and IDW groups increased in order. Compared to the NW group, the average compressive strength of the samples from the INW and IDW groups increased by 26% and 80%, respectively, indicating that the filling of the PVA–silica sol-modified impregnation could increase the compressive strength of the wood [46,47]. Although the DW group, which was only delignified–chilled, showed a sharp decrease in compressive strength due to the reduction of lignin and hemicellulose and the disappearance of the wood's internal support material, the IDW group obtained by impregnating it with PVA–silica sol was the strongest, still 43% higher than the INW group, which had the second strongest compressive strength, due to the fact that the delignification treatment provided more channels inside the wood and space within the wood, allowing the PVA–nano-silica sol to fill the pores and lamellar structures in the poplar wood more fully and more often. On the other hand, the pre-freezing treatment improves the permeability of the wood, making it easier for the PVA–nano-silica sol to infiltrate. During the impregnation process, as the modified impregnation solution is cross-linked with PVA and nano-silica sol, which has a large number of hydroxyl groups on the outside, the delignified wood is also able to expose more hydroxyl groups and therefore can form hydrogen bonds more tightly, thus increasing the compressive strength of the IDW. Compared with the literature data of similar systems, the samples prepared in this study have good mechanical properties [33,48,49].

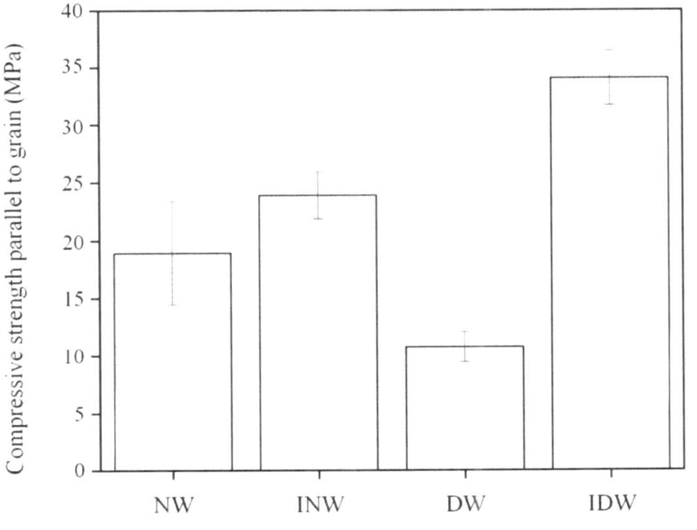

Figure 5. Mechanical properties of several groups of poplar wood samples.

3.5. Water Contact Angle, Hygroscopicity and Dry Shrinkage

Static wetting behavior was evaluated using 5 microliters of droplets. NW and DW were completely absorbed within 10 s and 2 s after the droplets contacted the surface

(Videos S1 and S2), respectively, showing strong hydrophilicity. The faster water absorption of DW may be due to delignification, which removes most hydrophobic lignin, exposes more hydroxyl groups and increases the porosity of wood. IDW tends to be stable after the droplet contacts the surface for 6 s, and finally, a static contact angle of 70° is obtained (Figure 6, Video S3); the improvement in hydrophobicity is mainly attributed to the increase in surface roughness and the barrier of surface film to water [50]. The silica nanoparticles on the wood surface will affect the wood surface roughness and capture more air, and PVA will coat the nanoparticles and form a film on the surface of delignified wood to slow down the water infiltration. In future experiments, it is necessary to use low-surface energy substances to modify the surface in order to further realize hydrophobicity [51–54].

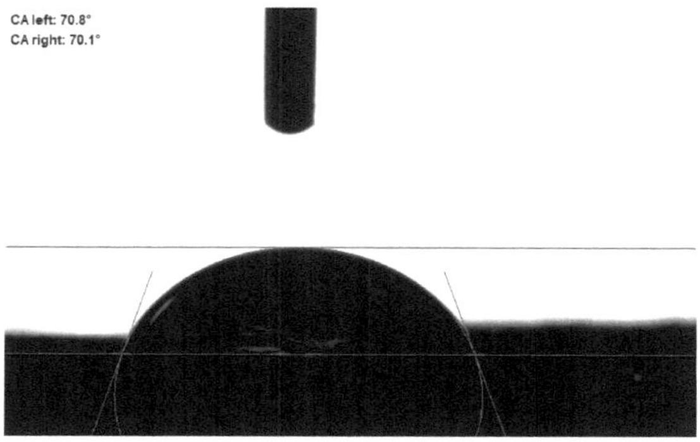

Figure 6. Static water contact angle of IDW surface.

The water absorption and dry shrinkage of DW and IDW are given in Table 1. Due to its rich pores and hydrophilicity, DW shows strong hygroscopicity, and its 24-h water absorption rate reaches 96%, while IDW is ~68%. After 2 weeks of continuous water absorption, the water absorption rate of DW is basically maintained at around 155%, while the IDW is still below 105%. PVA impregnation can effectively reduce the water absorption of wood, which may be due to the fact that PVA will form a film inside the wood during impregnation, blocking these pores, thus reducing the passage of water into the wood [55]. In addition, PVA impregnation can also reduce the water absorption by enhancing the structural stability of wood and reducing moisture absorption and expansion. For drying shrinkage, the radial full drying shrinkage of NW is 5.7%; the tangential full drying shrinkage is 7.1%. The shrinkage rate of volume full drying is 12.8%. Compared with NW, the radial shrinkage of IDW is slightly increased, while the tangential shrinkage and volume shrinkage are slightly decreased, which are 6.2%, 6.9% and 12.8%, respectively. It is speculated that PVA may play a role in filling and enhancing the structural stability of wood and reducing the deformation of wood in the tangential direction [56].

Table 1. Water absorption and full dry shrinkage of DW and IDW.

		NW	IDW
Water absorption (%)	1 day	96	68
	3 day	121	83
	7 day	142	94
	10 day	148	99
	12 day	152	101
	14 day	155	103
Full Dry Shrinkage (%)	Radial	5.7	6.2
	Tangential	7.1	6.9
	Volume	12.8	12.8

3.6. Thermogravimetric Properties

The thermal stability of various samples between 30 °C and 800 °C was investigated using thermogravimetric analysis (Figure 7). For delignified wood (DW), the initial mass loss from room temperature to 105 °C is attributed to water evaporation (~4%). A negligible weight loss is observed between 105 °C and 220 °C, during which the sample absorbs heat slowly, resulting in minimal mass change and a low weight loss rate. The highest weight loss rate, approximately 60%, occurs during the pyrolysis stage between 220 °C and 350 °C, due to the initial oxidative decomposition of hemicellulose and cellulose, as most lignin has been removed from DW. Hemicellulose begins to pyrolyze at 220 °C, with substantial decomposition products forming between 250 °C and 350 °C. Beyond 350 °C, the degradation rate decreases significantly, entering the carbonization stage where residual substances slowly pyrolyze until carbonization, resulting in little mass change. Post-experiment, only black carbon remains in the crucible. Pure PVA exhibits three distinct weight loss stages in its thermogravimetric analysis. From room temperature to 150 °C, the first stage shows a gentle downward trend due to the loss of physical water, with a mass loss of approximately 7%. The second stage, between 260 °C and 380 °C, shows a mass loss of up to 60%, primarily due to dehydration caused by the elimination of hydroxyl side groups. The final stage, from 400 °C to 460 °C, corresponds to the breakdown of the main PVA chain, leaving a low residual mass characteristic of polymeric materials [57]. The PVA-IDW sample, impregnated only with pure PVA, demonstrates overlapping degradation stages of cellulose and PVA from 200 °C to 450 °C, with the final mass residue falling between those of DW and pure PVA. Notably, the residual mass fraction of IDW at 800 °C is approximately 61%, significantly higher than that of DW and PVA-IDW, indicating enhanced thermal resistance. This improvement may be due to the excellent thermal properties of nano-silica, and with the increase in temperature, nano-silica remains stable without mass loss.

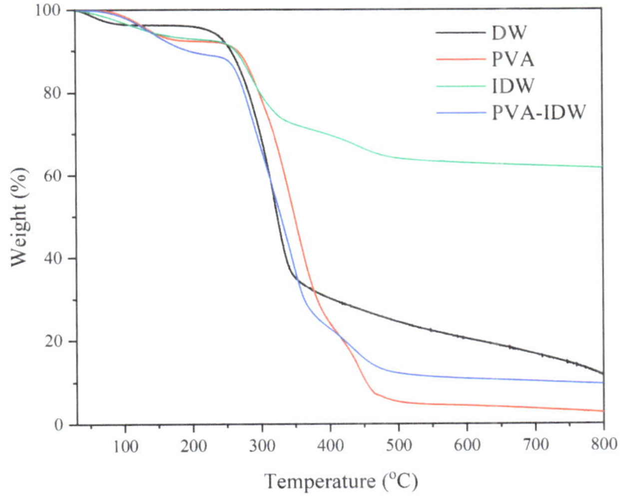

Figure 7. A typical TGA thermogram of DW, pure PVA, IDW and PVA-IDW.

4. Conclusions

In this paper, the fast-growing poplar wood is modified under vacuum impregnation using PVA–silica sol as an impregnation modifier, in combination with delignification–freezing treatment. The prepared PVA–silica sol poplar wood composites possess excellent smooth grain compressive strength, and the impregnation effect, structural modifications and compressive strength were investigated for analysis. The results showed that the removal of lignin and the increase in porosity in the delignified poplar samples facilitate the entry of PVA and silica sol into the wood, which is macroscopically expressed as an increase in weight gain and impregnation rate of IDW, and an 80.01% increase in compressive strength compared to NW. In summary, the method proposed in this study provides a research idea for the high-value utilization of low-quality wood and offers new possibilities for the application of wood in the construction industry. Additionally, given the unique adsorption characteristics of nano-silica, a promising future research direction involves modifying the silica before impregnation (such as introducing silver nanoparticles or natural antibacterial components) to enhance its antibacterial properties [58,59].

Supplementary Materials: The following supporting information can be downloaded at https://www.mdpi.com/article/10.3390/polym16131949/s1. Video S1: Absorption time of droplets after contacting NW surface, Video S2: Absorption time of droplets after contacting the DW surface, Video S3: Absorption time of droplets after contacting the IDW surface.

Author Contributions: Conceptualization, R.C. and T.C.; Data Curation, S.G.-Z.; Formal Analysis, R.C.; Investigation, R.C. and T.C.; Methodology, R.C. and T.C.; Project Administration, H.Y.; Supervision, S.G.-Z. and H.Y.; Validation, R.C., T.C. and S.G.-Z.; Visualization, T.C.; Writing—Original Draft, R.C. and T.C.; Writing—Review and Editing, T.C., S.G.-Z. and C.Z. All authors have read and agreed to the published version of the manuscript.

Funding: This research was funded by the Fundamental Research Funds for the Central Universities (No.2572020BC07).

Data Availability Statement: The original contributions presented in the study are included in the article/Supplementary Materials, further inquiries can be directed to the corresponding author/s.

Conflicts of Interest: The authors declare no conflicts of interest.

References

1. Rudd, L.; Kulshreshtha, S.; Belcher, K.; Amichev, B. Carbon life cycle assessment of shelterbelts in Saskatchewan, Canada. *J. Environ. Manag.* **2021**, *297*, 113400. [CrossRef] [PubMed]
2. Mora Rollo, A.; Rollo, A.; Mora, C. The tree-lined path to carbon neutrality. *Nat. Rev. Earth Environ.* **2020**, *1*, 332. [CrossRef]
3. Nian, V. The carbon neutrality of electricity generation from woody biomass and coal, a critical comparative evaluation. *Appl. Energy* **2016**, *179*, 1069–1080. [CrossRef]
4. Zhang, C.; Ge-Zhang, S.; Wang, Y.; Mu, H. A Wooden Carbon-Based Photocatalyst for Water Treatment. *Int. J. Mol. Sci.* **2024**, *25*, 4743. [CrossRef] [PubMed]
5. Dong, L.; Bettinger, P.; Liu, Z. Estimating the optimal internal carbon prices for balancing forest wood production and carbon sequestration: The case of northeast China. *J. Clean. Prod.* **2021**, *281*, 125342. [CrossRef]
6. Shakeel, U.; Li, X.; Wang, B.; Geng, F.; Zhang, Q.; Zhang, K.; Xu, X.; Xu, J. Structure and integrity of sequentially extracted lignin during poplar (alkaline) pretreatment. *Process Biochem.* **2022**, *117*, 198–208. [CrossRef]
7. Yang, H.; Wang, H.; Cai, T.; Ge-Zhang, S.; Mu, H. Light and wood: A review of optically transparent wood for architectural applications. *Ind. Crops Prod.* **2023**, *204*, 117287. [CrossRef]
8. Yang, L.; Liu, H. A review of Eucalyptus wood collapse and its control during drying. *BioResources* **2018**, *13*, 2171–2181. [CrossRef]
9. Missio, A.L.; Mattos, B.D.; De Cademartori, P.H.; Gatto, D.A. Effects of two-step freezing-heat treatments on Japanese raisintree (*Hovenia dulcis* thunb.) wood properties. *J. Wood Chem. Technol.* **2016**, *36*, 16–26. [CrossRef]
10. Liu, H.; Gao, J.; Chen, Y. Effects of pre-freezing prior to drying upon some physical and mechanical properties of Eucalyptus urophylla × Eucalyptus grandis wood. *BioResources* **2015**, *10*, 6417–6427. [CrossRef]
11. Farid, T.; Rafiq, M.I.; Ali, A.; Tang, W. Transforming wood as next-generation structural and functional materials for a sustainable future. *EcoMat* **2022**, *4*, e12154. [CrossRef]
12. Wang, Y.; Wu, Y.; Yang, F.; Yang, L.; Wang, J.; Zhou, J.; Wang, J. A highly transparent compressed wood prepared by cell wall densification. *Wood Sci. Technol.* **2022**, *56*, 669–686. [CrossRef]
13. Luo, H.; Si, R.; Liu, J.; Li, P.; Tao, Y.; Zhao, X.; Chen, H. Preparing the reinforced wood via embedding cellulose nanocrystals (CNC) into delignified fast-growing wood followed by densification. *Cellulose* **2022**, *29*, 7377–7396. [CrossRef]
14. Wang, J.; Liu, J.; Li, J.; Zhu, J.Y. Characterization of microstructure, chemical, and physical properties of delignified and densified poplar wood. *Materials* **2021**, *14*, 5709. [CrossRef] [PubMed]
15. Chen, X.; Ge-Zhang, S.; Han, Y.; Yang, H.; Ou-Yang, W.; Zhu, H.; Hao, J.; Wang, J. Ultraviolet-assisted modified delignified wood with high transparency. *Appl. Sci.* **2022**, *12*, 7406. [CrossRef]
16. Jakob, M.; Czabany, I.; Veigel, S.; Müller, U.; Gindl-Altmutter, W. Comparing the suitability of domestic spruce, beech, and poplar wood for high-strength densified wood. *Eur. J. Wood Wood Prod.* **2022**, *80*, 859–876. [CrossRef]

17. Spear, M.J.; Curling, S.F.; Dimitriou, A.; Ormondroyd, G.A. Review of functional treatments for modified wood. *Coatings* **2021**, *11*, 327. [CrossRef]
18. Wan, C.; Liu, X.; Huang, Q.; Cheng, W.; Su, J.; Wu, Y. A Brief Review of Transparent Wood: Synthetic Strategy, Functionalization and Applications. *Curr. Org. Synth.* **2021**, *18*, 615–623. [CrossRef] [PubMed]
19. Kango, S.; Kalia, S.; Celli, A.; Njuguna, J.; Habibi, Y.; Kumar, R. Surface modification of inorganic nanoparticles for development of organic–inorganic nanocomposites—A review. *Prog. Polym. Sci.* **2013**, *38*, 1232–1261. [CrossRef]
20. Yu, J.; Wang, A.C.; Zhang, M.; Lin, Z. Water treatment via non-membrane inorganic nanoparticles/cellulose composites. *Mater. Today* **2021**, *50*, 329–357. [CrossRef]
21. Mulla, M.Z.; Rahman, M.R.T.; Marcos, B.; Tiwari, B.; Pathania, S. Poly lactic acid (Pla) nanocomposites: Effect of inorganic nanoparticles reinforcement on its performance and food packaging applications. *Molecules* **2021**, *26*, 1967. [CrossRef] [PubMed]
22. Dong, X.; Zhuo, X.; Wei, J.; Zhang, G.; Li, Y. Wood-based nanocomposite derived by in situ formation of organic–inorganic hybrid polymer within wood via a sol–gel method. *ACS Appl. Mater. Interfaces* **2017**, *9*, 9070–9078. [CrossRef] [PubMed]
23. Chang, H.; Tu, K.; Wang, X.; Liu, J. Facile preparation of stable superhydrophobic coatings on wood surfaces using silica-polymer nanocomposites. *BioResources* **2015**, *10*, 2585–2596. [CrossRef]
24. De Hoyos-Martínez, P.L.; Issaoui, H.; Herrera, R.; Labidi, J.; Charrier-El Bouhtoury, F. Wood Fireproofing Coatings Based on Biobased Phenolic Resins. *ACS Sustain. Chem. Eng.* **2021**, *9*, 1729–1740. [CrossRef]
25. Ma, Y.; Wang, Q.; Zhou, X.; Hao, J.; Gault, B.; Zhang, Q.; Dong, C.; Nieh, T.G. A Novel Soft-Magnetic B2-Based Multiprincipal-Element Alloy with a Uniform Distribution of Coherent Body-Centered-Cubic Nanoprecipitates. *Adv. Mater.* **2021**, *33*, 2006723. [CrossRef] [PubMed]
26. Wang, M.; Dong, L.; Wu, J.; Shi, J.; Gao, Q.; Zhu, C.; Morikawa, H. Leaf-meridian bio-inspired nanofibrous electronics with uniform distributed microgrid and 3D multi-level structure for wearable applications. *NPJ Flex. Electron.* **2022**, *6*, 34. [CrossRef]
27. Shi, Y.C.; Ma, B.G.; Guan, J.P.; Ye, B.; Yu, L.C.; Wu, Z.M.; Shen, X.J.; Habaue, S. Rational design of a functionalized silicone polymer for modifying epoxy-based composites. *J. Mater. Res. Technol.* **2022**, *19*, 3867–3876. [CrossRef]
28. *GB/T 1935-2009*; Method of Testing in Compressivestrength Parallel to Grain of Wood. National Standards of People's Republic of China: Beijing, China, 2009.
29. *GB/T 1934.1-2009*; Method for Determination of the Water Absorption of Wood. National Standards of People's Republic of China: Beijing, China, 2009.
30. Al Arni, S. Extraction and isolation methods for lignin separation from sugarcane bagasse: A review. *Ind. Crops Prod.* **2018**, *115*, 330–339. [CrossRef]
31. Pejic, B.; Vukcevic, M.; Kostic, M.; Skundric, P. Biosorption of heavy metal ions from aqueous solutions by short hemp fibers: Effect of chemical composition. *J. Hazard. Mater.* **2009**, *164*, 146–153. [CrossRef]
32. Alqrinawi, H.; Ahmed, B.; Wu, Q.; Hai, L. Effect of partial delignification and densification on chemical, morphological, and mechanical properties of wood: Structural property evolution. *Ind. Crops Prod.* **2024**, *213*, 118430. [CrossRef]
33. Yang, H.; Wang, D.; Han, Y.; Tian, P.; Gao, C.; Yang, X.; Mu, H.; Zhang, M. Preparation and properties of modified poplar impregnated with PVA-nano silica sol composite dispersion system. *J. Wood Chem. Technol.* **2022**, *42*, 211–221. [CrossRef]
34. Ge-Zhang, S.; Yang, H.; Mu, H. Interfacial solar steam generator by MWCNTs/carbon black nanoparticles coated wood. *Alex. Eng. J.* **2023**, *63*, 1–10. [CrossRef]
35. Wu, S.S.; Tao, X.; Xu, W. Thermal conductivity of Poplar wood veneer impregnated with graphene/polyvinyl alcohol. *Forests* **2021**, *12*, 777. [CrossRef]
36. Vitas, S.; Segmehl, J.S.; Burgert, I.; Etienne, C. Porosity and pore size distribution of native and delignified beech wood determined by mercury intrusion porosimetry. *Materials* **2019**, *12*, 416. [CrossRef]
37. Liang, R.; Zhu, Y.H.; Wen, L.; Zhao, W.W.; Kuai, B.B.; Zhang, Y.L.; Cai, L.P. Exploration of effect of delignification on the mesopore structure in poplar cell wall by nitrogen absorption method. *Cellulose* **2020**, *27*, 1921–1932. [CrossRef]
38. Ilic, J. Advantages of prefreezing for reducing shrinkage-related degrade in eucalypts: General considerations and review of the literature. *Wood Sci. Technol.* **1995**, *29*, 277–285. [CrossRef]
39. Rooni, V.; Raud, M.; Kikas, T. The freezing pre-treatment of lignocellulosic material: A cheap alternative for Nordic countries. *Energy* **2017**, *139*, 1–7. [CrossRef]
40. Ji, Y.; Zhang, H.; Zhang, C.; Quan, Z.; Huang, M.; Wang, L. Fluorescent and mechanical properties of silicon quantum dots modified sodium alginate-carboxymethylcellulose sodium nanocomposite bio-polymer films. *Polymers* **2019**, *11*, 1476. [CrossRef] [PubMed]
41. Lv, Q.; Wang, W.; Liu, Y. Study on thermal insulation performance of cross-laminated bamboo wall. *J. Renew. Mater.* **2019**, *7*, 1231–1250. [CrossRef]
42. Yuan, T.; Zhang, T.; Huang, Y.; Wu, Y.; Wang, X.; Li, Y. Study on bamboo longitudinal flattening technology. *Polymers* **2022**, *14*, 816. [CrossRef]
43. Ding, L.; Han, X.; Jiang, S. Impregnation of poplar wood with multi-functional composite modifier and induction of in-situ polymerization by heating. *J. Wood Chem. Technol.* **2021**, *41*, 220–228. [CrossRef]
44. Xu, E.; Wang, D.; Lin, L. Chemical structure and mechanical properties of wood cell walls treated with acid and alkali solution. *Forests* **2020**, *11*, 87. [CrossRef]

45. Jiang, F.; Li, T.; Li, Y.; Zhang, Y.; Gong, A.; Dai, J.; Hitz, E.; Luo, W.; Hu, L. Wood-based nanotechnologies toward sustainability. *Adv. Mater.* **2018**, *30*, 1703453. [CrossRef]
46. Yona, A.M.C.; Žigon, J.; Matjaž, P.; Petrič, M. Potentials of silicate-based formulations for wood protection and improvement of mechanical properties: A review. *Wood Sci. Technol.* **2021**, *55*, 887–918. [CrossRef]
47. Cai, Y.; Wu, Y.; Yang, F.; Gan, J.; Wang, Y.; Zhang, J. Wood sponge reinforced with polyvinyl alcohol for sustainable oil–water separation. *ACS Omega* **2021**, *6*, 12866–12876. [CrossRef] [PubMed]
48. Zhang, Y.; Guan, P.; Ma, X.; Ping, L.; Zhenyu, S.; Xianjun, L.; Yingfeng, Z. Study on the Effect of Acrylic Acid Emulsion on the Properties of Poplar Wood Modified by Sodium Silicate Impregnation. *Forests* **2023**, *14*, 1221. [CrossRef]
49. Lu, Y.; Feng, M.; Zhan, H. Preparation of SiO_2–wood composites by an ultrasonic-assisted sol–gel technique. *Cellulose* **2014**, *21*, 4393–4403. [CrossRef]
50. Zhang, X.; Liu, W.; Liu, W.; Xueqing, Q. High performance PVA/lignin nanocomposite films with excellent water vapor barrier and UV-shielding properties. *Int. J. Biol. Macromol.* **2020**, *142*, 551–558. [CrossRef]
51. Jiang, J.; Zhou, Y.; Mei, C.; Cao, J. Polyethylene glycol and silica sol penetration improves hydrophobicity and dimensional stability of wood after a short-time treatment. *Eur. J. Wood Wood Prod.* **2021**, *79*, 1395–1404. [CrossRef]
52. Wang, Y.; Ge-Zhang, S.; Mu, P.; Wang, X.; Li, S.; Qiao, L.; Mu, H. Advances in Sol-Gel-Based Superhydrophobic Coatings for Wood: A Review. *Int. J. Mol. Sci.* **2023**, *24*, 9675. [CrossRef]
53. Ge-Zhang, S.; Cai, T.; Yang, H.; Ding, Y.; Song, M. Biology and nature: Bionic superhydrophobic surface and principle. *Front. Bioeng. Biotechnol.* **2022**, *10*, 1033514. [CrossRef] [PubMed]
54. Ge-Zhang, S.; Yang, H.; Ni, H.; Mu, H.; Zhang, M. Biomimetic superhydrophobic metal/nonmetal surface manufactured by etching methods: A mini review. *Front. Bioeng. Biotechnol.* **2022**, *10*, 958095. [CrossRef] [PubMed]
55. Zhang, J.; Koubaa, A.; Tao, Y.; Peng, L. The emerging development of transparent wood: Materials, characteristics, and applications. *Curr. For. Rep.* **2022**, *8*, 333–345. [CrossRef]
56. Ramage, M.H.; Burridge, H.; Busse-Wicher, M. The wood from the trees: The use of timber in construction. *Renew. Sustain. Energy Rev.* **2017**, *68*, 333–359. [CrossRef]
57. Bana, R.; Banthia, A.K. Mechanical and thermal analysis of poly (Vinyl-Alcohol) and modified wood dust composites. *J. Wood Chem. Technol.* **2011**, *31*, 218–232. [CrossRef]
58. Gankhuyag, S.; Bae, D.S.; Lee, K. One-pot synthesis of SiO_2@ Ag mesoporous nanoparticle coating for inhibition of Escherichia coli bacteria on various surfaces. *Nanomaterials* **2021**, *11*, 549. [CrossRef]
59. Cai, T.; Ge-Zhang, S.; Song, M. Anthocyanins in metabolites of purple corn. *Front. Plant Sci.* **2023**, *14*, 1154535. [CrossRef]

polymers

MDPI

Review

Wood Sponge for Oil–Water Separation

Chang Zhang [1], Taoyang Cai [2], Shangjie Ge-Zhang [1], Pingxuan Mu [1], Yuwen Liu [1] and Jingang Cui [1,*]

[1] College of Science, Northeast Forestry University, Harbin 150040, China; zc2021@nefu.edu.cn (C.Z.)
[2] Aulin College, Northeast Forestry University, Harbin 150040, China
* Correspondence: cjg@nefu.edu.cn

Abstract: In addition to filtering some sediments, hydrophobic wood sponges can also absorb many organic solvents, particularly crude oil. The leakage of crude oil poses a serious threat to the marine ecosystem, and oil mixed with water also generates great danger for its use. From the perspective of low cost and high performance, wood sponges exhibit great potential for dealing with crude oil pollution. Wood sponge is a renewable material. With a highly oriented layered structure and a highly compressible three-dimensional porous frame, wood sponges are extremely hydrophobic, making them ideal for oil–water separation. Currently, the most common approach for creating wood sponge is to first destroy the wood cell wall to obtain a porous-oriented layered structure and then enhance the oil–water separation ability via superhydrophobic treatment. Wood sponge prepared using various experimental methods and different natural woods exhibits distinctive properties in regards to robustness, compressibility, fatigue resistance, and oil absorption ability. As an aerogel material, wood sponge offers multi-action (absorption, filtration) and reusable oil–water separation functions. This paper introduces the advantages of the use of wood sponge for oil–water separation. The physical and chemical properties of wood sponge and its mechanism of adsorbing crude oil are explained. The synthesis method and the properties are discussed. Finally, the use of wood sponge is summarized and prospected.

Keywords: wood sponge; sewage treatment; crude oil adsorption; hydrophobic; adsorption mechanism

Citation: Zhang, C.; Cai, T.; Ge-Zhang, S.; Mu, P.; Liu, Y.; Cui, J. Wood Sponge for Oil–Water Separation. *Polymers* **2024**, *16*, 2362. https://doi.org/10.3390/polym16162362

Academic Editor: Antonios N. Papadopoulos

Received: 6 July 2024
Revised: 7 August 2024
Accepted: 19 August 2024
Published: 21 August 2024

1. Introduction

With the large-scale exploitation and use of oil, the technology regarding oil–water separation has been developing. A large number of water-driven crude oil production methods will inevitably mix water with oil (asphalt, resin, and naphthenate in oil act as emulsifiers, making it difficult for water to gather and separate) [1–4]. And the presence of this moisture will greatly affect the transportation and safe use of oil. In addition, the increase in oil spills also poses a great threat to the ecological environment and the safety of biological life. However, due to the increasing industrial demand, the scale of crude oil exploitation will only increase [3,5–12]. At present, the simplest and safest methods for oil–water separation include the use of hydrophobic and lipophilic oil-absorbing agents. However, most of the adsorption materials now include carbon-based aerogels such as activated carbon, graphene, and carbon nanotubes [13–15]. There are also bio-based aerogels such as chitosan aerogels [16], cellulose aerogels [17], and chitin aerogel [18]. In addition, there are artificial polymer aerogels such as commercial sponge, melamine, silicone sponge, etc., which have been employed in the field of oil–water separation [19–21]. However, due to the need for expensive equipment, its imperfect mechanical properties, and a complicated preparation process, its large-scale application is seriously hindered [22]. Other physical methods, such as centrifugal separation and air flotation [23], membrane separation [24–26], gravity separation [27,28], photoelectrocatalysis [29], and other schemes, offer poor separation selectivity, low reliability, and high cost [30–32]. Chemical methods such as the use of dispersants, combustion [33], and biodegradation not only exhibit difficulty in regards to crude oil recovery, but also cause secondary pollution [7,34–40] and other problems. In the study of the separation of water and organic solvents, the biomass-based adsorbent in the form of carbon aerogel has shown the ability to realize both a light weight and the recyclability of organic solvents, achieving good results [37–40]. However, the mixed water solution, with crude oil as an organic solvent, requires sufficient robustness of the adsorption materials in order to achieve success.

As a new adsorption material, wood sponge is a lightweight, degradable, and highly compressible 3D aerogel porous material [41–43]. At the same time, because the main skeleton structure is cellulose, and natural wood is rich in cellulose, it is also rich in raw materials, low in cost, and environmentally friendly. The structure of wood is directional and hierarchical. After the cell wall of the wood cells is chemically treated to remove lignin and hemicellulose, it will leave behind a hard cellulose skeleton, with good stability [44–48]. At this point, wood is basically composed of cellulose microfibers, which is a cellulose aerogel, to a certain extent, with a specific solvent adsorption capacity and extremely high porosity, with holes are closely arranged along the longitudinal direction of the wood fibers to form a series of relatively uniform low-curvature pipes [49–52]. Because the spring-like layered structure endows the wood sponge with sufficient mechanical compressibility and elasticity, it can withstand repeated extrusion without incurring structural damage. This type of wood displays a number of sponge-like characteristics, so it is called wood sponge.

It is obvious that wood sponge exhibits great potential for absorbing organic solvents, and the appearance of wood sponge is only the beginning of its applications in sewage treatment for the adsorption of crude oil. However, wood sponge displays natural hydrophilicity and lipophilicity. If it is to be used for oil–water separation and the absorption of oil or organic solvent, it must be modified to be hydrophobic. Generally, the chemical deposition method is used to modify wood sponge. After hydrophobic modification of wood sponge using a chemical modifier, the porous structure was still observed in the cross-sectional plane. Under the enlarged observation of a radial section, it was noted that the modifier was uniformly deposited on the large surface of the sample in the form of small particles [53–55]. In mainstream research articles about hydrophobic sponge, the water contact angle of the treated wood can reach levels great than 100, the density is generally only 50–60 mg cm^{-3}, and the porosity ranges from at least 90% up to 98% [56]. And under the same strain, the stress required by modified wood sponge is less than that of unmodified wood sponge. This light, hydrophobic, and oleophilic wood sponge can not only filter sewage containing crude oil but also absorb crude oil directionally, which also shows that targeted hydrophobic modification treatment has little effect on the structure of wood sponge. Generally, the oil absorption method is employed to directly place the hydrophobic sponge into the oil–water mixture, or to connect a sealed collection device through one end of a conduit; the collection device is connected to a vacuum pump, and the other end is connected to a hydrophobic wood sponge and placed into the oil–water mixture to filter out the water from the crude oil and to introduce the pollutant into the absorber, and it is then absorbed in the direction of the fiber bundle [53,54,57,58]. The device for testing the oil absorption capacity using a vacuum pump is shown in Figure 1.

This paper points out the necessity of using wood sponge for oil–water separation (part I), and introduces the mechanism of absorbing crude oil originating from the characteristics of wood sponge (Section 2). The preparation methods for wood sponge are summarized, and different preparation methods are compared (Section 3). Next, the limitations of wood sponge used in oil–water separation are briefly introduced (Section 4), and the future of wood sponge used in oil–water separation is prospected (Section 5).

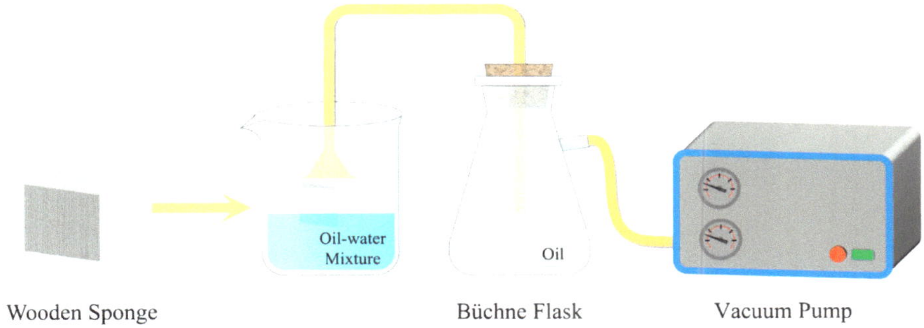

Figure 1. A hydrophobic wood sponge collects crude oil using a vacuum pump.

2. Sewage Treatment Mechanism of Adsorption of Crude Oil by Wood Sponge

2.1. Characteristics of Wood Sponge

Wood sponge, as a new material obtained by treating natural wood materials using chemical or physical methods, exhibits unique and diverse physical and chemical properties. The sewage treatment mechanism of wooden sponge for the adsorption of crude oil is mainly based on its unique material characteristics and structural advantages.

Thanks to its rich pore structure, the appearance of wood sponge is usually sponge-like or foam-like, offering excellent extrusion recovery and cutting ability, which can be adapted to the adsorption requirements of various shapes and surfaces, and it is not easily damaged during use [53]. Its density (40–80 kg m^{-3}) [54,59] is much lower than that of many other types of sponges or foam materials, making wood sponges more convenient to store, transport, and use. Its high porosity of more than 80% also means that it contains many tiny pores and channels, which not only provide abundant storage space for liquids (such as water or oil), but also enhance the capillary adsorption capacity of wood sponges [53,60]. Generally speaking, the higher the porosity, the better the adsorption performance of wood sponge [61].

Moreover, the three-dimensional porous network structure provides the wood sponge with a large specific surface area, increasing its surface contact with liquid, thus improving the adsorption efficiency [62–65]. Its three-dimensional network structure also helps to maintain the mechanical stability and durability of the wood sponge.

In the preparation process, lignin, hemicellulose, and other components in natural wood materials are usually stripped using chemical reagents, and the cellulose skeleton is retained, thus forming the unique structure of wood sponge [53]. Because cellulose is a natural polymer compound, wood sponge possesses good biodegradability [59,66]. This characteristic endows the porous polymer with incomparable advantages: it will not cause long-term pollution to the environment, and it conforms to the concept of sustainable development [67].

In addition, the wood sponge shows good stability in general chemical environments because the easily reactive substances have been stripped in advance using chemical reagent pretreatment [61]. However, it should be noted that some strong acids, strong bases, or organic solvents may destroy the structure and properties of wood sponges [68].

Most importantly, wood sponge exhibits great modification potential and can meet the needs of additional fields. Through further chemical or physical treatment, wood sponge can be endowed with more functional characteristics. For example, hydrophobic wood sponge can be prepared by hydrophobic treatment to improve its oil absorption performance [53,57], or superelastic sponge is prepared via superelastic treatment to increase its elasticity and ease of use [69,70].

The main criteria for evaluating the excellent properties of hydrophobic wood sponge are: internal water contact angle, porosity and recyclability [20,71], oil absorption capacity [72,73], separation efficiency [57], and compressive elasticity [63], along with other parameters. Recyclability refers to the ability of hydrophobic wood sponge to repeatedly absorb and squeeze crude oil from sewage containing petroleum, which reflects the compression fatigue resistance and repetitive work ability of hydrophobic wood sponge, largely determining whether it can actually be used to treat crude oil sewage. Oil absorption capacity is the ratio of the oil absorption capacity of hydrophobic wood sponge to its own mass, which focuses on the hydrophobic and oleophilic capacity of wood sponge at this stage. The separation efficiency is similar to the oil absorption capacity, but it is the ratio of the quality difference in the oil–water mixture, before and after absorption, to that of the original crude oil quality. Compressive elasticity is the maximum compression height ratio of hydrophobic wood sponge perpendicular to the fiber direction, which can restore its original state after compression; this is also a prominent feature of high porosity and deformability. Although wood itself exhibits high porosity and resilience, with the deepening of theoretical research, the methods for preparing hydrophobic wood sponge are becoming more and more mature and advanced, and the robustness, compression fatigue, hydrophobicity, and oil absorption ability of wood sponge are increasing. Table 1 shows the results of separating the oil–water mixture using wood sponge and other materials.

Table 1. Results of oil–water separation using wood sponge and other materials.

Ref.	Material	Oil	Absorption Capacity	Porosity	WCA	Cycles of Reuse
[59]	Modified wood sponge	Cuclohexane	17,300 mg g^{-1}	96.47%	121.37°	More than 10 times
[74]	Carbonaceous aerogel	Peanut oil	11,400 mg g^{-1}	Volume of 4.13 g cm^{-3}	144.2°	Not reported
[75]	Cabot thermal wrap	Dichloromethane	13,000 mg g^{-1}	92.1%	127.6°	More than 10 times
[75]	Aspen Aerogels Spaceloft	Dichloromethane	16,000 mg g^{-1}	86.1%	130.9°	More than 10 times
[76]	Popcorn-based carbon aerogel	Chloroethane	10,830 mg g^{-1}	Volume of 0.095 g cm^{-3}	151.6°	Not reported
[18]	Chitin sponge	Phenisin	58,000 mg g^{-1}	Pore sizes of 20–50 μm	148.7°	Above 93% after 10 cycles
[39]	Spongy graphene	Toluene	86,000 mg g^{-1}	Density of 12.5 mg cm^{-3}	114.2°	Basically unchanged after 10 cycles
[77]	Cellulose aerogel	Ruby	18,400 mg g^{-1}	97.3%	145°	Above 96.4% after 5 cycles
[78]	Polyurethane sponge	Chloroform	160,000 mg g^{-1}	99.3%	127°	Basically unchanged after 50 cycles
[79]	Butyl rubber	Crude oil	23,000 mg g^{-1}	Volume of 8.9 mL g^{-1}	Not reported	Above 70% after 8 cycles
[79]	Polypropylene	Fuel oil	15,700 mg g^{-1}	Pore sizes of 10 μm	Not reported	Above 10% after 8 cycles
[19]	Modified PU sponge	Lubricate oil	25,000 mg g^{-1}	Not reported	Not reported	More than 300 times
[39]	Graphene sponge	Castor oil	75,000 mg g^{-1}	Pore sizes of 570–620 μm	114.2°	More than 10 times
[80]	Graphene–CNT hybrid foam	Sesame oil	105,000 mg g^{-1}	Pore sizes of 100 μm	152.3°	More than 6 times
[81]	CNT sponge	Mineral oil	126,000 mg g^{-1}	98%	156°	Above 96% after 10 cycles
[82]	CNF/carbon foam	Wash oil	28,400 mg g^{-1}	95%	140°	Not reported
[83]	Exfoliate graphite	Heavy oil	75,000 mg g^{-1}	73–77%	Not reported	Not reported
[84]	Fir fibers	Grade-C oil	15,000 mg g^{-1}	Not reported	Not reported	Above 78% after 8 cycles
[85]	EV/CNT	Diesel oil	26,700 mg g^{-1}	Pore sizes of 5–10 nm	Not reported	Above 94% after 10 cycles
[81]	CNT sponges	Vegetable oil	130,100 mg g^{-1}	Density of 5.8 mg cm^{-3}	Not reported	More than 10 times
[81]	CNT sponges	Vegetable oil	32,300 mg g^{-1}	Density of 25.5 mg cm^{-3}	Not reported	More than 10 times
[86]	Corn stalk	Diesel oil	8600 mg g^{-1}	Not reported	Not reported	Not reported
[87]	Kapok	Diesel oil	36,700 mg g^{-1}	Pore sizes of 16.5 μm	102°	Basically unchanged after 15 cycles

2.2. Mechanism of Crude Oil Adsorption

The mechanism of adsorption of crude oil by wood sponge is mainly reflected in three aspects: capillary force, surface chemical action, and physical barrier action.

Capillary force is the force produced by the phenomenon that liquid rises or falls in a thin tube, and this action is derived from the interaction between the surface tension of liquid and the wall of the thin tube [88]. In wood sponge, because of its porous structure with a three-dimensional network [62], these pores can be regarded as countless tubules. When crude oil comes into contact with the wood sponge, the crude oil molecules will be attracted by the surface tension of the pore wall [89], thus being sucked into the pores. The pore size is widely distributed, with samples ranging from micropores to macropores. The pore size distribution is beneficial to the effective adsorption of crude oil molecules with different viscosities. Moreover, the pores are interconnected, forming a complex network structure, which enables crude oil molecules to diffuse and transport freely in the pores [62]. When the crude oil comes into contact with the wood sponge, the crude oil will spontaneously penetrate into the pores due to the surface tension of the pore wall and the interaction of crude oil molecules. The infiltration process does not require external pressure or energy input. Moreover, the adsorption speed of crude oil under the action of capillary force is increased. The diffusion and transport of crude oil molecules in the pores are driven by the surface tension of the pore walls, thus realizing rapid adsorption.

The surface chemical action of wood sponge to adsorb crude oil is one of the keys to its high-efficiency adsorption performance. This effect is mainly based on the interaction between the surface of the wood sponge and crude oil molecules. In the preparation process of wood sponge, hydrophobic groups (such as polysiloxane [90], fluoride, etc.) are usually introduced onto its surface through chemical modification. These hydrophobic groups can significantly reduce the hydrophilicity of the surface of wood sponge [57], making it more inclined to interact with the oil phase (such as crude oil). The existence of hydrophobic groups causes the wood sponge to repel water molecules and preferentially adsorb crude oil molecules when it comes into contact with oily wastewater. There will be intermolecular interactions, such as the van der Waals force and hydrogen bonding, between the crude oil molecules and the hydrophobic groups on the surface of wood sponge [90]. These interaction forces enable crude oil molecules to be firmly adsorbed on the surface of the wood sponge. Therefore, the modified wood sponge can adsorb crude oil more effectively. In addition to the hydrophobic groups, the cellulose skeleton of wood sponge itself exhibits a certain degree of lipophilicity [91]. The lipophilicity enables the wood sponge to more effectively combine with crude oil molecules, thus improving the adsorption efficiency.

The pores in the wood sponge not only provide abundant space for the adsorption of crude oil, but also constitute a physical barrier [92], which can prevent other impurities in the water phase from entering the pores. The pores of wood sponge have certain pore size selectivity. This means that molecules of different sizes will be hindered to varying degrees when passing through the pores. The physical barrier enables the wood sponge to achieve efficient oil–water separation without adding any chemical reagents. Besides separating the oil–water mixture, the physical barrier function of the wood sponge can also prevent other pollutants (such as heavy metal ions and organic pollutants) in the water phase from entering the pores [93], thus protecting the adsorption performance and service life of the sponge.

3. Preparation Methods of Hydrophobic Wood Sponge

The first step in making hydrophobic wood sponge is to soak raw materials (usually balsa wood) in acid (such as acetic acid and sodium chlorite) or alkaline solution (such as sodium hydroxide and sodium sulfite), allowing them to react for several hours at a constant temperature in a water bath to remove most of the lignin and hemicellulose in the raw materials before cleaning them with deionized water and finally subjecting them to freeze drying to obtain an anisotropic compressible wood aerogel via a directional tube. Lignin and hemicellulose display physical and chemical connections and are easily hydrolyzed by acid, while lignocellulose exhibits relatively stable chemical properties and only physical connections with lignin and hemicellulose. Therefore, the lignocellulose skeleton can be retained after the removal of lignin and cellulose [94–96]. We introduce mainly two basic methods for preparing wood sponge: chemical vapor deposition and chemical liquid deposition; on the basis of the two methods, solar energy and light energy are used to improve the preparation technique.

3.1. Chemical Vapor Deposition Method

In order to make the wood sponge hydrophobic, chemical vapor deposition (CVD) technology, employing a silane coupling agent, is generally used to embed a superhydrophobic coating onto the surface of the skeleton, i.e., polysiloxane is formed on the surface due to silanization [57]. This method has been widely used in recent years. The preparation of hydrophobic wood sponge is roughly divided into three steps, as follows: (1) the process of preparing wood sponge by removing lignin and hemicellulose using acidic or alkaline solution, leaving only the cellulose skeleton in the wood; (2) the gelation process of placing the wood aerogel into a freezing chamber for drying, removing the water to prepare a porous aerogel; (3) the hydrophobic modification process of placing the modifier and the wood sponge into a dryer for silanization, removing the excess modifier using a room-temperature vacuum oven [72]. However, there are many wood channels displaying a low curvature in the wood sponge itself, providing the treated hydrophobic aerogel with the ability to transport anisotropic liquid. The obtained wood aerogel not only displays oleophilic and hydrophobic separation characteristics, but also shows excellent mechanical compressibility. For this method, Guan et al. [53] used methyltrimethoxysilane (MTMS), a cheap and commonly used silylating agent, via chemical vapor deposition.

It is worth noting that in the wood sponge production process, in addition to the use of silane as a modifier, many other compounds can be used for surface modification or functionalization to improve performance or provide new characteristics. These compounds can include low molecular weight reagents and polymers.

As a low molecular weight reagent, fluorosilane, which is similar to silane, is also often used to enhance the hydrophobicity of materials [97]. The introduction of fluorine atoms can significantly reduce the surface energy of materials, thus improving their waterproof and oil-resistant properties. Moreover, there are some organic acids, such as acetic acid and propionic acid [59,67], which can react with hydroxyl groups on the surface of wood to form ester bonds, thus changing the surface properties of the materials. However, it should be noted that this method may not be as effective as silylation or fluorosilication. In addition, some natural polymers, such as chitosan [98] and cellulose derivatives [99,100], offer good biocompatibility and biodegradability and can be used as coatings or modifiers for wood sponges. They can enhance the mechanical properties, hydrophilicity, or hydrophobicity of materials. Yi et al. [98] used chitosan to prepare wood aerogels for the efficient removal of oil from water. Wen et al. [101] used polyacrylic acid (PAA) as a modifier of wood sponge and introduced PAA into the wood cellulose skeleton using graft copolymerization, thus promoting good water absorption and water retention in the wood sponge. Li et al. [102] studied the surface modification of wood sponge using polyethylene glycol (PEG). The introduction of PEG not only improved the hydrophilicity of the material, but also helped to reduce its surface tension and promote the oil–water separation process. MTCS [59], PMHS [54], and TCMS reagents [103] are also commonly used in chemical vapor deposition.

After CVD treatment, the porosity of the wood is basically unchanged, and its hydrophobic and oil-absorbing ability is enhanced. At present, this method has been widely used to create aerogels for separating oil–water mixtures [94–96]. Although CVD can significantly improve the hydrophobicity of the substrate, in this method, the reaction conditions, including reaction temperature, pressure and time, and initial reagent dosage, must be accurately controlled. With the increase in sample size, the diffusion time and reaction time also increase. In addition, the internal grafting distribution of the materials treated using this method is often uneven, and the adhesion between the superhydrophobic surface coating and the substrate is weak, which also limits its application in the actual treatment of oily wastewater [57].

3.2. Chemical Liquid Deposition Method

The liquid deposition (LPD) method was originally employed as a unique process for preparing various metal oxide films via the wet method [57]. It was also discovered that this method can be used for wood-function integration, mainly by soaking wood sponge in evenly mixed aqueous solution to complete the reaction. This method has been gradually applied to the hydrophobic modification of wood sponge. Using balsa wood as the matrix, balsa wood blocks were first transformed into cellulose wood sponge. The hydrophobic wood sponge prepared by liquid-phase silylation was more hydrophobic than that prepared by CVD, and the internal hydrophobic angle could reach more than 130 degrees, which indicated that it possessed a stronger oil absorption capacity and better compression fatigue resistance. The hydrophobic wood modified using the LPD method also shows excellent superhydrophobicity under acidic, alkaline, and salt water conditions, as well as in hot water, which proves its excellent durability [104,105]. These results mainly result from the fact that, compared with CVD method, this process is very simple, and the silanization reaction on the cellulose skeleton is more comprehensive and uniform. In addition, the size and morphology of the object are not relevant, special equipment, such as a vacuum system, is not needed, and most of the reactions can be completed under mild conditions. More importantly, the hydrophobic treatment process does not require catalysts or toxic substances, which solves the problems of complex processes and poor sustainability encountered in the related research. Through experiments, it was proven that this method can yield a cellulose aerogel structure exhibiting robustness, mechanical compressibility, anisotropy, and superhydrophobicity, and it can then be used for reusable oil–water separation [106–109].

Lignocellulose is rich in hydroxyl groups, providing opportunities for chemical functionalization and hybridization with other components [110]. Under the conditions of humidity and mild heating, a small amount of water and ethanol dispersant are used to hydrolyze the silylating agent to form silanol. These silanols further react with hydroxyl groups or other silanols on the surface of wood sponge to form covalently connected silane layers, ultimately forming a cross-linked polymer [18,111,112].

Compared with CVD, the liquid deposition method can also be accurately quantified, and the optimal proportion can be determined by controlling the amount of silylating agent. Liu et al. [113] used a liquid phase hydrophobic modification strategy to realize hydrophobic modification. Firstly, the obtained wood sponge is reacted in situ in a water system, and the surface is modified using low surface energy materials (MTMS). The hydrophobic angle is the highest, reaching 159°, which exceeds that of the hydrophobic and oleophilic wood aerogels used in most cellulose-based materials [114–116]. When the amount of MTMS continues to increase, the contact angle will decrease slightly, which may be due to the formation of larger polysiloxane particles on the WS surface at this ratio, resulting in uneven surface roughness [117,118]. In silane treatment, the silane monomer or oligomer can easily diffuse into the wood cell wall in the solution, thus providing the functional structure with better performance [98]. In this way, the superhydrophobic wood sponge is more likely to lose its superhydrophobicity in a corrosive environment. The chemical stability of superhydrophobic surfaces [65,119] is a key feature for its practical application. This also shows that LPD, as a hydrophobic treatment method, is very reliable, uniform, and effective.

3.3. Improving the Preparation Method by Utilizing Solar Energy and Light Energy

In order to improve the oil absorption capacity of wood sponge, many researchers have tried different hydrophobic modifiers in their own research. In addition, the researchers accurately mastered the best reaction conditions for lignocellulosic gels prepared using the CVD or LPD method and the best preparation conditions for different modifiers by controlling the variables. Through an in-depth understanding of the oil absorption mechanism of hydrophobic sponge, many researchers have put forward a new method for oil–water separation, improving the sponge preparation method using solar and electric energy.

Because of the high viscosity of crude oil, the separation efficiency is seriously limited. However, the viscosity of crude oil decreases with the increase in temperature. Thus, many researchers suggest that solar and electric energy can be used to realize heat conversion on hydrophobic wood sponge, thus increasing the temperature of crude oil and enhancing the effect of oil–water separation [73,120,121]. In recent years, hydrophobic/lipophilic porous adsorbents, with excellent Joule and solar heating properties, have attracted increasing attention because of their high adsorption speed, high electrothermal conversion, and photothermal conversion efficiency [122–126]. The three-dimensional porous wood sponge, with low density, high porosity, and compressibility with this heat-absorbing coating, can be used as a self-heating absorber to soak up oil [123,127]. These coatings, commonly used in many commercial sponges, include dopamine coatings [128], reduced graphene oxide coating [71,129,130], carbon nanotube modification [131,132], the CuO@CuS package [133], etc. However, compared with the serrated pores of commercial sponges, the radially oriented microchannels of wood sponges not only improve the performance of the self-heating absorber, but also improve the oil absorption capacity through low transfer resistance [134,135].

For example, Chao et al. [73] used solar and electric energy to realize the thermal conversion of hydrophobic wood sponge, choosing to add graphene coating on wood sponge to improve its self-heating ability. Graphene can effectively capture photons in the lattice and then convert solar energy into heat energy through lattice vibration to increase the in situ fluidity of crude oil, showing high separation efficiency [120]. The researchers first obtained graphene oxide using the improved Hummers method [136]. Then, the obtained wood sponge was soaked in a graphene oxide (GO) dispersion, dried under vacuum for 10 h at room temperature, and freeze-dried. Then, samples were taken from a drying tray, sealed under vacuum, dried at 100 °C for 6 h to obtain the wood sponge, and the graphene coating was reduced in situ. Then, a layer of transparent octadecyltrichlorosilane (OTS) was deposited in n-hexane via simple dip coating to maintain the sample's hydrophobicity and photothermal conversion ability [120].

Huang et al. [57] used electric energy to realize the thermal conversion of hydrophobic wood sponge. They developed a highly hydrophobic wood sponge coated with fluoroalkylsilane modified reduced graphene oxide (F-rGO@WS) for separating viscous crude oil from water. Firstly, chemical treatment was carried out to realize unobstructed longitudinal channels; then, a layer of graphene oxide was applied by simply immersing and coating the sponge, and then the sponge was reduced with ascorbic acid and grafted with perfluorooctyl triethoxysilane. The reduction of graphene oxide mainly endows WS with electrothermal ability [137]. The grafting of perfluorooctyl triethoxysilane aims at reducing the surface energy of reduced graphene oxide on the WS skeleton [73].

In addition, Zhang et al. [138] developed a polyurethane sponge coated with $Ti_3C_2T_x$ MXene ($Ti_3C_2T_x$@PU) to repair oil leakage; their results showed that the $Ti_3C_2T_x$ layer exhibited good Joule heating and a significant photothermal effect. Wang et al. [73] used a new type of $Ti_3C_2T_x$ MXene to wrap the wood sponge, and their results quickly cleaned up the crude oil leakage via excellent Joule heating and a good photothermal effect. $Ti_3C_2T_x$ MXene is obtained by the selective etching of Ti_3AlC_2, as reported in the etching method [122]. First, 1 g of lithium fluoride and 20 mL of 9 M hydrochloric acid were mixed at room temperature. Secondly, 1 g of $Ti_3C_2T_x$ powder was slowly poured into an HCl/LiF solution at 35 °C and allowed to react for 24 h, under continuous stirring. Thirdly, the obtained suspension was washed with ultrapure water several times until its pH value was greater than 5. Finally, the precipitate was ultrasonicated for 1 h and centrifuged to obtain $Ti_3C_2T_x$ MXene nanosheets [121]. $Ti_3C_2T_x$ MXene generates heat in situ, with the assistance of electricity and solar energy [139,140]. MXene comprises a large family of two-dimensional early transition metal carbides and/or carbonitrides [121], exhibiting high metal conductivity and a large specific surface area [141]. Li et al. [142] proved that the efficiency of $Ti_3C_2T_x$ MXene in regards to thermal optic conversion is 100%.

The material used by the all of above researchers is fir, which is a natural renewable material with obvious low density and high mechanical properties [142]. Researchers can generate heat in situ through electricity and solar energy. Nanomaterials can be combined with hydrophobic wood sponge, which can effectively capitalize on the Joule-assisted and solar-assisted effect to generate heat in situ to efficiently clean and recover high-viscosity crude oil. This new type of wood sponge adsorbent, with high mechanical superelasticity and hydrophobicity/lipophilicity, displays outstanding advantages for crude oil leakage repair, and the use of Joule and photothermal heat can realize all-weather operation. Compared with ordinary sponge-like adsorption materials, wood sponge possesses a rare radial pore structure and excellent compressibility, which significantly improves adsorption rate and recyclability. The application of hydrophobic wood sponge can be divided into four categories: chemical vapor deposition, chemical liquid deposition, improving the absorption rate of crude oil by wood sponge using using solar heat, and achieving this improvement via electric energy.

The change process, shown from top to bottom in the Figure 2, indicates that the research on the modification of hydrophobic wood sponge is gradually developing towards a more comprehensive and professional process. The preparation of wood sponge has gradually developed from a simple modification of hydrophobic and hydrophilic oil to the use of the thermal effect to accelerate the oil and water separation ability on the basis of the original high hydrophobic water, which also proved that this process resulted in the improved oil absorption ability of hydrophobic wood sponge. However, the results also show that hydrophobic wood sponges are becoming more expensive to make. Table 2 shows a comparison of the preparation methods of wood sponge.

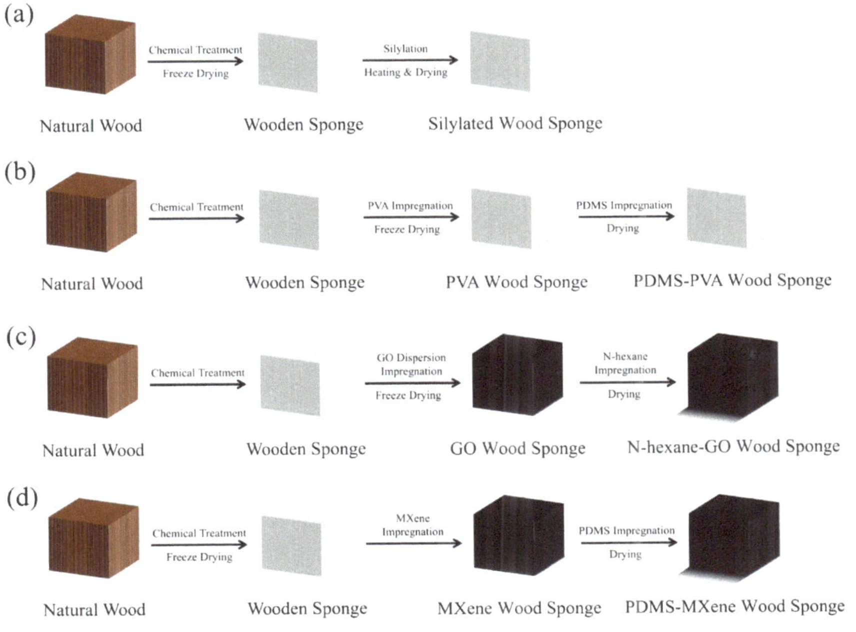

Figure 2. Development history of a hydrophobic wood sponge. (**a**) CVD, (**b**) LPD, (**c**) LPD and Joule heat collection using graphene nanocoatings; (**d**) LPD and added metal carbide nanosheet coatings utilize heat and electricity.

Table 2. Comparison of preparation methods of wood sponges.

Ref.	Preparation Method	Wood Sponge Species	WCA	Absorption Capacity	Compression	Advantages	Disadvantages
[94]	CVD	Cellulose/graphene aerogel	153°	80–197	Compressible to 90%	Strong hydrophobic ability, diverse materials, and good mechanical properties	Uneven deposition and harsh reaction conditions
[143]		Silylated cellulose fibers	142°	51–142.9	Compressible in diesel oil		
[53]		Silylated wood sponge	151°	16–41	Compressible to 60%		
[17]		Silylated nanocellulose sponge	136°	49–102	Compressible to 96%		
[113]	LPD	Superhydrophobic wood sponge	159°	23–60	Compressible	Simple treatment process, mild reaction conditions, uniform deposition; durability and accurate quantification	Slow reaction speed and low product purity
[96]		Superhydrophobic microfibrillated cellulose aerogel	151.8°	116–260	Compressible		
[73]	Utilizing solar energy and light energy	Graphene–wood sponge	134.2°	7.28	Compressible to 90%	Rare radial pore structure; excellent compressibility, recyclability, and high adsorption rate	Complex technology, stability problems, and high cost
[144]		Methyltrichlorosilane treated PVA-CNF aerogel	150.2°	44–96	Compressible to 80%		
[145]		CNT sponge	156°	87–176	Compressible in ethanol		
[144]		PVA/cellulose nanofibril aerogels	Not reported	44–96	Compressible		

4. Limitation of Wood Sponge Used in Oil–Water Separation

The wood sponge material shows both potential and advantages in the field of oil–water separation, but it also has some limitations. The following provides an analysis of the limitations of wood sponge for oil–water separation.

As a material itself, wood sponge exhibits some physical limitations. First of all, although wood sponge shows high oil absorption performance, its oil absorption capacity is still limited by the structure and preparation process of the material itself [146]. In practical application, a large number of wood sponges may be needed to achieve the ideal oil–water separation effect, which increases the cost and complexity of operation. In addition, the wood sponge material may be affected by physical wear, chemical corrosion, and other factors during use, resulting in its relatively poor durability [147]. This may affect its stability and reliability for long-term use. Moreover, due to the softness and elasticity of the wood sponge, it is easily affected by external forces and can be deformed during use [92], which may impact the effectiveness and accuracy of oil–water separation.

There are also some technological difficulties and challenges regarding the process of oil–water separation using wood sponge material. First of all, the oil–water mixture may contain various components and impurities, which may affect the oil absorption performance of the wood sponge [73]. For example, some chemicals may react with wood sponge, resulting in its performance degradation. Secondly, while pursuing high separation efficiency, we also need to consider the cost factor. The preparation cost, use cost, and subsequent treatment cost of wood sponge must be comprehensively considered. If the cost is too high, it may limit its popularization in practical applications [148]. In addition, the wood sponge will absorb a significant amount of oil and impurities in the process of oil–water separation. The accumulation of these oils, along with their residue and impurities, in wood sponge may cause difficulties during its regeneration and recovery [149]. Therefore, it is necessary to develop effective regeneration and recovery technologies to reduce treatment costs and decrease environmental pollution.

In order to resolve these limitations, further research and development work may be needed to optimize the design and performance of wood sponges specifically used for sewage treatment applications. This may include improving chemical treatment, freeze drying, and other preparation processes; introducing modifiers; and improving mechanical properties.

5. Summary and Prospect

Wood sponge is a type of wood aerogel with cellulose as the main skeleton; wood sponge is obtained by removing lignin and hemicellulose from natural wood. The surface energy of the sample can be changed via chemical modification, and sponges with different hydrophobic and oil-absorbing properties can be obtained [150–152]. There is a certain predictability between the preparation conditions and the properties of hydrophobic wood sponge, showing that its preparation process is controllable, and the most suitable preparation method can be identified by controlling the variables. After silylation modification, the adsorption capacity of hydrophobic wood sponge for oil and organic solvents is greatly improved, and this can also be recovered by simple extrusion. This recyclable green wood sponge exhibits high resilience, porosity, and oil absorption capacity, and the 3D aerogel material, displaying high porosity, light weight, and directional channels, is an attractive and promising candidate material for wastewater treatment. It is expected to be used in environmental protection fields, including sewage treatment and oil spill cleaning. At present, the research investigating the absorption of crude oil using wood sponge has achieved significant in-depth research results, reporting many breakthroughs regarding oil absorption capacity and reliability, e.g., Run et al. [21] proposed a superhydrophobic sponge with flame retardant properties that swells easily and can absorb 79–195 times its own weight in oil and organic solvents.

In the future, wood sponge preparation technology will tend to directly extract a biological matrix (such as chitosan, cellulose, chitin, and other polysaccharide structures) and then use directional freezing technology to achieve the special 3D structure, which is more suitable for large-scale production and application than are block-shaped wood materials. We anticipate that future research will focus on the practical application and commercialization of hydrophobic wood sponge materials, with an emphasis on strengthening their strong oil absorption, low cost, and high recyclability characteristics.

Author Contributions: Conceptualization, C.Z., T.C. and S.G.-Z.; formal analysis, C.Z. and T.C.; investigation, C.Z. and T.C.; methodology, C.Z. and T.C.; project administration, J.C.; resources, S.G.-Z.; supervision, J.C.; validation, C.Z., T.C. and P.M.; visualization, C.Z. and T.C.; writing—original draft, C.Z., T.C., S.G.-Z. and P.M.; writing—review and editing, C.Z., T.C., S.G.-Z. and Y.L. All authors have read and agreed to the published version of the manuscript.

Funding: This research was funded by the Fundamental Research Funds for the Central Universities (No. 2572022DJ03).

Data Availability Statement: Data availability is not applicable to this article, as no new data were created or analyzed in this study.

Conflicts of Interest: The authors declare no conflicts of interest.

References

1. Ventikos, N.P.; Vergetis, E.; Psaraftis, H.N.; Triantafyllou, G. A high-level synthesis of oil spill response equipment and countermeasures. *J. Hazard. Mater.* **2004**, *107*, 51–58. [CrossRef] [PubMed]
2. Kang, L.; Wang, B.; Zeng, J.; Cheng, Z.; Li, J.; Xu, J.; Gao, W.; Chen, K. Degradable dual superlyophobic lignocellulosic fibers for high-efficiency oil/water separation. *Green Chem.* **2020**, *22*, 504–512. [CrossRef]
3. Li, L.; Zhang, J.; Wang, A. Removal of Organic Pollutants from Water Using Superwetting Materials. *Chem. Rec.* **2018**, *18*, 118–136. [CrossRef]
4. Higaki, Y.; Hatae, K.; Ishikawa, T.; Takanohashi, T.; Hayashi, J.; Takahara, A. Adsorption and Desorption Behavior of Asphaltene on Polymer-Brush-Immobilized Surfaces. *ACS Appl. Mater. Interfaces* **2014**, *6*, 20385–20389. [PubMed]
5. Ge-Zhang, S.; Chen, X.; Zhu, H.; Song, Y.; Ding, Y.; Cui, J. Computer Simulation of Polyethylene Terephthalate Carbonated Beverage Bottle Bottom Structure Based on Manual–Automatic Double-Adjustment Optimization. *Polymers* **2022**, *14*, 2845. [CrossRef] [PubMed]
6. Tian, F.; Yang, Y.; Wang, X.-L.; An, W.-L.; Zhao, X.; Xu, S.; Wang, Y.-Z. From waste epoxy resins to efficient oil/water separation materials *via* a microwave assisted pore-forming strategy. *Mater. Horiz.* **2019**, *6*, 1733–1739. [CrossRef]

7. Ge-Zhang, S.; Song, M.; Huang, Z.; Li, M.; Mu, L. Comparison and Optimization: Research on the Structure of the PET Bottle Bottom Based on the Finite Element Method. *Polymers* **2022**, *14*, 3174. [CrossRef] [PubMed]
8. Wang, X.; Yu, J.; Sun, G.; Ding, B. Electrospun nanofibrous materials: A versatile medium for effective oil/water separation. *Mater. Today* **2016**, *19*, 403–414. [CrossRef]
9. Si, Y.; Fu, Q.; Wang, X.; Zhu, J.; Yu, J.; Sun, G.; Ding, B. Superelastic and Superhydrophobic Nanofiber-Assembled Cellular Aerogels for Effective Separation of Oil/Water Emulsions. *ACS Nano* **2015**, *9*, 3791–3799. [CrossRef] [PubMed]
10. Ge-Zhang, S.; Liu, H.; Song, M.; Wang, Y.; Yang, H.; Fan, H.; Ding, Y.; Mu, L. Advances in Polyethylene Terephthalate Beverage Bottle Optimization: A Mini Review. *Polymers* **2022**, *14*, 3364. [CrossRef]
11. Wang, X.; Pan, Y.; Yuan, H.; Su, M.; Shao, C.; Liu, C.; Guo, Z.; Shen, C.; Liu, X. Simple fabrication of superhydrophobic PLA with honeycomb-like structures for high-efficiency oil-water separation. *Chin. Chem. Lett.* **2020**, *31*, 365–368. [CrossRef]
12. Ma, W.; Ding, Y.; Zhang, M.; Gao, S.; Li, Y.; Huang, C.; Fu, G. Nature-inspired chemistry toward hierarchical superhydrophobic, antibacterial and biocompatible nanofibrous membranes for effective UV-shielding, self-cleaning and oil-water separation. *J. Hazard. Mater.* **2020**, *384*, 121476. [CrossRef]
13. Li, Z.; Wang, B.; Qin, X.; Wang, Y.; Liu, C.; Shao, Q.; Wang, N.; Zhang, J.; Wang, Z.; Shen, C.; et al. Superhydrophobic/Superoleophilic Polycarbonate/Carbon Nanotubes Porous Monolith for Selective Oil Adsorption from Water. *ACS Sustain. Chem. Eng.* **2018**, *6*, 13747–13755. [CrossRef]
14. Villacañas, F.; Pereira, M.F.R.; Órfão, J.J.; Figueiredo, J.L. Adsorption of simple aromatic compounds on activated carbons. *J. Colloid Interface Sci.* **2006**, *293*, 128–136. [CrossRef]
15. Chen, B.; Ma, Q.; Tan, C.; Lim, T.T.; Huang, L.; Zhang, H.J.S. Carbon-based sorbents with three-dimensional architectures for water remediation. *Small* **2015**, *11*, 3319–3336. [PubMed]
16. Li, Y.; Guo, C.; Shi, R.; Zhang, H.; Gong, L.; Dai, L. Chitosan/nanofibrillated cellulose aerogel with highly oriented microchannel structure for rapid removal of Pb (II) ions from aqueous solution. *Carbohydr. Polym.* **2019**, *223*, 115048. [CrossRef] [PubMed]
17. Zhang, Z.; Sèbe, G.; Rentsch, D.; Zimmermann, T.; Tingaut, P. Ultralightweight and Flexible Silylated Nanocellulose Sponges for the Selective Removal of Oil from Water. *Chem. Mater.* **2014**, *26*, 2659–2668. [CrossRef]
18. Duan, B.; Gao, H.; He, M.; Zhang, L. Hydrophobic Modification on Surface of Chitin Sponges for Highly Effective Separation of Oil. *ACS Appl. Mater. Interfaces* **2014**, *6*, 19933–19942. [CrossRef]
19. Zhu, Q.; Chu, Y.; Wang, Z.; Chen, N.; Lin, L.; Liu, F.; Pan, Q. Robust superhydrophobic polyurethane sponge as a highly reusable oil-absorption material. *J. Mater. Chem. A* **2013**, *1*, 5386–5393. [CrossRef]
20. Wu, L.; Li, L.; Li, B.; Zhang, J.; Wang, A. Magnetic, durable, and superhydrophobic polyurethane@ Fe_3O_4@ SiO_2@ fluoropolymer sponges for selective oil absorption and oil/water separation. *ACS Appl. Mater. Interfaces* **2015**, *7*, 4936–4946. [CrossRef]
21. Ruan, C.; Ai, K.; Li, X.; Lu, L. A Superhydrophobic Sponge with Excellent Absorbency and Flame Retardancy. *Angew. Chem.* **2014**, *126*, 5662–5666. [CrossRef]
22. Wu, M.-B.; Huang, S.; Liu, C.; Wu, J.; Agarwal, S.; Greiner, A.; Xu, Z.-K. Carboxylated wood-based sponges with underoil superhydrophilicity for deep dehydration of crude oil. *J. Mater. Chem. A* **2020**, *8*, 11354–11361. [CrossRef]
23. Al-Shamrani, A.; James, A.; Xiao, H. Separation of oil from water by dissolved air flotation. *Colloids Surf. A Physicochem. Eng. Asp.* **2002**, *209*, 15–26. [CrossRef]
24. Zhao, X.; Su, Y.-L.; Liu, Y.; Li, Y.; Jiang, Z. Free-Standing Graphene Oxide-Palygorskite Nanohybrid Membrane for Oil/Water Separation. *ACS Appl. Mater. Interfaces* **2016**, *8*, 8247–8256. [CrossRef]
25. Padaki, M.; Murali, R.S.; Abdullah, M.S.; Misdan, N.; Moslehyani, A.; Kassim, M.; Hilal, N.; Ismail, A.F. Membrane technology enhancement in oil–water separation. A review. *Desalination* **2015**, *357*, 197–207. [CrossRef]
26. Yang, C.; Zhang, G.; Xu, N.; Shi, J. Preparation and application in oil–water separation of $ZrO_2/\alpha\text{-}Al_2O_3$ MF membrane. *J. Membr. Sci.* **1998**, *142*, 235–243. [CrossRef]
27. Yoon, H.; Na, S.-H.; Choi, J.-Y.; Latthe, S.S.; Swihart, M.T.; Al-Deyab, S.S.; Yoon, S.S. Gravity-Driven Hybrid Membrane for Oleophobic–Superhydrophilic Oil–Water Separation and Water Purification by Graphene. *Langmuir* **2014**, *30*, 11761–11769. [CrossRef]
28. Tai, M.H.; Gao, P.; Tan, B.Y.L.; Sun, D.D.; Leckie, J.O. Highly Efficient and Flexible Electrospun Carbon–Silica Nanofibrous Membrane for Ultrafast Gravity-Driven Oil–Water Separation. *ACS Appl. Mater. Interfaces* **2014**, *6*, 9393–9401. [CrossRef]
29. Zhao, X.; Liu, W.; Fu, J.; Cai, Z.; O'Reilly, S.; Zhao, D. Dispersion, sorption and photodegradation of petroleum hydrocarbons in dispersant-seawater-sediment systems. *Mar. Pollut. Bull.* **2016**, *109*, 526–538. [CrossRef]
30. Fu, B.; Yang, Q.; Yang, F. Flexible Underwater Oleophobic Cellulose Aerogels for Efficient Oil/Water Separation. *ACS Omega* **2020**, *5*, 8181–8187. [CrossRef]
31. Wang, K.; Liu, X.; Tan, Y.; Zhang, W.; Zhang, S.; Li, J. Two-dimensional membrane and three-dimensional bulk aerogel materials via top-down wood nanotechnology for multibehavioral and reusable oil/water separation. *Chem. Eng. J.* **2019**, *371*, 769–780. [CrossRef]
32. Yue, X.; Li, Z.; Zhang, T.; Yang, D.; Qiu, F. Design and fabrication of superwetting fiber-based membranes for oil/water separation applications. *Chem. Eng. J.* **2019**, *364*, 292–309. [CrossRef]
33. Liang, H.; Guan, Q.; Chen, L.; Zhu, Z.; Zhang, W.; Yu, S. Macroscopic-Scale Template Synthesis of Robust Carbonaceous Nanofiber Hydrogels and Aerogels and Their Applications. *Angew. Chem. Int. Ed.* **2012**, *51*, 5101–5105. [CrossRef]

34. Zhang, W.; Liu, N.; Cao, Y.; Lin, X.; Liu, Y.; Feng, L. Superwetting Porous Materials for Wastewater Treatment: From Immiscible Oil/Water Mixture to Emulsion Separation. *Adv. Mater. Interfaces* **2017**, *4*, 29. [CrossRef]
35. Hayase, G.; Kanamori, K.; Fukuchi, M.; Kaji, H.; Nakanishi, K. Facile Synthesis of Marshmallow-like Macroporous Gels Usable under Harsh Conditions for the Separation of Oil and Water. *Angew. Chem. Int. Ed.* **2013**, *52*, 1986–1989. [CrossRef]
36. Zhang, A.; Chen, M.; Du, C.; Guo, H.; Bai, H.; Li, L. Poly(dimethylsiloxane) Oil Absorbent with a Three-Dimensionally Interconnected Porous Structure and Swellable Skeleton. *ACS Appl. Mater. Interfaces* **2013**, *5*, 10201–10206. [CrossRef]
37. Yu, C.; Yu, C.; Cui, L.; Song, Z.; Zhao, X.; Ma, Y.; Jiang, L. Facile Preparation of the Porous PDMS Oil-Absorbent for Oil/Water Separation. *Adv. Mater. Interfaces* **2017**, *4*, 1600862. [CrossRef]
38. Zhao, Y.; Hu, C.; Hu, Y.; Cheng, H.; Shi, G.; Qu, L. A versatile, ultralight, nitrogen-doped graphene framework. *Angew. Chem. Int. Ed.* **2012**, *51*, 11371–11375. [CrossRef]
39. Bi, H.; Xie, X.; Yin, K.; Zhou, Y.; Wan, S.; He, L.; Xu, F.; Banhart, F.; Sun, L.; Ruoff, R.S. Spongy Graphene as a Highly Efficient and Recyclable Sorbent for Oils and Organic Solvents. *Adv. Funct. Mater.* **2012**, *22*, 4421–4425. [CrossRef]
40. Zhang, C.; Ge-Zhang, S.; Wang, Y.; Mu, H. A Wooden Carbon-Based Photocatalyst for Water Treatment. *Int. J. Mol. Sci.* **2024**, *25*, 4743. [CrossRef] [PubMed]
41. Liu, H.; Chen, X.; Zheng, Y.; Zhang, D.; Zhao, Y.; Wang, C.; Pan, C.; Liu, C.; Shen, C. Lightweight, Superelastic, and Hydrophobic Polyimide Nanofiber/MXene Composite Aerogel for Wearable Piezoresistive Sensor and Oil/Water Separation Applications. *Adv. Funct. Mater.* **2021**, *31*, 2008006. [CrossRef]
42. Zhou, J.; Du, E.; He, Y.; Fan, Y.; Ye, Y.; Tang, B. Preparation of Carbonized Kapok Fiber/Reduced Graphene Oxide Aerogel for Oil-Water Separation. *Chem. Eng. Technol.* **2020**, *43*, 2418–2427. [CrossRef]
43. Yeh, S.K.; Tsai, Y.B.; Gebremedhin, K.F.; Chien, T.Y.; Chang, R.Y.; Tung, K.L. Preparation of polypropylene/high-melt-strength PP open-cell foam for oil absorption. *Polym. Eng. Sci.* **2021**, *61*, 1139–1149. [CrossRef]
44. Zhang, H.; Li, Y.; Shi, R.; Chen, L.; Fan, M. A robust salt-tolerant superoleophobic chitosan/nanofibrillated cellulose aerogel for highly efficient oil/water separation. *Carbohydr. Polym.* **2018**, *200*, 611–615. [CrossRef]
45. Gong, X.; Wang, Y.; Zeng, H.; Betti, M.; Chen, L. Highly porous, hydrophobic, and compressible cellulose nanocrystals/poly(vinyl alcohol) aerogels as recyclable absorbents for oil–water separation. *ACS Sustain. Chem. Eng.* **2019**, *7*, 11118–11128. [CrossRef]
46. Parmar, K.R.; Dora, D.; Pant, K.; Roy, S. An ultra-light flexible aerogel-based on methane derived CNTs as a reinforcing agent in silica-CMC matrix for efficient oil adsorption. *J. Hazard. Mater.* **2019**, *375*, 206–215. [CrossRef]
47. Zhu, H.; Yang, X.; Cranston, E.D.; Zhu, S. Flexible and Porous Nanocellulose Aerogels with High Loadings of Metal–Organic-Framework Particles for Separations Applications. *Adv. Mater.* **2016**, *28*, 7652–7657. [CrossRef]
48. Peng, H.; Wang, H.; Wu, J.; Meng, G.; Wang, Y.; Shi, Y.; Liu, Z.; Guo, X. Preparation of Superhydrophobic Magnetic Cellulose Sponge for Removing Oil from Water. *Ind. Eng. Chem. Res.* **2016**, *55*, 832–838. [CrossRef]
49. Xu, Y.; Ma, S.; Liu, C.; Wang, L.; Wang, H.; Xu, W.; Zhuang, Y.; Yang, H. Wood-inspired polyacrylonitrile foam with hierarchically aligned porous structure for application in water purification. *J. Appl. Polym. Sci.* **2021**, *138*, 50870. [CrossRef]
50. Zhu, M.; Li, Y.; Chen, G.; Jiang, F.; Yang, Z.; Luo, X.; Wang, Y.; Lacey, S.D.; Dai, J.; Wang, C. Tree-inspired design for high-efficiency water extraction. *Adv. Mater.* **2017**, *29*, 1704107. [CrossRef]
51. Che, W.; Xiao, Z.; Wang, Z.; Li, J.; Wang, H.; Wang, Y.; Xie, Y. Wood-Based Mesoporous Filter Decorated with Silver Nanoparticles for Water Purification. *ACS Sustain. Chem. Eng.* **2019**, *7*, 5134–5141. [CrossRef]
52. Li, X.; Jin, X.; Wu, Y.; Zhang, D.; Sun, F.; Ma, H.; Pugazhendhi, A.; Xia, C. A comprehensive review of lignocellulosic biomass derived materials for water/oil separation. *Sci. Total Environ.* **2023**, *876*, 162549. [CrossRef]
53. Guan, H.; Cheng, Z.; Wang, X. Highly Compressible Wood Sponges with a Spring-like Lamellar Structure as Effective and Reusable Oil Absorbents. *ACS Nano* **2018**, *12*, 10365–10373. [CrossRef]
54. Cai, Y.; Wu, Y.; Yang, F.; Gan, J.; Wang, Y.; Zhang, J. Wood Sponge Reinforced with Polyvinyl Alcohol for Sustainable Oil–Water Separation. *ACS Omega* **2021**, *6*, 12866–12876. [CrossRef]
55. Yang, R.; Liang, Y.; Hong, S.; Zuo, S.; Wu, Y.; Shi, J.; Cai, L.; Li, J.; Mao, H.; Ge, S.; et al. Novel Low-Temperature Chemical Vapor Deposition of Hydrothermal Delignified Wood for Hydrophobic Property. *Polymers* **2020**, *12*, 1757. [CrossRef]
56. Ruan, C.; Shen, M.; Ren, X.; Ai, K.; Lu, L. A Versatile and Scalable Approach toward Robust Superhydrophobic Porous Materials with Excellent Absorbency and Flame Retardancy. *Sci. Rep.* **2016**, *6*, 31233. [CrossRef]
57. Huang, W.; Zhang, L.; Lai, X.; Li, H.; Zeng, X. Highly hydrophobic F-rGO@ wood sponge for efficient clean-up of viscous crude oil. *Chem. Eng. J.* **2020**, *386*, 123994. [CrossRef]
58. Li, T.; Chen, C.; Brozena, A.H.; Zhu, J.Y.; Xu, L.; Driemeier, C.; Dai, J.; Rojas, O.J.; Isogai, A.; Wågberg, L.; et al. Developing fibrillated cellulose as a sustainable technological material. *Nature* **2021**, *590*, 47–56. [CrossRef]
59. Wang, Z.; Lin, S.; Li, X.; Zou, H.; Zhuo, B.; Ti, P.; Yuan, Q. Optimization and absorption performance of wood sponge. *J. Mater. Sci.* **2021**, *56*, 8479–8496. [CrossRef]
60. Ge-Zhang, S.; Yang, H.; Mu, H. Interfacial solar steam generator by MWCNTs/carbon black nanoparticles coated wood. *Alex. Eng. J.* **2023**, *63*, 1–10. [CrossRef]
61. Du, B.; Li, B.; Yang, K.; Chao, Y.; Luo, R.; Zhou, S.; Li, H. Superhydrophobic wood sponge with intelligent pH responsiveness for efficient and continuous oil-water separation. *Mater. Res. Express* **2023**, *10*, 055101. [CrossRef]
62. Chen, Y.; Zhang, L.; Mei, C.; Li, Y.; Duan, G.; Agarwal, S.; Greiner, A.; Ma, C.; Jiang, S. Wood-Inspired Anisotropic Cellulose Nanofibril Composite Sponges for Multifunctional Applications. *ACS Appl. Mater. Interfaces* **2020**, *12*, 35513–35522. [CrossRef]

63. Guan, H.; Zhang, C.; Tu, K.; Dai, X.; Wang, X.; Wang, X. Wet-Stable Lamellar Wood Sponge with High Elasticity and Fatigue Resistance Enabled by Chemical Cross-Linking. *ACS Appl. Mater. Interfaces* **2024**, *16*, 18173–18183. [CrossRef]
64. He, W.; Qiang, H.; Liang, S.; Guo, F.; Wang, R.; Cao, J.; Guo, Z.; Pang, Q.; Wei, B.; Sun, J. Hierarchically porous wood aerogel/polypyrrole(PPy) composite thick electrode for supercapacitor. *Chem. Eng. J.* **2022**, *446*, 137331. [CrossRef]
65. Yu, Y.; Li, N.; Lu, X.; Yan, B.; Chen, G.; Wang, Y.; Duan, X.; Cheng, Z.; Wang, S. Co/N co-doped carbonized wood sponge with 3D porous framework for efficient peroxymonosulfate activation: Performance and internal mechanism. *J. Hazard. Mater.* **2022**, *421*, 126735. [CrossRef]
66. Credou, J.; Berthelot, T. Cellulose: From biocompatible to bioactive material. *J. Mater. Chem. B* **2014**, *2*, 4767–4788. [CrossRef]
67. Sun, J.; Guo, H.; Ribera, J.; Wu, C.; Tu, K.; Binelli, M.; Panzarasa, G.; Schwarze, F.W.; Wang, Z.L.; Burgert, I. Sustainable and biodegradable wood sponge piezoelectric nanogenerator for sensing and energy harvesting applications. *ACS Nano* **2020**, *14*, 14665–14674.
68. Bi, W.; Li, H.; Hui, D.; Gaff, M.; Lorenzo, R.; Corbi, I.; Corbi, O.; Ashraf, M. Effects of chemical modification and nanotechnology on wood properties. *Nanotechnol. Rev.* **2021**, *10*, 978–1008. [CrossRef]
69. Farid, T.; Rafiq, M.I.; Ali, A.; Tang, W. Transforming wood as next-generation structural and functional materials for a sustainable future. *EcoMat* **2022**, *4*, e12154. [CrossRef]
70. Suppanucroa, N. Development and Characterization of Natural Rubber-Cellulose Sponge. Master's Thesis, Chulalongkorn University, Bangkok, Thailand, 2019.
71. Chang, J.; Shi, Y.; Wu, M.; Li, R.; Shi, L.; Jin, Y.; Qing, W.; Tang, C.; Wang, P. Solar-assisted fast cleanup of heavy oil spills using a photothermal sponge. *J. Mater. Chem. A* **2018**, *6*, 9192–9199. [CrossRef]
72. Shang, Y.; Gao, R.; Wang, Y.; Dong, X.; Li, X.; Tang, J.; Wang, X.; Liu, J.; Xie, Y.; Li, J.; et al. Scalable fabrication of efficient and recycling wood residue-derived sponge for crude oil adsorption. *J. Hazard. Mater.* **2023**, *441*, 129979. [CrossRef]
73. Chao, W.; Wang, S.; Li, Y.; Cao, G.; Zhao, Y.; Sun, X.; Wang, C.; Ho, S.-H. Natural sponge-like wood-derived aerogel for solar-assisted adsorption and recovery of high-viscous crude oil. *Chem. Eng. J.* **2020**, *400*, 125865. [CrossRef]
74. Wang, Z.; Jin, P.; Wang, M.; Wu, G.; Dong, C.; Wu, A. Biomass-Derived Porous Carbonaceous Aerogel as Sorbent for Oil-Spill Remediation. *ACS Appl. Mater. Interfaces* **2016**, *8*, 32862–32868. [CrossRef]
75. Karatum, O.; Steiner, S.A., III; Griffin, J.S.; Shi, W.; Plata, D.L. Flexible, Mechanically Durable Aerogel Composites for Oil Capture and Recovery. *ACS Appl. Mater. Interfaces* **2016**, *8*, 215–224. [CrossRef]
76. Dai, J.; Zhang, R.; Ge, W.; Xie, A.; Chang, Z.; Tian, S.; Zhou, Z.; Yan, Y. 3D macroscopic superhydrophobic magnetic porous carbon aerogel converted from biorenewable popcorn for selective oil-water separation. *Mater. Des.* **2018**, *139*, 122–131. [CrossRef]
77. Nguyen, S.T.; Feng, J.; Le, N.T.; Le, A.T.T.; Hoang, N.; Tan, V.B.C.; Duong, H.M. Cellulose Aerogel from Paper Waste for Crude Oil Spill Cleaning. *Ind. Eng. Chem. Res.* **2013**, *52*, 18386–18391. [CrossRef]
78. Liu, Y.; Ma, J.; Wu, T.; Wang, X.; Huang, G.; Qiu, H.; Li, Y.; Wang, W.; Gao, J. Cost-Effective Reduced Graphene Oxide-Coated Polyurethane Sponge As a Highly Efficient and Reusable Oil-Absorbent. *ACS Appl. Mater. Interfaces* **2013**, *5*, 10018–10026. [CrossRef] [PubMed]
79. Ceylan, D.; Dogu, S.; Karacik, B.; Yakan, S.D.; Okay, O.S.; Okay, O. Evaluation of Butyl Rubber as Sorbent Material for the Removal of Oil and Polycyclic Aromatic Hydrocarbons from Seawater. *Environ. Sci. Technol.* **2009**, *43*, 3846–3852. [CrossRef]
80. Dong, X.; Chen, J.; Ma, Y.; Wang, J.; Chan-Park, M.B.; Liu, X.; Wang, L.; Huang, W.; Chen, P. Superhydrophobic and superoleophilic hybrid foam of graphene and carbon nanotube for selective removal of oils or organic solvents from the surface of water. *Chem. Commun.* **2012**, *48*, 10660–10662. [CrossRef]
81. Gui, X.; Li, H.; Wang, K.; Wei, J.; Jia, Y.; Li, Z.; Fan, L.; Cao, A.; Zhu, H.; Wu, D. Recyclable carbon nanotube sponges for oil absorption. *Acta Mater.* **2011**, *59*, 4798–4804. [CrossRef]
82. Xiao, N.; Zhou, Y.; Ling, Z.; Qiu, J. Synthesis of a carbon nanofiber/carbon foam composite from coal liquefaction residue for the separation of oil and water. *Carbon* **2013**, *59*, 530–536. [CrossRef]
83. Zheng, Y.-P.; Wang, H.-N.; Kang, F.-Y.; Wang, L.-N.; Inagaki, M. Sorption capacity of exfoliated graphite for oils-sorption in and among worm-like particles. *Carbon* **2004**, *42*, 2603–2607. [CrossRef]
84. Inagaki, M.; Kawahara, A.; Konno, H. Sorption and recovery of heavy oils using carbonized fir fibers and recycling. *Carbon* **2002**, *40*, 105–111. [CrossRef]
85. Zhao, M.-Q.; Huang, J.-Q.; Zhang, Q.; Luo, W.-L.; Wei, F. Improvement of oil adsorption performance by a sponge-like natural vermiculite-carbon nanotube hybrid. *Appl. Clay Sci.* **2011**, *53*, 1–7. [CrossRef]
86. Husseien, M.; Amer, A.; El-Maghraby, A.; Hamedallah, N. A comprehensive characterization of corn stalk and study of carbonized corn stalk in dye and gas oil sorption. *J. Anal. Appl. Pyrolysis* **2009**, *86*, 360–363. [CrossRef]
87. Abdullah, M.; Rahmah, A.U.; Man, Z. Physicochemical and sorption characteristics of Malaysian *Ceiba pentandra* (L.) Gaertn. as a natural oil sorbent. *J. Hazard. Mater.* **2010**, *177*, 683–691. [CrossRef] [PubMed]
88. Butt, H.-J.; Kappl, M. Normal capillary forces. *Adv. Colloid Interface Sci.* **2009**, *146*, 48–60. [CrossRef] [PubMed]
89. Perez-Sanchez, J.F.; Palacio-Perez, A.; Suarez-Dominguez, E.J.; Diaz-Zavala, N.P.; Izquierdo-Kulich, E. Evaluation of surface tension modifiers for crude oil transport through porous media. *J. Pet. Sci. Eng.* **2020**, *192*, 107319. [CrossRef]
90. Sam, E.K.; Liu, J.; Lv, X. Surface Engineering Materials of Superhydrophobic Sponges for Oil/Water Separation: A Review. *Ind. Eng. Chem. Res.* **2021**, *60*, 2353–2364. [CrossRef]

91. Sjahro, N.; Yunus, R.; Abdullah, L.C.; Rashid, S.A.; Asis, A.J.; Akhlisah, Z.N. Recent advances in the application of cellulose derivatives for removal of contaminants from aquatic environments. *Cellulose* **2021**, *28*, 7521–7557. [CrossRef]
92. Li, X.; Yang, Z.; Peng, Y.; Zhang, F.; Lin, M.; Zhang, J.; Lv, Q.; Dong, Z. Wood-Inspired Compressible Superhydrophilic Sponge for Efficient Removal of Micron-Sized Water Droplets from Viscous Oils. *ACS Appl. Mater. Interfaces* **2022**, *14*, 11789–11802. [CrossRef] [PubMed]
93. Abu-Thabit, N.Y.; Uwaezuoke, O.J.; Abu Elella, M.H. Superhydrophobic nanohybrid sponges for separation of oil/water mixtures. *Chemosphere* **2022**, *294*, 133644. [CrossRef] [PubMed]
94. Mi, H.-Y.; Jing, X.; Politowicz, A.L.; Chen, E.; Huang, H.-X.; Turng, L.-S. Highly compressible ultra-light anisotropic cellulose/graphene aerogel fabricated by bidirectional freeze drying for selective oil absorption. *Carbon* **2018**, *132*, 199–209. [CrossRef]
95. Zhang, X.; Liu, M.; Wang, H.; Yan, N.; Cai, Z.; Yu, Y. Ultralight, hydrophobic, anisotropic bamboo-derived cellulose nanofibrils aerogels with excellent shape recovery via freeze-casting. *Carbohydr. Polym.* **2019**, *208*, 232–240. [CrossRef]
96. Zhou, S.; Liu, P.; Wang, M.; Zhao, H.; Yang, J.; Xu, F. Sustainable, Reusable, and Superhydrophobic Aerogels from Microfibrillated Cellulose for Highly Effective Oil/Water Separation. *ACS Sustain. Chem. Eng.* **2016**, *4*, 6409–6416. [CrossRef]
97. Li, L.; Li, B.; Dong, J.; Zhang, J. Roles of silanes and silicones in forming superhydrophobic and superoleophobic materials. *J. Mater. Chem. A* **2016**, *4*, 13677–13725. [CrossRef]
98. Yi, L.; Yang, J.; Fang, X.; Xia, Y.; Zhao, L.; Wu, H.; Guo, S. Facile fabrication of wood-inspired aerogel from chitosan for efficient removal of oil from Water. *J. Hazard. Mater.* **2020**, *385*, 121507. [CrossRef]
99. Oprea, M.; Voicu, S.I. Recent advances in composites based on cellulose derivatives for biomedical applications. *Carbohydr. Polym.* **2020**, *247*, 116683. [CrossRef]
100. Hon, D.N.-S. Cellulose and its derivatives: Structures, reactions, and medical uses. In *Polysaccharides in Medicinal Applications*; Routledge: Abingdon, UK, 2017; pp. 87–105.
101. He, W.; Wei, B.; Liang, S.; Wang, R.; Ji, Q.; Hu, G.; Li, W.; He, L.; Yu, J.; Zhu, H.; et al. Highly nanostructured and carboxylated wood aerogel-based adsorption membrane reconstructed by grafting of polyacrylic acid for efficient removal of heavy-metal ions. *Chem. Eng. J.* **2024**, *493*, 152411. [CrossRef]
102. Li, Y.; Wang, B.; Zhang, W.; Zhao, J.; Fang, X.; Sun, J.; Xia, R.; Guo, H.; Liu, Y. Processing wood into a phase change material with high solar-thermal conversion efficiency by introducing stable polyethylene glycol-based energy storage polymer. *Energy* **2022**, *254*, 124206. [CrossRef]
103. Ramakrishnan, R.K.; Padil, V.V.; Škodová, M.; Wacławek, S.; Černík, M.; Agarwal, S. Hierarchically porous bio-based sustainable conjugate sponge for highly selective oil/organic solvent absorption. *Adv. Funct. Mater.* **2021**, *31*, 2100640.
104. Deki, S.; Béléké, A.B.; Kotani, Y.; Mizuhata, M. Synthesis of tungsten oxide thin film by liquid phase deposition. *Mater. Chem. Phys.* **2010**, *123*, 614–619. [CrossRef]
105. Janesch, J.; Arminger, B.; Gindl-Altmutter, W.; Hansmann, C. Superhydrophobic coatings on wood made of plant oil and natural wax. *Prog. Org. Coat.* **2020**, *148*, 105891. [CrossRef]
106. Lin, W.; Huang, Y.; Li, J.; Liu, Z.; Yang, W.; Li, R.; Chen, H.; Zhang, X. Preparation of highly hydrophobic and anti-fouling wood using poly(methylhydrogen)siloxane. *Cellulose* **2018**, *25*, 7341–7353. [CrossRef]
107. Zhang, M.; Zheng, R.; Chen, J.; Huang, H. Investigation on the determination of lignocellulosics components by NREL method. *Chin. J. Anal. Lab.* **2010**, *29*, 15–18.
108. Fan, J.; Zhao, L.; Yu, J.; Liu, G. The effect of calcination temperature on the microstructure and photocatalytic activity of TiO2-based composite nanotubes prepared by an in situ template dissolution method. *Nanoscale* **2012**, *4*, 6597–6603. [CrossRef]
109. Chen, C.; Kuang, Y.; Zhu, S.; Burgert, I.; Keplinger, T.; Gong, A.; Li, T.; Berglund, L.; Eichhorn, S.J.; Hu, L. Structure–property–function relationships of natural and engineered wood. *Nat. Rev. Mater.* **2020**, *5*, 642–666. [CrossRef]
110. Artus, G.R.J.; Jung, S.; Zimmermann, J.; Gautschi, H.; Marquardt, K.; Seeger, S. Silicone Nanofilaments and Their Application as Superhydrophobic Coatings. *Adv. Mater.* **2006**, *18*, 2758–2762. [CrossRef]
111. Korhonen, J.T.; Kettunen, M.; Ras, R.H.A.; Ikkala, O. Hydrophobic Nanocellulose Aerogels as Floating, Sustainable, Reusable, and Recyclable Oil Absorbents. *ACS Appl. Mater. Interfaces* **2011**, *3*, 1813–1816. [CrossRef]
112. Rohrbach, K.; Li, Y.; Zhu, H.; Liu, Z.; Dai, J.; Andreasen, J.; Hu, L. A cellulose based hydrophilic, oleophobic hydrated filter for water/oil separation. *Chem. Commun.* **2014**, *50*, 13296–13299. [CrossRef]
113. Liu, Y.; Huang, Y.; Huang, Q.; Li, F.; Liu, X. Liquid-phase deposition functionalized wood sponges for oil/water separation. *J. Mater. Sci.* **2021**, *56*, 19075–19092. [CrossRef]
114. Biermann, O.; Hädicke, E.; Koltzenburg, S.; Müller-Plathe, F. Hydrophilicity and Lipophilicity of Cellulose Crystal Surfaces. *Angew. Chem. Int. Ed.* **2001**, *40*, 3822–3825. [CrossRef]
115. Kettunen, M.; Silvennoinen, R.J.; Houbenov, N.; Nykänen, A.; Ruokolainen, J.; Sainio, J.; Pore, V.; Kemell, M.; Ankerfors, M.; Lindström, T.; et al. Photoswitchable Superabsorbency Based on Nanocellulose Aerogels. *Adv. Funct. Mater.* **2011**, *21*, 510–517. [CrossRef]
116. Yang, Y.; Deng, Y.; Tong, Z.; Wang, C. Renewable Lignin-Based Xerogels with Self-Cleaning Properties and Superhydrophobicity. *ACS Sustain. Chem. Eng.* **2014**, *2*, 1729–1733. [CrossRef]
117. Yang, Y.; Yi, H.; Wang, C. Oil Absorbents Based on Melamine/Lignin by a Dip Adsorbing Method. *ACS Sustain. Chem. Eng.* **2015**, *3*, 3012–3018. [CrossRef]
118. Donath, S.; Militz, H.; Mai, C. Wood modification with alkoxysilanes. *Wood Sci. Technol.* **2004**, *38*, 555–566. [CrossRef]

119. Cheng, H.; Gu, B.; Pennefather, M.P.; Nguyen, T.X.; Phan-Thien, N.; Duong, H.M. Cotton aerogels and cotton-cellulose aerogels from environmental waste for oil spillage cleanup. *Mater. Des.* **2017**, *130*, 452–458. [CrossRef]
120. Wang, P.-L.; Ma, C.; Yuan, Q.; Mai, T.; Ma, M.-G. Novel $Ti_3C_2T_x$ MXene wrapped wood sponges for fast cleanup of crude oil spills by outstanding Joule heating and photothermal effect. *J. Colloid Interface Sci.* **2022**, *606*, 971–982. [CrossRef]
121. Shahzad, F.; Alhabeb, M.; Hatter, C.B.; Anasori, B.; Hong, S.M.; Koo, C.M.; Gogotsi, Y. Electromagnetic interference shielding with 2D transition metal carbides (MXenes). *Science* **2016**, *353*, 1137–1140. [CrossRef]
122. Niu, H.; Li, J.; Wang, X.; Luo, F.; Qiang, Z.; Ren, J. Solar-assisted, fast, and in situ recovery of crude oil spill by a superhydrophobic and photothermal sponge. *ACS Appl. Mater. Interfaces* **2021**, *13*, 21175–21185. [CrossRef]
123. Xu, G.; Zhang, L.; Yu, W.; Sun, Z.; Guan, J.; Zhang, J.; Lin, J.; Zhou, J.; Fan, J.; Murugadoss, V.; et al. Low optical dosage heating-reduced viscosity for fast and large-scale cleanup of spilled crude oil by reduced graphene oxide melamine nanocomposite adsorbents. *Nanotechnology* **2020**, *31*, 225402. [CrossRef]
124. Wang, X.; Peng, G.; Chen, M.; Zhao, M.; He, Y.; Jiang, Y.; Zhang, X.; Qin, Y.; Lin, S. Reduced graphene oxide composites and its real-life application potential for in-situ crude oil removal. *Chemosphere* **2020**, *249*, 126141. [CrossRef] [PubMed]
125. Luo, Z.; Wang, X.; Yang, D.; Zhang, S.; Zhao, T.; Qin, L.; Yu, Z.-Z. Photothermal hierarchical carbon nanotube/reduced graphene oxide microspherical aerogels with radially orientated microchannels for efficient cleanup of crude oil spills. *J. Colloid Interface Sci.* **2020**, *570*, 61–71. [CrossRef]
126. Ge, J.; Shi, L.-A.; Wang, Y.-C.; Zhao, H.-Y.; Yao, H.-B.; Zhu, Y.-B.; Zhang, Y.; Zhu, H.-W.; Wu, H.-A.; Yu, S.-H. Joule-heated graphene-wrapped sponge enables fast clean-up of viscous crude-oil spill. *Nat. Nanotechnol.* **2017**, *12*, 434–440. [CrossRef]
127. Zhang, C.; Wu, M.-B.; Wu, B.-H.; Yang, J.; Xu, Z.-K. Solar-driven self-heating sponges for highly efficient crude oil spill remediation. *J. Mater. Chem. A* **2018**, *6*, 8880–8885. [CrossRef]
128. Dong, X.; Cui, M.; Huang, R.; Su, R.; Qi, W.; He, Z. Polydopamine-Assisted Surface Coating of MIL-53 and Dodecanethiol on a Melamine Sponge for Oil–Water Separation. *Langmuir* **2020**, *36*, 1212–1220. [CrossRef] [PubMed]
129. Wang, K.; Wang, D.Y.; Wang, M.Z.; Dan, X.X.; Che, L.M.; Xu, H.H.; Zhou, H.; Liu, H.; Singh, L.; Wu, X.E. Functional photothermal sponges for efficient solar steam generation and accelerated cleaning of viscous crude-oil spill. *Sol. Energy Mater. Sol. Cells* **2020**, *204*, 110203. [CrossRef]
130. Fan, T.; Su, Y.; Fan, Q.; Li, Z.; Cui, W.; Yu, M.; Ning, X.; Ramakrishna, S.; Long, Y. Robust Graphene@PPS Fibrous Membrane for Harsh Environmental Oil/Water Separation and All-Weather Cleanup of Crude Oil Spill by Joule Heat and Photothermal Effect. *ACS Appl. Mater. Interfaces* **2021**, *13*, 19377–19386. [CrossRef] [PubMed]
131. Wu, X.; Lei, Y.; Li, S.; Huang, J.; Teng, L.; Chen, Z.; Lai, Y. Photothermal and Joule heating-assisted thermal management sponge for efficient cleanup of highly viscous crude oil. *J. Hazard. Mater.* **2021**, *403*, 124090. [CrossRef]
132. Li, Q.; Sun, Q.; Li, Y.; Wu, T.; Li, S.; Zhang, H.; Huang, F. Solar-heating crassula perforata-structured superoleophilic CuO@CuS/PDMS nanowire arrays on copper foam for fast remediation of viscous crude oil spill. *ACS Appl. Mater. Interfaces* **2020**, *12*, 19476–19482. [CrossRef]
133. Wu, M.; Huang, S.; Liu, T.; Wu, J.; Agarwal, S.; Greiner, A.; Xu, Z. Compressible Carbon Sponges from Delignified Wood for Fast Cleanup and Enhanced Recovery of Crude Oil Spills by Joule Heat and Photothermal Effect. *Adv. Funct. Mater.* **2021**, *31*, 2006806. [CrossRef]
134. Kuang, Y.; Chen, C.; Chen, G.; Pei, Y.; Pastel, G.; Jia, C.; Song, J.; Mi, R.; Yang, B.; Das, S.; et al. Bioinspired Solar-Heated Carbon Absorbent for Efficient Cleanup of Highly Viscous Crude Oil. *Adv. Funct. Mater.* **2019**, *29*, 1900162. [CrossRef]
135. Zhang, L.; Li, H.; Lai, X.; Su, X.; Liang, T.; Zeng, X. Thiolated graphene-based superhydrophobic sponges for oil-water separation. *Chem. Eng. J.* **2017**, *316*, 736–743. [CrossRef]
136. Chang, H.; Jia, Y.; Xiao, L.; Chen, H.; Zhao, K.; Chen, Y.; Ma, Y. Three dimensional cross-linked and flexible graphene composite paper with ultrafast electrothermal response at ultra-low voltage. *Carbon* **2019**, *154*, 150–155. [CrossRef]
137. Gong, C.; Lao, J.; Wang, B.; Li, X.; Li, G.; Gao, J.; Wan, Y.; Sun, X.; Guo, R.; Luo, J.J. Fast and all-weather cleanup of viscous crude-oil spills with Ti 3 C 2 TX MXene wrapped sponge. *J. Mater. Chem. A* **2020**, *8*, 20162–20167.
138. Zhang, Q.; Yi, G.; Fu, Z.; Yu, H.; Chen, S.; Quan, X. Vertically Aligned Janus MXene-Based Aerogels for Solar Desalination with High Efficiency and Salt Resistance. *ACS Nano* **2019**, *13*, 13196–13207. [CrossRef] [PubMed]
139. Ma, Z.; Kang, S.; Ma, J.; Shao, L.; Zhang, Y.; Liu, C.; Wei, A.; Xiang, X.; Wei, L.; Gu, J. Ultraflexible and mechanically strong double-layered aramid nanofiber–$Ti_3C_2T_x$ mxene/silver nanowire nanocomposite papers for high-performance electromagnetic interference shielding. *ACS Nano* **2020**, *14*, 8368–8382. [CrossRef]
140. Naguib, M.; Kurtoglu, M.; Presser, V.; Lu, J.; Niu, J.; Heon, M.; Hultman, L.; Gogotsi, Y.; Barsoum, M.W. Two-dimensional nanocrystals produced by exfoliation of Ti_3AlC_2. In *MXenes*; Jenny Stanford Publishing: London, UK, 2011; pp. 15–29.
141. Li, R.; Zhang, L.; Shi, L.; Wang, P.J. MXene Ti_3C_2: An effective 2D light-to-heat conversion material. *ACS Nano* **2017**, *11*, 3752–3759.
142. Song, J.; Chen, C.; Yang, Z.; Kuang, Y.; Li, T.; Li, Y.; Huang, H.; Kierzewski, I.; Liu, B.; He, S.; et al. Highly Compressible, Anisotropic Aerogel with Aligned Cellulose Nanofibers. *ACS Nano* **2018**, *12*, 140–147. [CrossRef]
143. Laitinen, O.; Suopajärvi, T.; Österberg, M.; Liimatainen, H. Hydrophobic, Superabsorbing Aerogels from Choline Chloride-Based Deep Eutectic Solvent Pretreated and Silylated Cellulose Nanofibrils for Selective Oil Removal. *ACS Appl. Mater. Interfaces* **2017**, *9*, 25029–25037. [CrossRef]
144. Zheng, Q.; Cai, Z.; Gong, S. Green synthesis of polyvinyl alcohol (PVA)–cellulose nanofibril (CNF) hybrid aerogels and their use as superabsorbents. *J. Mater. Chem. A* **2014**, *2*, 3110–3118. [CrossRef]

145. Chen, Z.; Ren, W.; Gao, L.; Liu, B.; Pei, S.; Cheng, H.-M. Three-dimensional flexible and conductive interconnected graphene networks grown by chemical vapour deposition. *Nat. Mater.* **2011**, *10*, 424–428. [CrossRef] [PubMed]
146. Fang, Y.; Jing, C.; Li, G.; Ling, S.; Wang, Z.; Lu, P.; Li, Q.; Dai, C.; Gao, S.; Chen, B.; et al. Wood-Derived Systems for Sustainable Oil/Water Separation. *Adv. Sustain. Syst.* **2021**, *5*, 2100039. [CrossRef]
147. Spear, M.J. Preservation, protection and modification of wood composites. In *Wood Composites*; Woodhead Publishing: Sawston, UK, 2015; pp. 253–310.
148. Li, M.; Xu, M.; Shi, Q.; Wang, H.; Guo, H.; Wang, L.; James, T.D. An upcycled wood sponge adsorbent for drinking water purification by solar steam generation. *Environ. Sci. Nano* **2022**, *9*, 2559–2571. [CrossRef]
149. Zhang, T.; Li, Z.; Lü, Y.; Liu, Y.; Yang, D.; Li, Q.; Qiu, F. Recent progress and future prospects of oil-absorbing materials. *Chin. J. Chem. Eng.* **2019**, *27*, 1282–1295. [CrossRef]
150. Ge-Zhang, S.; Yang, H.; Ni, H.; Mu, H.; Zhang, M. Biomimetic superhydrophobic metal/nonmetal surface manufactured by etching methods: A mini review. *Front. Bioeng. Biotechnol.* **2022**, *10*, 958095. [CrossRef]
151. Ge-Zhang, S.; Cai, T.; Yang, H.; Ding, Y.; Song, M. Biology and nature: Bionic superhydrophobic surface and principle. *Front. Bioeng. Biotechnol.* **2022**, *10*, 1033514. [CrossRef]
152. Wang, Y.; Ge-Zhang, S.; Mu, P.; Wang, X.; Li, S.; Qiao, L.; Mu, H. Advances in Sol-Gel-Based Superhydrophobic Coatings for Wood: A Review. *Int. J. Mol. Sci.* **2023**, *24*, 9675. [CrossRef]

polymers MDPI

Article

Characterization and Application of Non-Formaldehyde Binder Based Citric Acid, Maleic Acid, and Molasses Adhesive for Plywood Composite

Jajang Sutiawan [1,2], Alifah Syahfitri [3], Deni Purnomo [2], Sudarmanto [2], Narto [2], Fazhar Akbar [2], Dimas Triwibowo [2], Ismadi [2], Putri Amanda [2], Sukma Surya Kusumah [2], Muhammad Adly Rahandi Lubis [2], Dede Hermawan [3], Ignasia Maria Sulastiningsih [2], Arif Nuryawan [1] and Luthfi Hakim [1,*]

[1] Department of Forest Products, Faculty of Forestry, Universitas Sumatera Utara, Medan 20155, Indonesia; jajang.sutiawan@brin.go.id (J.S.); arif5@usu.ac.id (A.N.)
[2] Research Center for Biomass and Bioproducts, National Research and Innovation Agency, Cibinong 16911, Indonesia; deni007@brin.go.id (D.P.); suda015@brin.go.id (S.); nart001@brin.go.id (N.); fazh001@brin.go.id (F.A.); dima018@brin.go.id (D.T.); isma011@brin.go.id (I.); putr011@brin.go.id (P.A.); sukm002@brin.go.id (S.S.K.); muha142@brin.go.id (M.A.R.L.); ignasiasulastin@gmail.com (I.M.S.)
[3] Forest Product Department, Faculty of Forestry and Environment, IPB University, Bogor 16680, Indonesia; alifahsyahfitri@apps.ipb.ac.id (A.S.); dedehe@apps.ipb.ac.id (D.H.)
* Correspondence: luthfi@usu.ac.id

Abstract: Emissions of formaldehyde from wood-based panels, such as plywood, are gaining increased attention due to their carcinogenic impact on human health and detrimental effects on the environment. Plywood, which is primarily bound with a urea-formaldehyde adhesive, releases formaldehyde during hot pressing and gradually over time. Therefore, this study aims to analyze the impact of non-formaldehyde adhesive types on plywood performance. In addition, plywood performance was assessed by comparing Jabon wood (*Anthocephalus cadamba* Miq) veneer with other Indonesian wood veneers such as Mempisang (*Alphonse* spp.) and Mahogany (*Swietenia mahagoni*). To manufacture a three-layer plywood panel, a two-step manufacturing process was devised. The first step involved the use of Jabon veneers treated with citric acid (CA), maleic acid (MA), and molasses (MO), and another step was carried out for various wood veneers such as Jabon, Mempisang, and Mahogany using CA. The performance of plywood was examined using JAS 233:2003. The performance of plywood bonded with CA was better than that of plywood bonded with MA and MO. The Jabon wood veneer resulted in a lower density of plywood than other wood veneers. The water absorption, thickness swelling, modulus of elasticity, and tensile shear strength of plywood from Jabon wood veneer were similar to those of plywood from Mahogany wood veneer and lower than those of Mempisang wood veneer. The ester linkages of plywood bonded with CA were greater than those of plywood bonded with MA and MO because plywood bonded with CA has better performance than plywood bonded with MA and MO.

Keywords: eco-friendly composite; Indonesian wood; non-formaldehyde adhesive; plywood

check for updates

Citation: Sutiawan, J.; Syahfitri, A.; Purnomo, D.; Sudarmanto; Narto; Akbar, F.; Triwibowo, D.; Ismadi; Amanda, P.; Kusumah, S.S.; et al. Characterization and Application of Non-Formaldehyde Binder Based Citric Acid, Maleic Acid, and Molasses Adhesive for Plywood Composite. *Polymers* **2023**, *15*, 3897. https://doi.org/10.3390/polym15193897

Academic Editor: Ming Zhang

Received: 9 August 2023
Revised: 23 August 2023
Accepted: 28 August 2023
Published: 27 September 2023

1. Introduction

The expansion of the home construction and furniture industries is driving global plywood production and consumption, which has experienced a significant increase from 158 million m^3 to 162 million m^3 [1]. Additionally, to meet the needs of importers such as Japan, the Republic of Korea, and the United States, Indonesia exports 3.85 million m^3 of plywood annually [2]. Plywood can be used in various applications, including furniture, musical instruments, modes of transportation, packaging, sporting goods, and construction [3]. Notably, the manufacturing of plywood has witnessed a transformation in the use of fast-growing wood species. These species, which are prized for their short rotation cycles and suitable harvest diameters, have emerged as an alternative source for veneer-based products [4]. This shift signifies a sustainable method to meet the ever-expanding demand for plywood while mitigating the impact on natural wood forests.

Plywood is typically bonded with urea-formaldehyde (UF) adhesive due to its cost-effectiveness, rapid curing, transparent glue line, and ability to produce high-quality panels that meet the required standards [5]. Additionally, plywood adhesive such as phenol-formaldehyde (PF), melamine-urea-formaldehyde (MF), and polyurethane adhesive can be substituted [6,7]. However, the use of UF, MF, and PF resins in plywood manufacturing, both during the hot-pressing process and in the panel's lifespan, poses environmental concerns and health risks [8] including conditions like nasopharyngeal cancer, leukaemia, respiratory tract irritation, genotoxicity, and skin sensitization [9–12]. Several effective techniques have been devised to mitigate formaldehyde emissions from wood-based panels joined with formaldehyde-based resins. [13–19]. Kristak et al. [8] reported that the emission of formaldehyde from wood-based panels joined with formaldehyde-based resins can be reduced using several techniques. The primary ones include (i) lowering the formaldehyde-urea molar ratio; (ii) changing the hot-pressing parameters, such as the temperature and duration used during pressing; (iii) including formaldehyde scavengers like tannins, lignin, starch, wheat flour, and rice husk flour; and (iv) the use of alternative adhesive systems based on carbon materials, modified amide-containing biopolymer, carboxymethyl cellulose, and soy flour, post-treatment of the wood-based goods, surface treatment, or combining UF resin with other resins. In addition, non-formaldehyde adhesives have been developed to reduce formaldehyde emissions [20]. Non-formaldehyde adhesives were successfully fabricated as wood adhesives [21–27]. The latter have been considered the cheapest non-formaldehyde adhesives (CA) with remarkable adhesion properties [28–30].

In 2012, a significant advancement emerged with the development of citric acid (CA) as a composite adhesive. Some of these developments include adhesive for molding [31–33], plywood [34–37], laminated veneer lumber [38], oriented strand board [39], composite plywood [40], and particleboard [24,41]. Notably, CA adhesive has gained widespread recognition and use in the domain of particleboard adhesives. Some of these developments include particleboard from bamboo [42,43], Nypa [44], Salacca [45], Imperata cylindrica [46], rice biomass [47], Washingtonia palm caches [48], giant reed [49], cardoon leaf [50], and rubberwood [51]. Sutiawan et al. [37] successfully used non-formaldehyde adhesives such as CA adhesive for Jabon (*Anthocephalus cadamba* Miq), a fast-growing species of plywood. The results showed that plywood pressed at 190 °C for 10 min had less delamination and higher tensile shear strength (TSS). Additionally, some properties of plywood complied with the requirements of JAS 233:2003. In this study, the Jabon and CA adhesive combination was compared to other Indonesian wood veneers and other non-formaldehyde adhesives such as maleic acid (MA) and molasses (MO).

Sutiawan et al. [52] reported that JIS A 5908-2003 [53] type 8 was satisfied by the performance of a sorghum bagasse composite particleboard bonded with MA. In addition, Sutiawan et al. [54] highlighted the application of MA in bonding table tennis blades fabricated from sorghum bagasse particleboard. The optimum condition of the previous study was an MA content of 15 wt%, particle size of 4–20 mesh, and pressing process conditions of 200 °C and 20 min [55]. Expanding on this investigation, Syahfitri et al. [56] achieved favorable outcomes by combining sorghum biomass with MO to produce particleboards. According to JIS A 5908:2003, the performance of the particleboard fulfilled the required standards. The optimum condition of the previous study was an MO content of 20 wt% and particle size of 4–20 mesh [56]. Therefore, this study aims to analyze the influence of non-formaldehyde adhesive on the performance of plywood. In addition, plywood performance was assessed by comparing Jabon wood veneer with other Indonesian wood veneers such as Mempisang (*Alphonse* spp.) and Mahogany (*Swietenia mahagoni*).

2. Material and Methods

2.1. Material

Three Indonesian wood veneers—Jabon, Mempisang, and Mahogany—of dimensions of 350 mm × 350 mm × 2 mm, were obtained from the Center for Standardization of Sustainable Forest Management Instruments, Bogor, Indonesia. The respective densities of the Jabon wood, Mempisang, and Mahogany used were 0.40 g/cm^3, 0.67 g/cm^3, 0.58 g/cm^3. For consistency, the veneers were dried at 60 °C for 24 h, resulting in an MC of 5%, which was used throughout the study. The CA and MA were obtained from MERCK and PT Telagasakti Sakatautama, respectively. In addition, MO PTPN XII East Java, Indonesia provided the materials. According to previous investigations, the CA, MA, and MO were set at concentrations of 59 wt%, 44 wt%, and 59 wt%, respectively [37,52,56]. The adhesives used in this study have a solids content of 44.23–58.8%, gel time of 4.6–10.3 min, pH of 1.17–4.79, and viscosity of 6.5–152 mPa·s.

2.2. Characterization of Adhesive

The adhesive characteristics tested consisted of solids content, gelation time, pH, and viscosity.

2.2.1. Solids Content

The solids content of the adhesive identifies the number of particles in the adhesive. The more adhesive particles that react with wood in the gluing process, the stronger the bond strength. An adhesive sample of 1 g was added on aluminum foil and then placed in an oven (Memmert, Germany) at 103 ± 3 °C for 3 h. After the sample dried, the aluminum foil was transferred to a desiccator and weighed. The solids content was calculated using the formula below:

$$\text{Solids content (\%)} = (\text{Oven-dried weight}/\text{Initial weight}) \times 100$$

2.2.2. Gelation Time

To evaluate gelation time, the adhesive was placed in a test tube. The gel time meter (Techne GT-6, Coleparmer, Vernon, IL, USA) was positioned to submerge the needle in the sample. Dimethyl sulfoxide (DMSO) was used in a water bath, and the temperature was raised to 135 °C. After that, the time required for the adhesive to gelatinate was observed. The adhesive gelation time limit was obtained when the timer stopped automatically and showed the gelation time number marked "gel" on the screen.

2.2.3. Viscosity

Approximately 20 mL of the adhesive samples were introduced into a glass and mounted on a rotational rheometer (RheolabQC, AntonPaar, Graz, Austria). Viscosity measurements were performed using a concentric cylinder (cc)-type spindle no. 27 with a rotation speed of 100/s. Tests were carried out at 25 °C to determine the viscosity, and dynamic viscosity was measured for 120 s.

2.2.4. pH Value

The pH value of the adhesive was determined using a pH meter (Laqua pH 1200, Horiba, Kyoto, Japan). The pH value was shown on the screen a few moments after the electrode probe of the pH meter was dipped into the adhesive sample placed in a container.

2.2.5. Curing Behaviors

The curing behaviors of the adhesive were examined using differential scanning calorimetry (DSC) and thermogravimetric analysis (TGA). The adhesive samples were analyzed using TGA, and all samples were freeze-dried for one hour and then pulverized to less than 60 mesh. TGA was conducted using a TGA 4000 instrument (PerkinElmer 4000, Waltham, MA, USA). The sample adhesive was analyzed using DSC. The samples were scanned from 25 °C to 500 °C at a 10 °C/min rate under nitrogen purging. DSC measurements were completed using a DSC 4000 instrument (PerkinElmer 4000, United States). Under nitrogen purging, the samples were scanned from 25 °C to 400 °C at 10 °C/min.

2.3. Manufacture of Plywood

A layer of 134 g/m^2 based on the solids content of non-formaldehyde adhesive with a single glue line was spread on the wood veneer surface [38]. According to a previous study, a three-layer plywood panel was created using pressing conditions of 190 °C, for 10 min, under 1.3 MPa [37]. In this study, a two-step manufacturing process was devised (Table 1). In the first step, Jabon veneer was created using several non-formaldehydes adhesive such as CA, MA, and MO. Then, in the second step, another plywood panel was created using CA adhesive for various wood veneers such as Jabon, Mempisang, and Mahogany.

Table 1. Conditions of manufacture of plywood.

Steps	Type of Adhesive	Type of Wood Veneer
	CA	Jabon
1	MA	Jabon
	MO	Jabon
	CA	Jabon
2	CA	Mempisang
	CA	Mahogany

2.4. Determination of Plywood Performance

The performance of plywood, including density, moisture content (MC), water absorption (WA), thickness swelling (TS), modulus of elasticity (MOE), modulus of rupture (MOR), and TSS, was examined according to the Japanese Agricultural Standard No. 233 (JAS 2003) [57].

2.4.1. Density

Density testing was carried out using samples measuring $5 \times 5 \times 0.6$ cm^3 in length, width, and thickness. The determination of density was expressed in the results of the comparison between the weight and volume of the board. The density of plywood was calculated using an equation based on the JAS 234-2003 standard [58].

$$\text{Density } (g/cm^3) = M/V$$

where:

M is the weight of plywood (g)
V is the volume of plywood (cm^3)

2.4.2. Moisture Content (MC)

An MC test was performed on samples measuring $5 \times 5 \times 0.6$ cm^3 in length, width, and thickness. Subsequently, the MC value was calculated using the difference between the initial weight and the final weight after 24 h of drying in an oven at 103 °C. The MC value was calculated using an equation based on the JAS 234-2003 standard.

$$MC\ (\%) = (BB - BKT)/BKT \times 100$$

where:

BB is the weight of the sample before drying (g)
BKT is the weight of the sample after drying (g)

2.4.3. Water Absorption (WA)

WA tests were performed on samples measuring $5 \times 5 \times 0.6$ cm^3. The difference in weight before soaking and weight after immersion in water for 24 h was measured. The WA value was derived using a calculation based on the JAS 234-2003 standard.

$$\text{WA}(\%) = (\text{B2} - \text{B1}) / \text{B1} \times 100$$

where:

B1 is the weight of the sample before soaking (g)
B2 is the weight of the sample after soaking (g)

2.4.4. Thickness Swelling (TS)

TS tests were performed on samples of $5 \times 5 \times 0.6$ in length, width, and thickness. The measurements were carried out by measuring the difference in initial thickness before and after soaking for 24 h. The TS value was calculated using an equation based on the JAS 234-2003 standard.

$$\text{TS}(\%) = (\text{T2} - \text{T1}) / \text{T1} \times 100$$

where:

T1 is the thickness of the sample before soaking (mm)
T2 is the thickness of the sample after soaking (mm)

2.4.5. Modulus of Elasticity (MOE) and Modulus of Rupture (MOR)

MOE and MOR tests were performed on samples measuring $20 \times 5 \times 0.6$ cm^3. A universal testing machine (Shimadzu AG-IS 50 kN, Japan) was then used to test the results. MOE and rupture tests were performed at 10 mm/minute loading speeds. The MOE and MOR were determined using an equation based on the JAS 234-2003 standard.

$$\text{MOE (MPa)} = (\Delta \text{PL3}) / (4\Delta \text{Ybh})$$

$$\text{MOR (MPa)} = (3\text{PmaxL}) / (2\text{bh})$$

where:

P max is the maximum load (N)
P is the load below the limit of proportion (N)
Y is the deflection at load P (mm)
L is the pacing distance (mm)
b is the width of the test sample (mm)
h is the thickness of the test sample (mm)

2.4.6. Tensile Shear Strength (TSS)

The test was carried out using samples measuring $8 \times 2.5 \times 0.6$ cm^3. The samples were sheared using a universal testing machine (Shimadzu AG-IS 50 kN, Kyoto, Japan) at a loading speed of 2 mm/min to a maximum load. Adhesive strength was calculated using the JAS 234-2003 standard.

$$\text{Shear strength (MPa)} = \text{P} / (\text{b} \times \text{h})$$

where:

P is the maximum load (N),
b is the width of the specimen (mm)
h is the distance between notches (mm)

2.5. Functional Groups Analysis

A Fourier transform infrared (FTIR) instrument was used to measure changes in functional groups (PerkinElmer, USA). An FTIR spectra range from 4000 to 400 with a 4 cm^{-1} resolution was captured in absorbance mode. The spectra were normalized for baseline using Perkin Elmer software (Spectrum, Version 10.5.1) [52].

2.6. Statistical Analysis

A simple completely randomized design was employed, featuring two factors: the type of non-formaldehyde adhesive (CA, MA, and MO) and the type of wood veneer (Jabon, Mempisang, and Mahogany). The difference in plywood properties was analyzed using analysis of variance (ANOVA) and Duncan's multi-range test (Duncan) at 0.05.

3. Result and Discussion

3.1. Characteristics of Adhesive

The results regarding adhesive characterization (refer to Table 2) reveal that the MA adhesive has lower solids content, pH, and viscosity than the CA and MO adhesives. The lower pH and viscosity contributed to the lower quality of plywood bonded with MA adhesive. The short gelation time indicates that the adhesive no longer required a long setting time during hot pressing in the manufacture of composite products [59]. The average viscosity of the CA, MA, and MO adhesives is quite low compared to conventional formaldehyde-based plywood adhesives, such as UF resins with an average viscosity of around 250–400 mPa·s [59]. The viscosity value affects the ability of the adhesive to penetrate the pores of the wood and the storage life of the adhesive. Adhesives with high viscosity have a short storage life because they harden faster and the quality of the adhesive is low [59]. Notably, the MA and MO adhesives have lower thermal degradation (145 °C and 150 °C) compared to the CA adhesive (165 °C) (Figure 1A). However, the MO adhesive also has a higher first endothermic peak (200 °C) compared to the MA and CA adhesives (140 °C and 159 °C) (Figure 1B). These phenomena resulted in low TSS in plywood bonded with MO adhesive.

Table 2. Characteristics of adhesives used in this study.

Type of Adhesive	Solids Content (%)	Gel Time (min)	pH	Viscosity (mPa·s)
CA	56.20	4.6	2.3	7.3
MA	44.23	10.3	1.17	6.5
MO	58.8	5.9	4.79	152

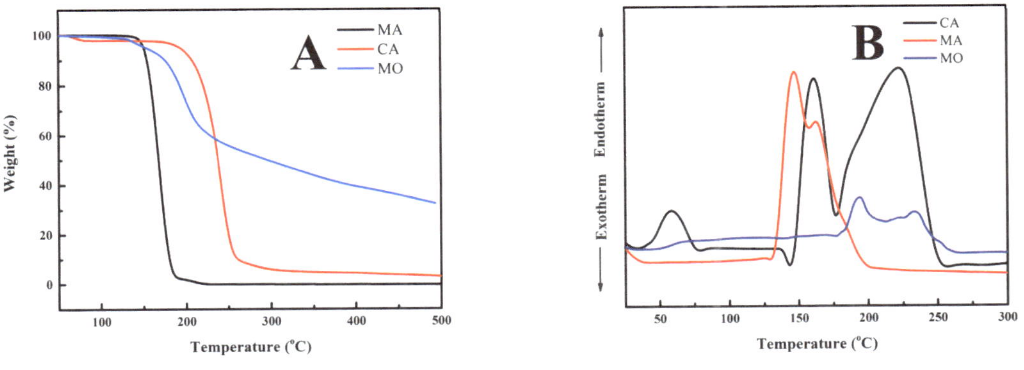

Figure 1. Thermal characteristics of adhesives used in this study (**A**) TGA and (**B**) DSC.

3.2. Variation in Type of Non-Formaldehyde Adhesive

Figure 2 shows the influence of non-formaldehyde adhesive type on the performance of plywood observed using Jabon plywood. The density and MC of plywood ranged between 0.43 and 0.50 g/cm^3 and between 5.92 and 9.70%, respectively (Figure 2A). The ANOVA revealed that there was no significant difference between plywood densities (Table 3). However, the MC of plywood bonded with CA was slightly higher than that of plywood bonded with MA and MO (Table 4). The samples' WA and TS ranged from 60.93 to 95.39% and from 4.35 to 7.74%, respectively (Figure 2B). The WA of plywood bonded with MO was slightly higher than that of plywood bonded with CA and MA. This phenomenon was affected by a cross-linker in CA and MA, and lignocellulose material was higher than in MO. Sutiawan et al. [37] reported that CA adhesive resulted in a higher concentration of cross-linkers in ester linkage than other adhesives, as detected by FTIR analysis.

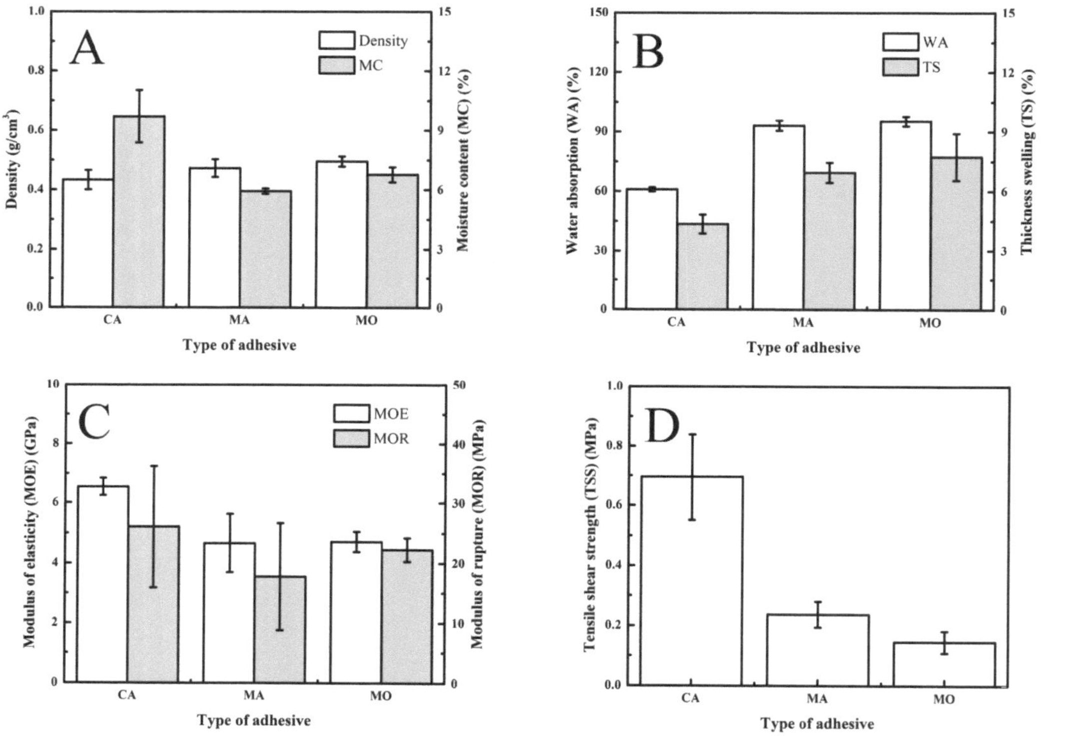

Figure 2. Influence of non-formaldehyde adhesive type on the properties of plywood. Note: error bars represent the standard deviation. (**A**) Density and moisture content, (**B**) water absorption and thickness swelling, (**C**) modulus of elasticity and rupture, and (**D**) Tensile shear strength.

Table 3. ANOVA of plywood properties with variation by type of non-formaldehyde adhesive.

Parameter	ANOVA
MOE	0.015 **
MOR	0.497 ns
TSS	0.001 **
Density	0.084 ns
MC	0.003 **
WA	0.000 **
TS	0.005 **

ns not difference, ** difference.

Table 4. Duncan test results of plywood properties with variation by type of non-formaldehyde adhesive.

Type of Adhesive	Density	MC	WA	TS	MOE	MOR	TSS
CA	0.43 a	9.70 c	60.93 a	4.35 a	6.54 b	26.06 a	0.70 b
MA	0.47 ab	5.92 a	93.08 b	6.94 b	4.66 a	17.74 a	0.24 a
MO	0.50 b	6.76 b	95.39 b	7.74 b	4.72 a	22.26 a	0.15 a

There is no difference between values with the same letter in a row.

As shown in Table 4, the MOE and MOR of plywood bonded with CA were marginally greater than in plywood bonded with MA and MO at a p-value of 0.05. The TSS of plywood bonded with CA was marginally greater than that of plywood bonded with MA and MO at the same level of p-value, which was the same as the MOE and MOR (Table 4). According to previous studies, these occurrences are caused by the presence of hydroxyl groups (OH), which are necessary for reacting with carboxyl groups (COOH) of CA to produce ester linkages (R-COO-R) [52].

3.3. Variation of Wood Veneer

The impact of wood veneer type on plywood performance was observed using CA adhesive, as illustrated in Figure 3. The average density of plywood ranged from 0.43 to 0.59 g/cm^3 and the MC ranged from 6.85 to 9.70%, as shown in Figure 3A. The Jabon wood veneer has a lower density of plywood than other wood veneers ($p < 0.05$, Tables 5 and 6). These phenomena are due to the density of Jabon wood (0.40 g/cm^3) being lower than that of Mempisang (0.67 g/cm^3) and Mahogany (0.58 g/cm^3) [60–62]. Karliati et al. [60] reported that the manufacture of wood-derived products, such as plywood, could increase the product's density compared to solid wood. According to the JAS 233: 2003 standard (JAS 2003), the MC of all plywood samples was inappropriate according to the standard (MC < 14%).

Table 5. ANOVA of plywood properties with variation by type of wood veneer.

Parameter	ANOVA
MOE	0.002 **
MOR	0.166 ns
TSS	0.051 ns
Density	0.000 **
MC	0.012 **
WA	0.000 **
TS	0.004 **

ns not difference, ** difference.

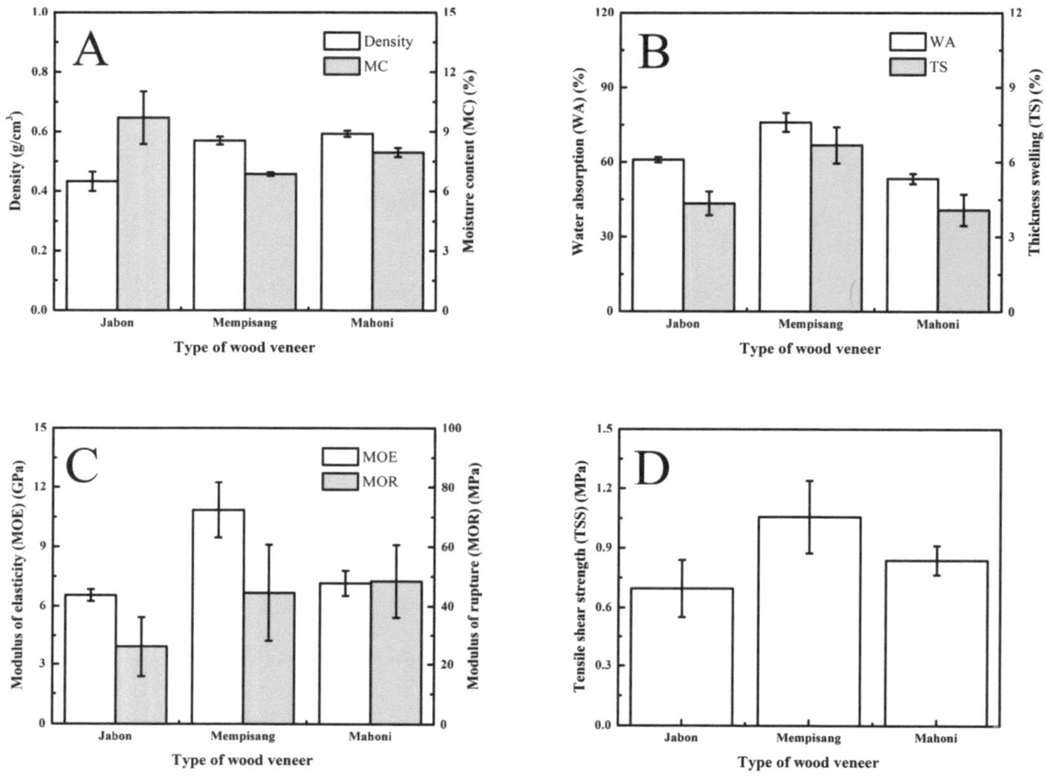

Figure 3. Influence of type of wood veneer on the performance of plywood. Note: error bars represent the standard deviation. (**A**) Density and moisture content, (**B**) water absorption and thickness swelling, (**C**) modulus of elasticity and rupture, and (**D**) Tensile shear strength.

Table 6. Duncan test results of plywood properties with variation by type of adhesive.

Type of Wood Veneer	Density	MC	WA	TS	MOE	MOR	TSS
Jabon	0.43 a	9.70 b	60.93 b	4.35 a	6.54 a	26.06 a	0.70 a
Mempisang	0.57 b	6.85 a	75.97 c	6.68 b	10.85 b	44.37 a	1.06 b
Mahogany	0.59 b	7.95 a	53.27 a	4.08 a	7.16 a	48.42 a	0.84 ab

There is no difference between values with the same letter in a row.

The WA and TS of plywood from Jabon wood veneer (60.93% and 4.35%) were similar to those from Mahogany wood veneer (53.27% and 4.08%) and were lower than those from Mempisang (75.97% and 6.68%) wood veneer ($p < 0.05$, Tables 5 and 6) (Figure 3B). A previous study reported that Jabon plywood bonded with PF adhesive has a WA of 107.6% [60]. Therefore, the plywood in this current study showcases commendable dimensional stability. The formation of ester linkages in wood-derived products has resulted in good dimensional stability [24,43,63].

The MOE and MOR of plywood from Jabon wood veneer (6.54 GPa and 26.06 MPa) were lower than those from Mahogany (7.16 GPa and 48.42 MPa) wood veneer and Mempisang (10.85 GPa and 44.37 MPa) wood veneer ($p < 0.05$, Tables 5 and 6) (Figure 3C). However, the TSS of plywood made with Jabon wood veneer (0.70 MPa) was similar to that of Mahogany wood veneer (0.84 MPa) and was lower than that of Mempisang (1.06 MPa) wood veneer ($p < 0.05$, Tables 5 and 6) (Figure 3C,D). The density of wood-derived products affected mechanical properties such as bending and bonding quality [24]. As an example, Sutrisno et al. [64] reported that wood-derived products, specifically laminated veneer lumber bonded with PF resins, exhibited an MOR of 43.14 MPa, which surpasses the MOR value observed in this study. In addition, Sun et al. [34] reported that Poplar plywood bonded with CA has a TSS value of 0.35 MPa, making all TSS values higher.

3.4. Functional Groups Analysis

The FTIR spectrum of plywood using different types of non-formaldehyde adhesive and wood veneers is shown in Figure 4. The difference is visible at approximately 1725 cm^{-1}. The peak of plywood bonded with CA, approximately 1725 cm^{-1} (ester linkages), was greater (Figure 4A) than that of plywood bonded with MA and MO [52]. This distinction in peaks contributes to the superior performance of CA-bonded plywood when contrasted with MA- and MO-bonded plywood. In addition, the peaks at 1725 cm^{-1} (ester linkages) and 1040 cm^{-1} (hemiacetal's C-O-C) in plywood from Jabon wood veneer were higher than those from Mahogany and Mempisang wood veneer (Figure 4B). This observation explains the similarity in TSS between plywood from Jabon wood veneer and plywood from Mahogany wood veneer, despite the lower density of Jabon wood. The peak at 1725 cm^{-1} signifies ester linkages, and the peak at 1040 cm^{-1} corresponds to hemiacetal's C-O-C stretching vibration, resulting in the esterification of CA with lignocellulose materials [63,65].

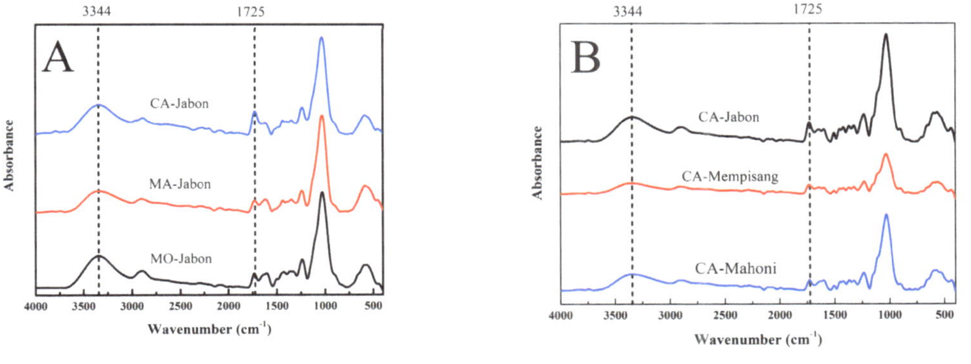

Figure 4. FTIR spectrum of plywood with different types of non-formaldehyde adhesive (**A**) and wood veneer (**B**).

4. Conclusions

In conclusion, the type of non-formaldehyde adhesive and the type of wood veneer affected the performance of plywood. Plywood bonded with CA showed superior performance compared to plywood bonded with MA and MO adhesives. The Jabon wood veneer resulted in a lower density of plywood than other wood veneers. The water absorption, thickness swelling, modulus of elasticity, and tensile shear strength of plywood crafted from Jabon wood veneer showed similarities to plywood made from Mahogany wood veneer and were lower than those derived from Mempisang wood veneer. The ester linkages of plywood bonded with CA were greater than those of plywood bonded with MA and MO because plywood bonded with CA has better performance than plywood bonded with MA and MO. Further study is still needed to determine solid content and hardener optimization factors when applying CA, MA, and MO in adhesive plywood.

Author Contributions: Data curation, J.S., A.S., D.P., S., N. and F.A.; methodology, J.S. and A.S.; supervision, D.H., S.S.K. and I.M.S.; writing—original draft, J.S.; writing—review and editing, L.H.

and A.N.; Formal analysis, D.T., I., P.A. and M.A.R.L. All authors have read and agreed to the published version of the manuscript.

Funding: This study was funded by World Class University (WCU) Universitas Sumatera Utara scheme with contract number 62/UN5.2.3.1/PPM/KP-WCU/2022.

Institutional Review Board Statement: Not applicable.

Informed Consent Statement: Not applicable.

Data Availability Statement: The data presented in this study are available on request from the corresponding author.

Acknowledgments: The authors express gratitude to the World Class University (WCU) Universitas Sumatera Utara scheme with contract number 62/UN5.2.3.1/PPM/KP-WCU/2022.

Conflicts of Interest: The authors declare no conflict of interest.

References

1. FAO. *Forest Products Annual Market Review, 2018–2019*; Food and Agriculture Organization: Rome, Italy, 2019.
2. BPS. *Statistics of Forestry Production*; The Central Bureau of Statistics: Jakarta, Indonesia, 2019.
3. Bekhta, P.; Salca, E.A. Influence of veneer densification on the shear strength and temperature behavior inside the plywood during hot press. *Constr. Build. Mater.* **2018**, *162*, 20–26. [CrossRef]
4. Mangurai, S.U.N.M.; Massijaya, M.Y.; Hadi, Y.S.; Hermawan, D. The physical characteristics of oil palm trunk and fast growing species veneer for composite-plywood. *IOP Conf. Ser. Earth Environ. Sci.* **2018**, *196*, 012025. [CrossRef]
5. Dunky, M. Urea–formaldehyde (UF) adhesive resins for wood. *Int. J. Adhes. Adhes.* **1998**, *18*, 95–107. [CrossRef]
6. Jiang, J.; Lu, X. Improving characteristics of melamine–urea–formaldehyde resin by addition of blocked polyurethane prepolymer. *Eur. J. Wood Wood Prod.* **2017**, *75*, 185–191. [CrossRef]
7. Bekhta, P.; Ortynska, G.; Sedliacik, J. Svojstva modificiranoga fenol-formaldehidnog ljepila za furnirske ploče proizvedene od furnira s visokim sadržajem vode. *Drv. Ind.* **2015**, *65*, 293–301. [CrossRef]
8. Kristak, L.; Antov, P.; Bekhta, P.; Lubis, M.A.R.; Iswanto, A.H.; Reh, R.; Sedliacik, J.; Savov, V.; Taghiyari, H.R.; Papadopoulos, A.N.; et al. Recent progress in ultra-low formaldehyde emitting adhesive systems and formaldehyde scavengers in wood-based panels: A review. *Wood Mater. Sci. Eng.* **2022**, *18*, 763–782. [CrossRef]
9. Böhm, M.; Salem, M.Z.M.; Srba, J. Formaldehyde emission monitoring from a variety of solid wood, plywood, blockboard and flooring products manufactured for building and furnishing materials. *J. Hazard. Mater.* **2012**, *221–222*, 68–79. [CrossRef]
10. Meyer, B. Formaldehyde exposure from building products. *Environ. Int.* **1986**, *12*, 283–288. [CrossRef]
11. Myers, G. *Resin Hydrolysis and Mechanisms of Formaldehyde Release from Bonded Wood Products*; Forest Products Laboratory: Madison, WI, USA, 1987; pp. 119–156.
12. Roffael, E.; Behn, C.; Dix, B. On the formaldehyde release of wood particles. *Eur. J. Wood Wood Prod.* **2012**, *70*, 911–912. [CrossRef]
13. Réh, R.; Igaz, R.; Krišt'ák, L.; Ružiak, I.; Gajtanska, M.; Božíková, M.; Kučerka, M. Functionality of beech bark in adhesive mixtures used in plywood and its effect on the stability associated with material systems. *Materials* **2019**, *12*, 1298. [CrossRef]
14. Mao, A.; Hassan, E.B.; Kim, M.G. Investigation of low mole ratio UF and UMF resins aimed at lowering the formaldehyde emission potential of wood composite boards. *BioResources* **2013**, *8*, 2453–2469. [CrossRef]
15. Barry, A.; Corneau, D. Effectiveness of barriers to minimize VOC emissions including formaldehyde. *For. Prod. J.* **2006**, *56*, 38–42.
16. Wang, Z.; Zhao, S.; Kang, H.; Zhang, W.; Li, J.; Zhang, S.; Huang, A. Reduction of energy consumption of green plywood production by implementing high-efficiency thermal conductive bio-adhesive: Assessment from pilot-scaled application. *J. Clean. Prod.* **2019**, *210*, 1366–1375. [CrossRef]
17. Faria, D.L.; Scatolino, M.V.; de Oliveira, J.E.; Gonçalves, F.G.; Soriano, J.; de Paula Protásio, T.; Lelis, R.C.C.; de Carvalho, L.M.H.; Mendes, L.M.; Junior, J.B.G. Cardanol-based adhesive with reduced formaldehyde emission to produce particleboards with waste from bean crops. *Environ. Sci. Pollut. Res.* **2023**, *30*, 48270–48287. [CrossRef]
18. Yadav, S.M.; Lubis, M.A.R.; Wibowo, E.S.; Park, B.D. Effects of nanoclay modification with transition metal ion on the performance of urea–formaldehyde resin adhesives. *Polym. Bull.* **2020**, *78*, 2375–2388. [CrossRef]
19. Nakanishi, E.Y.; Cabral, M.R.; de Souza Gonçalves, P.; dos Santos, V.; Savastano Junior, H. Formaldehyde-free particleboards using natural latex as the polymeric binder. *J. Clean. Prod.* **2018**, *195*, 1259–1269. [CrossRef]
20. He, Z. *Bio-Based Wood Adhesives*; CRC Press: Boca Raton, FL, USA, 2017.
21. Ghaffar, S.H.; Fan, M. Lignin in straw and its applications as an adhesive. *Int. J. Adhes. Adhes.* **2014**, *48*, 92–101. [CrossRef]
22. Mati-Baouche, N.; Elchinger, P.H.; De Baynast, H.; Pierre, G.; Delattre, C.; Michaud, P. Chitosan as an adhesive. *Eur. Polym. J.* **2014**, *60*, 198–213. [CrossRef]
23. Norström, E.; Fogelström, L.; Nordqvist, P.; Khabbaz, F.; Malmström, E. Xylan—A green binder for wood adhesives. *Eur. Polym. J.* **2015**, *67*, 483–493. [CrossRef]
24. Umemura, K.; Sugihara, O.; Kawai, S. Investigation of a new natural adhesive composed of citric acid and sucrose for particleboard II: Effects of board density and pressing temperature. *J. Wood Sci.* **2014**, *61*, 40–44. [CrossRef]

25. Yi, Z.; Wang, W.; Zhang, W.; Li, J. Preparation of tannin-formaldehyde-furfural resin with pretreatment of depolymerization of condensed tannin and ring opening of furfural. *J. Adhes. Sci. Technol.* **2016**, *30*, 947–959. [CrossRef]
26. Zhang, H.; Liu, P.; Musa, S.M.; Mai, C.; Zhang, K. Dialdehyde cellulose as a bio-based robust adhesive for wood bonding. *ACS Sustain. Chem. Eng.* **2019**, *7*, 10452–10459. [CrossRef]
27. Gumowska, A.; Robles, E.; Kowaluk, G. Evaluation of functional features of lignocellulosic particle composites containing biopolymer binders. *Materials* **2021**, *14*, 7718. [CrossRef]
28. Ando, D.; Umemura, K. Chemical structures of adhesive and interphase parts in sucrose/citric acid type adhesive wood-based molding derived from japanese cedar (*Cryptomeria japonica*). *Polymers* **2021**, *13*, 4224. [CrossRef] [PubMed]
29. Li, J.; Lei, H.; Xi, X.; Li, C.; Hou, D.; Song, J.; Du, G. A sustainable tannin-citric acid wood adhesive with favorable bonding properties and water resistance. *Ind. Crops Prod.* **2023**, *201*, 116933. [CrossRef]
30. Widyorini, R.; Umemura, K.; Isnan, R.; Putra, D.R.; Awaludin, A.; Prayitno, T.A. Manufacture and Properties of Citric Acid-Bonded Particleboard Made from Bamboo Materials. *Eur. J. Wood Wood Prod.* **2016**, *74*, 57–65. [CrossRef]
31. Ando, D.; Umemura, K. Bond structures between wood components and citric acid in wood-based molding. *Polymers* **2021**, *13*, 58. [CrossRef]
32. Umemura, K.; Ueda, T.; Kawai, S. Characterization of wood-based molding bonded with citric acid. *J. Wood Sci.* **2012**, *58*, 38–45. [CrossRef]
33. Widodo, E.; Kusumah, S.S.; Subyakto; Umemura, K. Development of moulding using sweet sorghum bagasse and citric acid: Effects of application method and citric acid content. *For. Prod. J.* **2020**, *70*, 151–157. [CrossRef]
34. Sun, S.; Zhao, Z.; Umemura, K. Further exploration of sucrose-citric acid adhesive: Synthesis and application on plywood. *Polymers* **2019**, *11*, 1875. [CrossRef]
35. Zhao, Z.; Sakai, S.; Wu, D.; Chen, Z.; Zhu, N.; Huang, C.; Sun, S.; Zhang, M.; Umemura, K.; Yong, Q. Further exploration of sucrose-citric acid adhesive: Investigation of optimal hot-pressing conditions for plywood and curing behavior. *Polymers* **2019**, *11*, 1996. [CrossRef]
36. Li, C.; Lei, H.; Wu, Z.; Xi, X.; Du, G.; Pizzi, A. Fully Biobased Adhesive from Glucose and Citric Acid for Plywood with High Performance. *ACS Appl. Mater. Interfaces* **2022**, *14*, 23859–23867. [CrossRef] [PubMed]
37. Sutiawan, J.; Hermawan, D.; Massijaya, M.Y.; Kusumah, S.S.; Lubis, M.A.R.; Marlina, R.; Purnomo, D.; Sulastiningsih, I.M. Influence of different hot-pressing conditions on the performance of eco-friendly jabon plywood bonded with citric acid adhesive. *Wood Mater. Sci. Eng.* **2021**, *17*, 400–409. [CrossRef]
38. Del Menezzi, C.; Amirou, S.; Pizzi, A.; Xi, X.; Delmotte, L. Reactions with wood carbohydrates and lignin of citric acid as a bond promoter of wood veneer panels. *Polymers* **2018**, *10*, 833. [CrossRef]
39. Hakim, L.; Widyorini, R.; Nugroho, W.D.; Prayitno, T.A. Performance of citric acid-bonded oriented board from modified fibrovascular bundle of salacca (*Salacca zalacca* (gaertn.) voss) frond. *Polymers* **2021**, *13*, 4090. [CrossRef]
40. Choowang, R.; Luengchavanon, M. Oil palm trunk powder blended with citric acid directly used as wood adhesive by one step liquefaction reaction. *Ind. Crops Prod.* **2021**, *170*, 113809. [CrossRef]
41. Umemura, K.; Sugihara, O.; Kawai, S. Investigation of a new natural adhesive composed of citric acid and sucrose for particleboard. *J. Wood Sci.* **2013**, *59*, 203–208. [CrossRef]
42. Widyorini, R.; Yudha, A.P.; Adifandi, Y.; Umemura, K.; Kawai, S. Characteristic of bamboo particleboard bonded with citric acid. *Wood Res. J.* **2013**, *4*, 31–35. [CrossRef]
43. Widyorini, R.; Nugraha, P.A.; Rahman, M.Z.A.; Prayitno, T.A. Bonding ability of a new adhesive composed of citric acid-sucrose for particleboard. *BioResources* **2016**, *11*, 4526–4535. [CrossRef]
44. Widyorini, R.; Prayitno, T.A.; Puspa Yudha, A.; Setiawan, B.A.; Wicaksono, B.H. Pengaruh konsentrasi asam sitrat dan suhu pengempaan terhadap kualitas papan partikel dari pelepah nipah. *Has. Penelit. J. Ilmu Kehutan.* **2012**, *6*, 61–70. [CrossRef]
45. Widyorini, R.; Umemura, K.; Soraya, D.K.; Dewi, G.K.; Nugroho, W.D. Effect of citric acid content and extractives treatment on the manufacturing process and properties of citric acid-bonded Salacca frond particleboard. *BioResources* **2019**, *14*, 4171–4180. [CrossRef]
46. Syamani, F.A.; Kusumah, S.S.; Astari, L.; Prasetiyo, K.W.; Wibowo, E.S.; Subyakto. Effect of pre-drying time and citric acid content on Imperata cylindrica particleboards properties. *IOP Conf. Ser. Earth Environ. Sci.* **2018**, *209*, 012034. [CrossRef]
47. Sejati, P.S.; Kusumah, S.S.; Dwianto, W.; Surjasa, D. Modification of rice biomass wastes for eco-friendly particleboard. *IOP Conf. Ser. Earth Environ. Sci.* **2020**, *572*, 012005. [CrossRef]
48. Ferrandez-Garcia, M.T.; Ferrandez-Garcia, A.; Garcia-Ortuño, T.; Ferrandez-Garcia, C.E.; Ferrandez-Villena, M. Influence of particle size on the properties of boards made from Washingtonia palm rachis with citric acid. *Sustainability* **2020**, *12*, 4841. [CrossRef]
49. Ferrandez-Garcia, M.T.; Ferrandez-Garcia, C.E.; Garcia-Ortuño, T.; Ferrandez-Garcia, A.; Ferrandez-Villena, M. Experimental evaluation of a new giant reed (*Arundo donax* L.) composite using citric acid as a natural binder. *Agronomy* **2019**, *9*, 882. [CrossRef]
50. Santos, J.; Pereira, J.; Ferra, J.; Magalhães, F.D.; Martins, J.M.; Carvalho, L.H. New cardoon (*Cynara cardunculus* L.) particleboards using cardoon leaf extract and citric acid as bio-adhesive. *Mater. Circ. Econ.* **2021**, *3*, 1–7. [CrossRef]
51. Huaxu, Z.; Hua, L.S.; Tahir, P.M.; Ashaari, Z.; Al-Edrus, S.S.O.; Ibrahim, N.A.; Abdullah, L.C.; Mohamad, S.F. Physico-mechanical and biological durability of citric acid-bonded rubberwood particleboard. *Polymers* **2021**, *13*, 98. [CrossRef]

52. Sutiawan, J.; Hadi, Y.S.; Nawawi, D.S.; Abdillah, I.B.; Zulfiana, D.; Lubis, M.A.R.; Nugroho, S.; Astuti, D.; Zhao, Z.; Handayani, M.; et al. The properties of particleboard composites made from three sorghum (*Sorghum bicolor*) accessions using maleic acid adhesive. *Chemosphere* **2022**, *290*, 133163. [CrossRef]
53. *JIS A 5908-2003*; Particleboards. Japanese Standards Association (JSA): Tokyo, Japan, 2003.
54. Sutiawan, J.; Hermawan, D.; Hadi, Y.S.; Nawawi, D.S.; Kusumah, S.S.; Ningrum, R.S.; Amanda, P.; Ismayati, M.; Abdillah, I.B. Properties of sorghum (Sorghum bicolor) biomass particleboard at different maleic acid content and particle size as potential materials for table tennis blade. *Biomass Convers. Biorefinery* **2022**, *in press*. [CrossRef]
55. Sutiawan, J.; Hartono, R.; Hermawan, D.; Hadi, Y.S.; Nawawi, D.S.; Abdillah, I.B.; Syahfitri, A.; Kusumah, S.S.; Adi, D.S.; Kusumaningrum, W.B.; et al. Properties of Table Tennis Blade from Sorghum Bagasse Particleboard Bonded with Maleic Acid Adhesive at Different Pressing Temperatures and Times. *Polymers* **2023**, *15*, 166. [CrossRef]
56. Syahfitri, A.; Hermawan, D.; Kusumah, S.S.; Ismadi; Lubis, M.A.R.; Widyaningrum, B.A.; Ismayati, M.; Amanda, P.; Ningrum, R.S.; Sutiawan, J. Conversion of agro-industrial wastes of sorghum bagasse and molasses into lightweight roof tile composite. *Biomass Convers. Biorefinery* **2022**, *in press*. [CrossRef]
57. JAS. *Japanese Agricultural Standard for Plywood*; JAS No. 233; Japanese Agricultural Standard Association: Tokyo, Japan, 2003.
58. *JAS 234-2003*; Japanese Agricultural Standard for Glued Laminated Timber. Japan Plywood Inspection Corporation: Tokyo, Japan, 2003.
59. Lubis, M.A.R.; Park, B.D. Modification of urea-formaldehyde resin adhesives with oxidized starch using blocked pMDI for plywood. *J. Adhes. Sci. Technol.* **2018**, *32*, 2667–2681. [CrossRef]
60. Karliati, T.; Sumardi, I.; Darwis, A.; Rumidatul, A. Incorporation of phenol-formaldehyde-based black liquor as an adhesive on the performance of plywood. *J. Biol. Sci.* **2018**, *18*, 346–353. [CrossRef]
61. Supriadi, A.; Abdurachman, A. Sifat pemesinan lima jenis kayu asal riau. *J. Penelit. Has. Hutan* **2018**, *36*, 85–100. [CrossRef]
62. Sulastiningsih, I.M.; Anggraini Indrawan, D.; Balfas, J. Karakteristik papan sandwich dengan inti papan partikel. *J. Penelit. Has. Hutan* **2020**, *38*, 161–172. [CrossRef]
63. Liao, R.; Xu, J.; Umemura, K. Low Density Sugarcane Bagasse Particleboard Bonded and additive content. *BioResources* **2016**, *11*, 2174–2185. [CrossRef]
64. Sutrisno; Alamsyah, E.M.; Sulistyawati, E.; Suheri, A. The potential use of wood waste ash nanofiller for improvement of laminated veneer lumber production made from Jabon (*Anthocephalus cadamba*). *J. Indian Acad. Wood Sci.* **2018**, *15*, 126–131. [CrossRef]
65. Amini, M.H.M.; Hashim, R.; Sulaiman, N.S.; Mohamed, M.; Sulaiman, O. Citric acid-modified starch as an environmentally friendly binder for wood composite making. *BioResources* **2020**, *15*, 4234–4248. [CrossRef]

polymers

Review

Progress in the Preparation of Stimulus-Responsive Cellulose Hydrogels and Their Application in Slow-Release Fertilizers

Zhenghui Li [1] and Ming Zhang [1,2,*]

1 School of Material Science and Engineering, Beihua University, Jilin City 132013, China; lizhenghui202304@163.com
2 Key Laboratory of Wooden Materials Science and Engineering of Jilin Province, Beihua University, Jilin City 132013, China
* Correspondence: mattzhming@163.com

Abstract: Agriculture is facing challenges such as water scarcity, low fertilizer utilization, food security and environmental sustainability. Therefore, the development of slow-release fertilizer (SRF) with controlled water retention and release is particularly important. Slow-release fertilizer hydrogel (SRFH) has a three-dimensional (3D) network structure combined with fertilizer processing, displaying excellent hydrophilicity, biocompatibility and controllability. Cellulose has abundant hydroxyl groups as well as outstanding biodegradability and special mechanical properties, which make it a potential candidate material for the fabrication of hydrogels. This work would analyze and discuss various methods for preparing stimulus-responsive cellulose hydrogels and their combinations with different fertilizers. Moreover, the application and release mechanism of stimulus-responsive cellulose hydrogels in SRF have been summarized as well. Finally, we would explore the potential issues of stimulus-responsive cellulose hydrogels serving as an SRF, propose reasonable solutions and give an outlook of the future research directions.

Keywords: slow-release fertilizers; hydrogel; cellulose; stimulus-responsive; release mechanism

Citation: Li, Z.; Zhang, M. Progress in the Preparation of Stimulus-Responsive Cellulose Hydrogels and Their Application in Slow-Release Fertilizers. *Polymers* **2023**, *15*, 3643. https://doi.org/10.3390/polym15173643

Academic Editor: Fengwei (David) Xie

Received: 31 July 2023
Revised: 28 August 2023
Accepted: 29 August 2023
Published: 4 September 2023

1. Introduction

Since the global population is increasing and food shortage issues are becoming more and more serious, fast-acting fertilizers such as nitrogen (N) phosphorus (P) and potassium (K) are essential in modern agriculture to improve food production. In the last 50 years, global agriculture has been improved due to the introduction of N fertilizers. The current data show that 2.50×10^6 tons of fertilizers will be needed annually to meet global food demand, an 800% increase in N fertilizer inputs, 35% of which has been used in China [1]. However, 50–70% of N fertilizer is usually lost to the environment through volatilization, denitrification and leaching. Instead, further overuse of N fertilizer is not effective in improving crop yields, but will increase the nitrate and N_2O (greenhouse gases) levels in groundwater [2], leading to severe environmental pollution and low fertilizer efficiency [3]. Nowadays, almost 87% of the globally consumed water was applied in irrigating the lands to increasing yield, the contribution of which has reached more than 40% [4]. However, the high evaporation, low rainfall, high variability, uneven seasonal distribution, high frequency of droughts, and serious water scarcity in arid and semiarid regions greatly limit the cultivation and production of crops [5]. Apparently, the fertilizer and water resources are not only the two main factors affecting crop growth, but also the key factors for sustainable agricultural development [6]. Therefore, it is of great importance to develop a novel fertilizer with good water-holding and slow-release properties. Slow-release fertilizer hydrogel not only can retain more water in the soil, but can also meet the various nutrient requirements towards the crops at different growth stages.

Due to the porous structure and intermolecular interactions, hydrogel is able to provide the loading sites for fertilizers. Its 3D hygroscopic network structure has excellent water absorption and retention capacities, but it is insoluble in water, and can retain large amounts of water/liquid fertilizer for longer periods of time, which plays amazing roles in SRF by slowly releasing in stressful environments. Stimulus-responsive hydrogels can sense a little change or stimulus from the external environment (e.g., temperature, pH, light, etc.), and subsequently undergo variations in physical structure or chemical properties [7]. After embedding fertilizers into stimulus-responsive hydrogels through dissolution adsorption, in situ embedding and coating envelopes, the as-obtained SRF can change its release behavior according to the variations of surrounding environment and the fluctuations in soil. It can provide a more efficient path for fertilizer delivery, which facilitates the proper amount of fertilizer applied at the optimal time, and synchronizes with the plant's nutrient requirement more precisely [8]. Despite the above attractiveness, the hydrogels have some drawbacks, such as the following: some physical hydrogels are soft and fragile (with weak mechanical properties, e.g., strength~100 kPa, rigidity~10 kPa and toughness~10 J/m^{-2}) [9], which are hardly expected to retain their shape and robustness; although the petroleum-based hydrogels represented by polyacrylamide (PAM) and polyacrylic acid have high water absorption and retention, they are expensive, poor biodegradability, and have potential environmental hazards (e.g., their degradation products lead to the function disorders of normal soil), which are not conducive to the sustainable development of agriculture [10]. Therefore, the incorporations of natural polymers and nanomaterials (e.g., cellulose nanofiber, etc.) into the matrix not only enhance the mechanical properties of the hydrogels, but also solves the high-cost and low-degradability problems.

Nowadays, the green, biodegradable and renewable polymers (e.g., cellulose [11], lignin [12], starch [13], chitosan [14], etc.) have been used to prepare SRF under the "carbon peaking and carbon neutrality" strategy. They greatly reduce the production costs, simplify the manufacturing process, and solve the environmental problems caused by excess byproducts. Some factors should be considered during the design procedure as follows: (a) to load sufficient nutrients; (b) the safety, efficiency and biodegradability of designed material; (c) no interference towards vegetation germination [15]; and (d) to release the fertilizer according to the stimulation of plant growth stages [3]. As a polymeric compound widely found in plant cell walls, cellulose has been considered as the most abundant and renewable material on earth [16]. It contains abundant functional groups, exceptional mechanical properties and high chemical modifiability. Moreover, cellulose could be produced into fibrous forms at the micro/nanometer size scale, which becomes a promising candidate for the fabrication of hydrogels in recent years [17–21]. Cellulose-based hydrogels can be prepared by physically crosslinking the cellulose molecules, or by chemically and/or physically crosslinking the cellulose derivatives [22]. The attractive properties of cellulose and its derivatives (e.g., ethylcellulose, cellulose acetate, methylcellulose, etc.) contain properties of biodegradability, biocompatibility, nontoxicity and functionality, which have motivated scientists and researchers worldwide to develop the cellulose hydrogels for a variety of applications. Notably, cellulose also improves the mechanical strength and biodegradability of the composite hydrogel product, which has high potential as an agricultural-grade reinforcing material in the manufacture of environmentally friendly stimulus-responsive hydrogel, indicating its promising application in SRF [23].

To control the release of cellulose hydrogels serving as the SRF, an in-depth understanding of the slow-release mechanism is necessary for further research and following applications [24]. The release mechanism of cellulose hydrogel is a hot research topic, and many reasonable release models have been proposed by many scholars. However, its release mechanism of nutrients cannot be determined uniquely due to the influence of original materials, soil and other factors [25–27]. This work would introduce and analyze various methods for preparing stimulus-responsive cellulose hydrogels and their combinations with different fertilizers. From Figure 1 and Table 1, the application and release mechanism of stimulus-responsive cellulose hydrogels in SRF would be discussed as well. Moreover, the potential issues of stimulus-responsive cellulose hydrogels serving as the

SRF would be summarized. Finally, we also propose the reasonable solutions and outlook the future research directions of stimulus-responsive cellulose hydrogels in SRF.

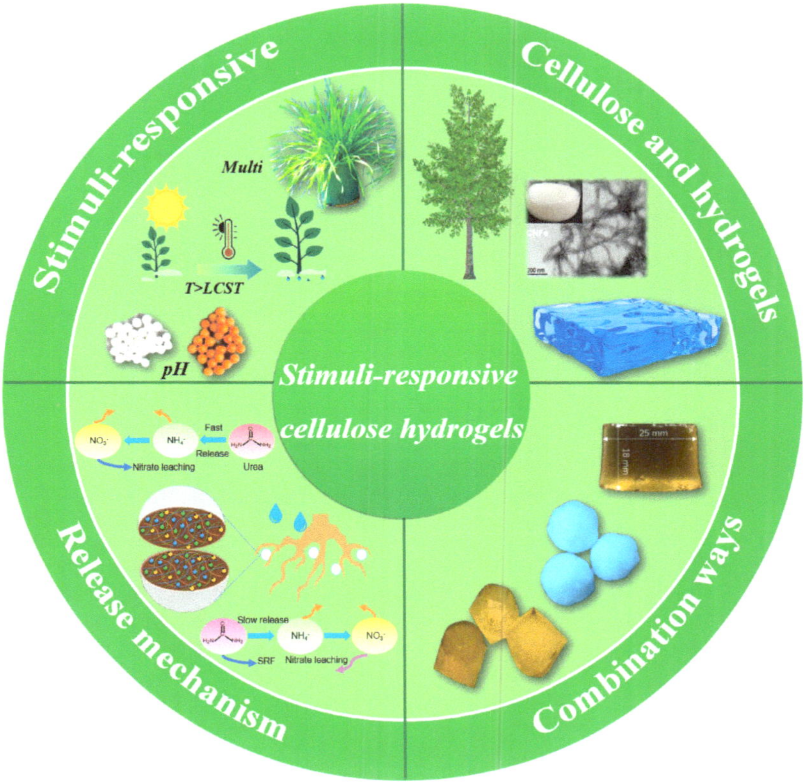

Figure 1. Schematic diagram showing the main topics of this review.

Table 1. Release mechanism, breakthrough and challenge of cellulose SRFHs.

Release Mechanism	Breakthrough	Challenge	Ref.
Stage I: First-order release kinetics model $\frac{M_t}{M_\infty} = 1 - e^{tk_1}$ Stage II: Zero-order release kinetics model $\frac{M_t}{M_\infty} = k_0 t$ Stage III: Higuchi model $\frac{M_t}{M_\infty} = k_H t^{1/2}$	Sustained AN release over 90 days; N fertilizer delivery pattern of pH-stimulus cellulose hydrogel has been synchronized with plant demands in the critical growth periods; Root activation can be achieved under short-term pH variation of soils during wetting/drying irrigation intervals.	Water retention and holding capacity of hydrogel need more systematic evaluations; Deep investigation of nutritional requirements for crops at different growth stages; Detailed and systematic research on the sustained release mechanism of fertilizers is still scarce.	[3]
Korsmeyer–Peppas model $\frac{M_t}{M_\infty} = kt^n$	Total release was 15% on the 1st day and no more than 75% on the 30th day; Hydrogels are in line with the SRF standard of the European Committee for Standardization;	Adjusting composition to increase the solubility, crosslinking degree, hydrophilicity, positive and negative charge of product.	[28]
	Hydrogel exhibited a swelling ratio exceeding 1400% at the optimal pH and temperature.	Study the composition of hydrogels after biodegradation; Ensure the nontoxicity, safety and reliability of hydrogels.	[29]

Table 1. *Cont.*

Release Mechanism	Breakthrough	Challenge	Ref.
Higuchi model followed by Korsmeyer–Peppas model	Nanocomposite hydrogel maintained 17.36% of soil moisture within 30 days; High urea loading capacity of 1.47 g/g; Cumulative urea release capacity of 60% within 30 days.	Consideration should be given to the stimulus-responsive cellulose SRFHs for smart regulation of water and fertilizer.	[30]
	Hydrogels exhibited a better swelling-capacity (37 g/g), water-retention (22.78%) and slow-release performance (40.84%).	To achieve large-scale manufacture, the preparation process should be simplified;	[31]
	Hydrogel with high surface area (45.25 m^2/g) showed high water adsorption (100 g/g); Lengthened period of totally losing the soil moisture by 18 days than pure soil.	Slow-release mechanisms should be studied in depth; Need economic evaluation of hydrogels, including raw materials, operation, fixed capital investment, etc.	[15]

2. Preparation of Hydrogels Based on Cellulose and Its Derivatives

2.1. Cellulose and Its Derivatives

Cellulose, first discovered and isolated by Anselme Payen in 1838, is a mixture of crystalline and amorphous forms consisting of glucose units linked by β-(1,4) glycosidic bonds, which play an important role in maintaining the structure of plant cell walls. As a natural polymer, cellulose is resourceful, biocompatible, renewable and degradable, which is why it is used in various applications such as food packaging [32], electromagnetic shielding [33], drug carriers [34], 3D printing [35], lithium–sulfur and zinc batteries [36,37], etc. Cellulose with an annual production of over 7.5~10.0 × 10^{10} tons is distributed in higher plants and marine animals (e.g., tuna) widely, and in algae, fungi, bacteria, invertebrates and even amoebae (protozoa) to a lesser extent [38]. In general, cellulose is a fibrous, tough, water-insoluble substance with the advantages of high mechanical strength, high Young's modulus, large specific surface area, hydrophilicity, and chemical modification potential, which has become a potential candidate for manufacturing cellulose-based hydrogels in recent years [39].

At the nanoscale level ($\leq$100 nm), nanocellulose is a class of flexible and elongated nanostructured cellulose, which can be usually separated from wood and other plants by specific pretreatments (e.g., deep eutectic solvent, carboxymethylation, TEMPO oxidation, homogenization, defibrillation, etc.) and the mechanical shearing procedure [10]. The chemical and physical features of nanocellulose usually vary depending on its source and extraction method (e.g., physical, chemical and biological methods, etc.). Nanocellulose fibers possess the merits of natural cellulose such as low density, nontoxicity and biodegradability, as well as high mechanical strength, high crystallinity, high specific surface area and self-assembly in aqueous media due to their nanoscale shape and size [40]. According to size and structure characteristics, nanocellulose is usually classified into cellulose nanocrystals (CNCs) [41], cellulose nanofibrils (CNFs) [42], and bacterial cellulose (BC) [43]. Notably, the intermolecular and intramolecular hydrogen bonds in cellulose and its high crystallinity greatly limit its water solubility and reactivity as well as its respective application. For the effective utilization of cellulose, a series of cellulose derivatives have been obtained after modification and functionalization through the chemical reactions (e.g., amination [3], esterification [44], etherification [45], oxidation [46], etc.) of the hydroxyl groups in the cellulose molecule as presented in Figure 2. These derivatives (e.g., cellulose ethers, cellulose esters and cellulose phosphates, etc.) have been widely used in material science, biomedicine and other fields.

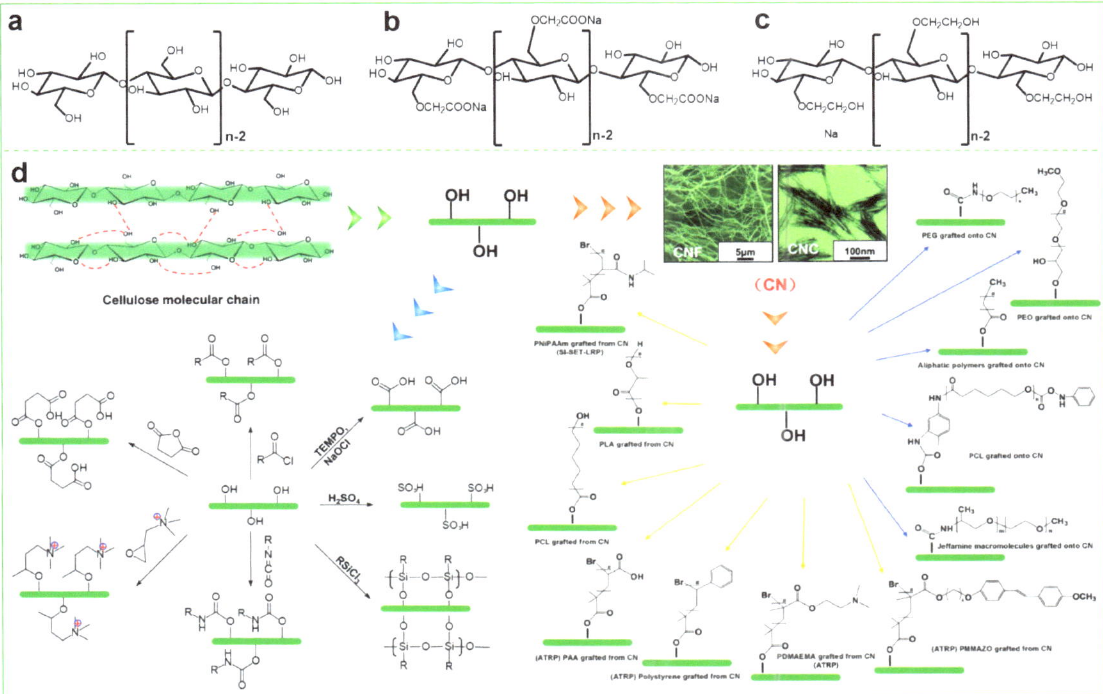

Figure 2. Chemical structures of (**a**) cellulose, (**b**) carboxymethyl cellulose (CMC), and (**c**) hydroxyethyl cellulose (HEC). (**d**) Schematic illustration of modification and functionalization of cellulose and its derivatives.

Cellulose ethers, as common cellulose derivatives including carboxymethyl cellulose (CMC), hydroxyethyl cellulose (HEC) and hydroxypropyl cellulose (HPMC), are the compounds obtained by reacting cellulose with alcohols, phenols, and amines through alkylation and etherification processes. They have good dispersion, water retention, and emulsification and thickening properties, which are commonly used in food, cosmetics, pharmaceuticals, building materials, etc. [47]. CMC is mainly used as a thickener, suspending aid, binder, gelling agent, stabilizer, water-retention agent, etc., displaying an outstanding ability to stabilize emulsions, absorb moisture from the atmosphere, and suspend solids in aqueous media. Sanosh et al. [48] obtained a sodium carboxymethylcellulose (CMC) hydrogel from the etherification between BC with chloroacetic acid in an alkaline medium. The as-prepared CMC without stirring (CMC-0 h) after dehydrothermal treatment generated the crosslinked hydrogel, which has a water absorption of 35 times its original weight. This work paves the path to artificial ultrapure hydrogels for their use in the healthcare and pharmaceutical industries. In an acidic environment, the hydroxyl group of cellulose can be esterified with the acyl halides and acid anhydrides to produce the cellulose esters with good solubility and film-formation properties. Cellulose esters have been commonly used in coatings, plastics, films, etc. Cellulose phosphate is a compound obtained by the reaction of cellulose and phosphoric acid. Due to its excellent flame retardancy and biocompatibility, cellulose phosphate has been widely used in textiles, medical materials, and other fields [49].

In addition, enolic acid, amide, olefin and other molecules containing unsaturated bonds can be used to modify cellulose's molecular structure for obtaining the cellulose derivatives with various functional groups by the grafting reaction. These cellulose derivatives not only retain the green and environmental advantages of natural cellulose, but also provide a large number of functional groups such as hydroxyl, carboxyl and amino groups, which contribute to the preparation of hydrogels with 3D porous structure, and endow them with new functional properties and application values. Furthermore, numerous hydroxyl groups in nanocellulose can adsorb the charged contaminants [50,51]. The products designed based on nanocellulose display many amazing properties such as excellent mechanical properties, good thermal stability, 3D porous structure, strong adsorption capacity towards metal ions and dyes, being an ideal carrier for loading nutrients, being fluorescent probes and other functional molecular properties [50,52].

2.2. Hydrogel and Its Crosslinking Strategy

Hydrogel is made from the combination of synthetic or/and natural polymers. As a polymeric material with a highly porous 3D network structure and a large number of hydrophilic groups (e.g., amines, hydroxyl and carboxylic) [53], hydrogel can absorb and hold substantial water or biological fluids (higher than 20%) without disintegration. In 1960, Wichterle and Lim co-polymerized glycolmonomothacrylate and glycoldimethacrylate to achieve the first generation of hydrogel, which greatly improved the compatibility between the materials and live tissues. Due to the outstanding biocompatibility, physical and chemical properties, hydrogel rapidly triggered a research boom in scientific fields since then, which has been widely used in tissue engineering [54], controlled drug release [55], agricultural production [56], hygiene products [57], and electronic devices [58]. Moreover, hydrogels have high permeability for water-soluble drugs, tunable mechanical properties, predictable degradation rates, and sensitivity to external stimuli such as pH and temperature [59].

The physical and chemical properties of hydrogels (swelling property, mechanical strength, elasticity, release kinetics, etc.) highly depend on their composition, stimulus condition (e.g., pH, temperature, light, magnetic field, etc.), nature and extent of crosslinking [60]. Physical crosslinked hydrogels, also called non-permanent hydrogels, are caused by macromolecular entanglement, ion attraction, electrostatic interaction, hydrogen bonds, van der Waals forces and hydrophobic interactions. Physical hydrogels are usually reversible (sol–gel transition) and preferable for applications where toxic crosslinkers are undesirable, and are often formed under modest conditions [16]. Although they have the advantages of easy preparation and operation, they also have the disadvantages of low mechanical strength and poor stability [61]. Common crosslinked hydrogels prepared via physical strategy include freeze-dried hydrogels (freeze-drying technology) [62], filament hydrogels (self-assembly and other methods) [63] and nanogels (nanoparticle formation) [64]. In conclusion, physical hydrogels have diverse structures, but poor mechanical properties and can easily be lost in soil, which are unfavorable to water retention and the slow release of fertilizer, greatly limiting their application in agriculture.

Chemical crosslinked hydrogels, also known as permanent hydrogels, are thermally irreversible (differ from physical crosslinked hydrogels), which are more stable in soil serving as the SRF for agriculture due to its rigid networks, mechanical robustness, high water absorption, and excellent physical and viscoelastic properties [16,65]. Usually, chemical crosslinked hydrogels are achieved by forming covalent bonds among the polymer chains in the presence of crosslinking agents or under specific external conditions. Such procedures might involve covalent bond graft, free-radical polymerization, click chemistry, enzymatic reactions, heat dehydration, and ionic and radiation (e.g., gamma and ultraviolet rays, or electron beam) [3,15,31,59] crosslinking [66–70]. Common chemical crosslinked hydrogels include PAM hydrogels [71], polyvinyl alcohol hydrogels [72], acrylic hydrogels [73] and so on. Although chemical crosslinked hydrogel has the mature preparing techniques and wide range of applications, its crosslinking process is usually tedious, and the introduced crosslinking agent may be toxic [74]. The chemical hydrogels synthesized via radiation without toxic crosslinking agent are relatively pure and safe, which are suitable for agriculture [70], biomedical and pharmaceutical applications [75].

Dual-network crosslinked hydrogels are formed by interweaving two identical or different networks, which are generally realized by free-radical copolymerization, water-soluble polymer crosslinking, and polymer interpenetrating networks [76]. One network is strongly covalently crosslinked (via covalent bond graft, free-radical polymerization, click chemistry, heat dehydration, etc.), which usually has excellent mechanical strength and rigidity [77]. Meanwhile, the other one is loosely physically crosslinked (via van der Waals forces, electrostatic interactions, hydrogen bonding, etc.), which usually has high ductility and reversibility [78]. These two networks can be entangled with each other by physical or chemical crosslinking. The hydrogel with dual networks is obviously superior to the single network hydrogel, showing excellent ductility, high plasticity, good biocompatibility, and outstanding self-healing property, which can withstand large tensile and compressive forces [79]. However, the radiation and dual-network crosslinking processes are usually complicated and expensive, which should be further simplified and lower the cost. Therefore, it is significant to develop a simple and low-cost process for preparing the water-retaining, structurally stable, biodegradable and nontoxic crosslinked hydrogels by using natural biomass materials.

2.3. Hydrogel Based on Cellulose and Its Derivatives

The mechanical properties of hydrogels are vital to preserving the shape, stiffness, and robustness of the hydrogel during its practical applications [54]. However, many pure hydrogels are very soft, displaying a weak mechanical property. The hydrogels based on cellulose and its derivatives, so-called cellulose hydrogels, can be obtained by the esterification or etherification of cellulose, and the chemical or dual-network crosslinking with cellulose derivatives [16]. Due to their extraordinary mechanical properties, high absorption efficiency and stimulation reactivity, the cellulose hydrogels are increasingly attractive in multidisciplinary fields (e.g., biomedicine [34], food [80] and agriculture [81]) [16].

Due to its high strength, crystallinity, surface area and aspect ratio, nanocellulose (e.g., CNF and CNC) has attracted increasing attention in the preparation of nanocomposite hydrogel as the reinforcing filler. Moreover, cellulose nanocomposite hydrogel can be more qualified to serve as the reservoir for improving irrigation efficiency and plant growth. Xu et al. [82] reported a novel nanocellulose hydrogel based on TEMPO-oxidized CNF and gelatin methacrylate (GelMA) for 3D printing the scaffold. The product can uptake water up to around 90 times its own weight. The hydrogel scaffold with high CNF concentration after dual crosslinking treatment achieved a further improvement of its mechanical strength. Its compressive Young's modulus and local surface stiffness could be adjusted by changing the CNF and GelMA contents. Specifically, the introduction of nanocellulose and its derivatives not only improves the mechanical strength of the hydrogel, but also enhances its biodegradable capability, which is conducive to green, environmentally friendly and sustainable development.

Das et al. [83] synthesized the hydrogel based on the carboxymethyl cellulose sodium (NaCMC), HEC and CNF with citric acid (CA) as the crosslinking agent (Figure 3a). The HEC greatly promoted the interaction between the molecular chains for forming a strong hydrogel network, and the CNFs (with a length of ~600 nm, 0.7%) served as the reinforcing agents in the hydrogel matrix, which largely improved the mechanical strength of hydrogel (16.27 MPa, Figure 3e). Such a mechanical improvement is essential to maintain the structural integrity of the hydrogel in agricultural applications. The hydrophilic groups (e.g., −OH and −COOH) attached to the cellulose chains are responsible for the water-absorbing capability of hydrogel. The water absorption of NaCMC/HEC/CNF hydrogel could reach up to ~1070% (Figure 3b,d). Moreover, the buried soil tests (60 days) showed appearance and color variations, structural decomposition and chemical changes of the hydrogels due to microorganism degradation (Figure 3c), confirming its outstanding biodegrading capability.

In summary, hydrogel based on cellulose (or nanocellulose) and its derivatives can be prepared by different crosslinking methods. Its outstanding overall performance makes it fit in absorbing and retaining water in the hydrogel networks, and releasing water and nutrients to the plants in a controlled manner, which is very essential for agricultural production in arid and semiarid regions.

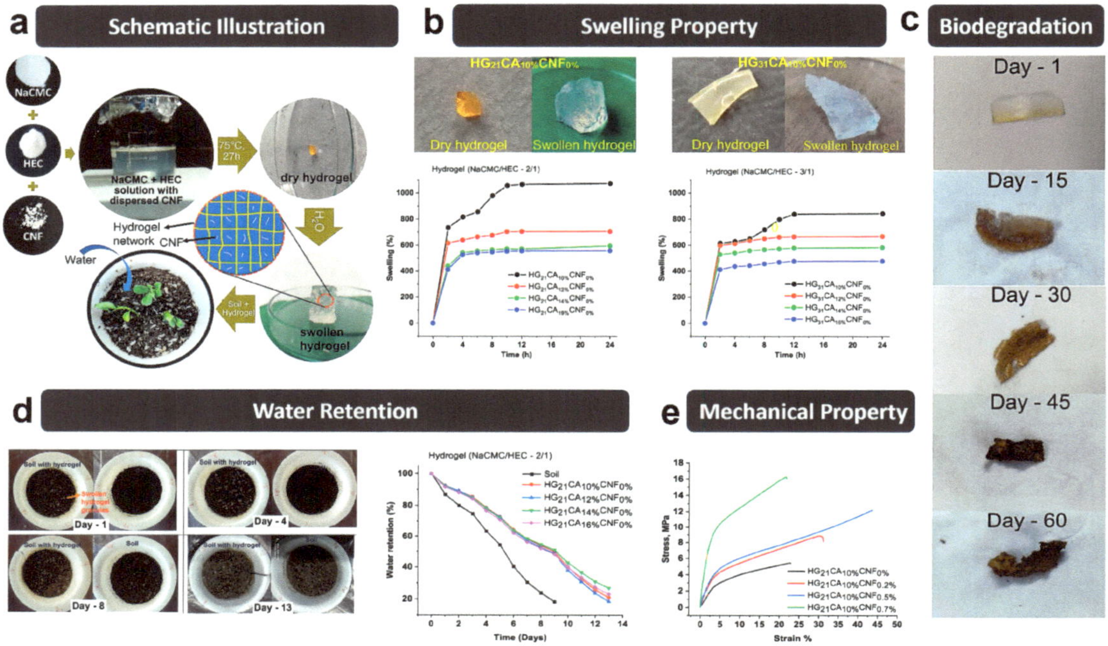

Figure 3. (**a**) Schematic illustration of NaCMC/HEC/CNF hydrogels. (**b**) Swelling performance of hydrogels: NaCMC/HEC:2/1 and NaCMC/HEC:3/1. (**c**) Hydrogel images during the buried soil test. (**d**) Water retention test of soil with and without hydrogel. (**e**) Stress−strain curves of the hydrogel and hydrogel nanocomposites. Reprinted with permission from ref. [83]. Copyright 2023 American Chemical Society.

3. Combination of Fertilizer and Cellulose Hydrogel

In the growth of crops, nitrogen, phosphorus, potassium, and some other trace elements (such as iron, zinc and molybdenum) are the essential nutrients [84]. The pore structure of cellulose hydrogels can provide suitable sites for loading the urea [85]. It is the recent research focus to combine cellulose hydrogel with nutrient fertilizer for obtaining a novel hydrogel with slow-release function. This paper summarizes three main strategies for loading the fertilizer into cellulose hydrogel in the following section.

3.1. Solubilization and Adsorption Method

Generally, the solubilization and adsorption method involves three steps (Figure 4). The first step is immersing the as-prepared aerogel (freeze-dried hydrogel) in a solution containing fertilizer. The second step is the adsorption of nutrients into the 3D network by the swelling of hydrogel. The third step is waiting for the equilibrium swelling ratio (ESR) of hydrogel with ultimate nutrient and water retention. In addition, the releasing speed of the hydrogel can be controlled by changing its physical shape after drying [86].

Liu et al. [31] used MIL-100(Fe) (metal–organic frameworks, MOF), CNF and sodium alginate (SA) to prepare the MC hydrogels by free-radical polymerization in the presence of APS (initiator) and MBA (crosslinking agent). Subsequently, the resulting hydrogel was dried using a freeze dryer, and further immersed into the urea solution (25 mg/mL) for

72 h at 25 °C, as presented in Figure 4a. Results of the fertilizer slow-releasing test show that the cumulative release rate of MC-10% (MOF to CNF: 10%) hydrogel after loading urea reached 40.84% in soil at 30 d.

Do et al. [87] prepared an alginate-coated gelatin/CNC hydrogel by a simple layer-by-layer process (Figure 4c). Then, NH_4NO_3, P_2O_5 and K_2O (NPK fertilizer) were dissolved in the distilled water to prepare NPK aqueous solution (10 mg·mL^{-1}). After immersing in NPK aqueous solution (100 mL), the pre-dried hydrogels gradually swelled to a steady state. Results showed that the alginate-coated gelatin/CNC hydrogel exhibited a slower NPK releasing speed than the gelatin/CNC hydrogel. Compression tests showed that the addition of CNCs improved the mechanical properties, water-retention capacity, and sustained fertilizer release of the hydrogel.

In summary, the solubilization and adsorption method can reduce the fertilizer's influence on the hydrogel-synthesis process, and better control the formation of SRFHs. However, the fertilizer loading in hydrogel is lower than the other methods, and the multifarious operation steps (including hydrogel synthesis, drying, fertilizer absorption and re-drying) leads to a long preparation period [65]. Moreover, the hydrogel must be crosslinked sufficiently to envelop the fertilizer very well.

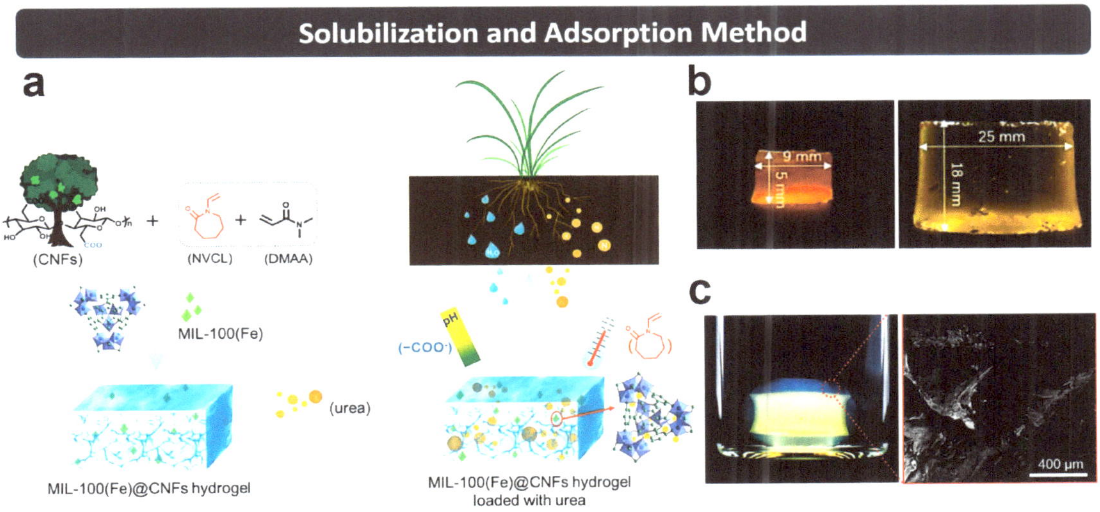

Figure 4. (**a**) Simplified scheme for preparing MC hydrogel. (**b**) Hydrogels before and after swelling equilibrium. Reprinted with permission from ref. [31]. Copyright 2021 Elsevier Ltd. (**c**) Alginate-coated gelatin/CNC hydrogel and its SEM image. Reprinted with permission from ref. [87]. Copyright 2022 Elsevier Ltd.

3.2. In Situ Encapsulation Method

The in situ encapsulation method (Figure 5) involves directly mixing the fertilizer and raw materials together. During the formation of hydrogel, the fertilizer can be encapsulated in the hydrogel matrix [65]. Wang et al. [15] prepared the CNF from pulp residue, which was further mixed with SA and urea in the presence MIL-100 (Fe) under stirring condition. Then, the ball-shaped CNF/SA/MOFs hydrogel (CAM) with good water- and fertilizer-regulating functions was prepared successfully by dropping the above mixture uniformly in $CaCl_2$ solution (200 mL, 2.5 wt%) by using a syringe (Figure 5a,b).

Liu et al. [28] prepared the CNF suspension (0.5 wt%), which was further mixed with SA under stirring condition to obtain a homogeneous suspension. Then, polyvinyl alcohol (PVA) and NPK fertilizer (1.5 wt%) were added to the above suspension, which was loaded into two plastic syringes and then pushed into a beaker containing $CaSO_4$ solution (400 mL 0.1 wt%). Then, slow-release fertilizer cellulose hydrogels with a macroporous flexible core and microporous semi-interpenetrating polymer network shell were successfully prepared (Figure 5c). Results showed that the water of the soils containing composite hydrogel was beyond 80% on the 60th day, and the N/P/K releases of composite hydrogel was 14.03%/11.95%/7.33% (1st day) and 67.90%/70.78%/71.12% (30th day), respectively.

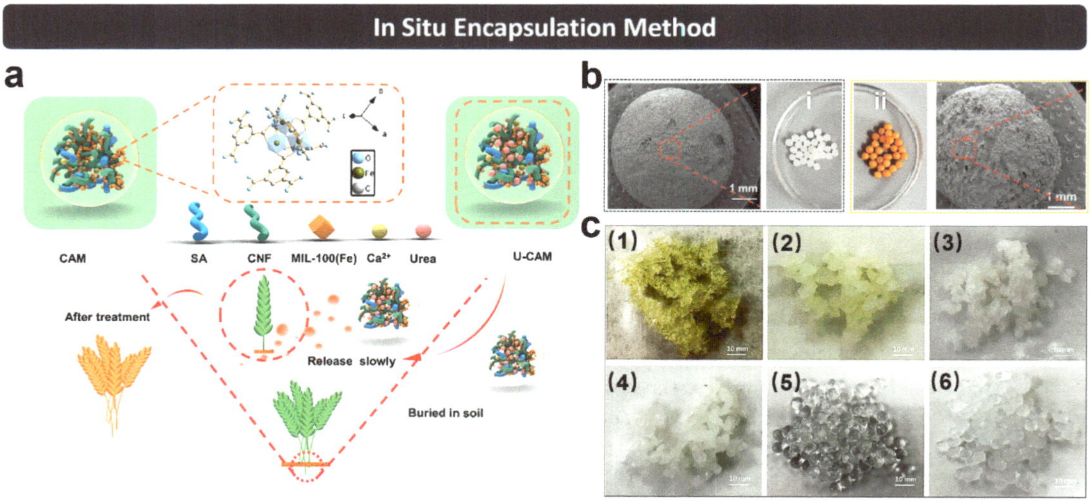

Figure 5. (**a**) Simplified scheme for preparing the CAM hydrogel and its application in the slow release of urea. (**b**) SEM and photograph of (i) CAM-0% (without MIL-100 (Fe)); (ii) CAM-10% (MIL-100 (Fe) to CNF at 10%). Reprinted with permission from ref. [15]. Copyright 2021 Elsevier Ltd. (**c**) Different hydrogels after swelling for 200 h. (1) SA, (2) SA/NPK, (3) SA/CNF/NPK, (4) SA/CNF, (5) SA/CNF/PVA and (6) SA/CNF/PVA/NPK hydrogels. Reprinted with permission from ref. [28]. Copyright 2021 Elsevier Ltd.

Compared with the solubilization and adsorption method, the in situ encapsulation method can load more fertilizer into the hydrogel, largely increasing the fertilizer loading capacity of cellulose hydrogel. Furthermore, this method is much simpler than the solubilization and adsorption method, which only requires one step to load the fertilizer in hydrogel, largely reducing the preparing cost and time for SRFH. Notably, the presence of fertilizer sometimes might adversely affect the formation and properties of the hydrogel. Therefore, the polymerization process must be strictly controlled until its completion in order to avoid the unreacted materials and the impurities which are difficult to remove [65].

3.3. Coating and Wrapping Method

Generally, the coating and wrapping method (Figure 6) involves (1) wrapping the fertilizer as a core in a hydrogel with water-absorption and retention capacity, and (2) forming a core–shell structure to achieve the effect of water retention and SRF. The product core can be pure urea or NPK composite fertilizer; in the meantime, some polymers and inorganic substances can be added to regulate its performance as well. The shell can be a water-absorbing and water-retaining hydrogel with one or multiple layers, or a layer of hydrophobic polymer which can delay the release of water and fertilizer.

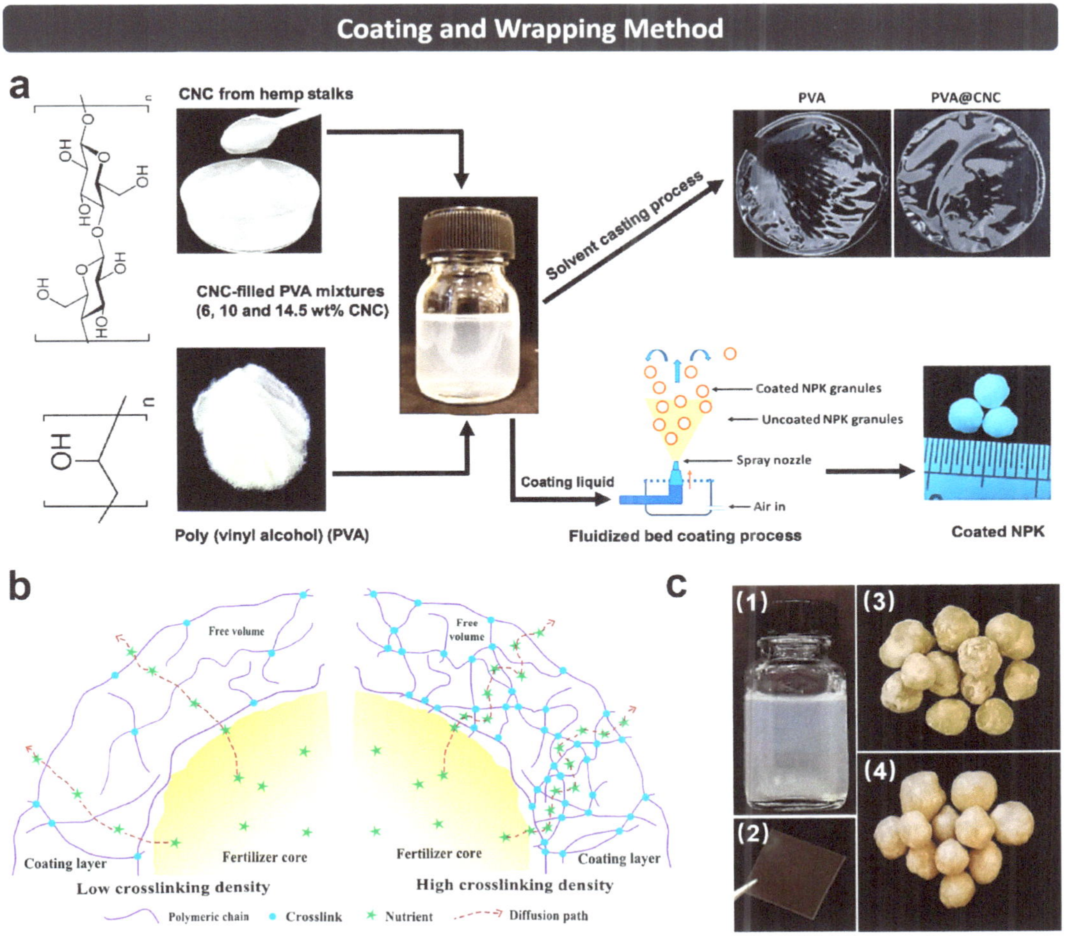

Figure 6. (**a**) Schematic illustration for preparing PVA@CNC nanocomposites and further coating on the NPK fertilizer granules. Reprinted with permission from ref. [88]. Copyright 2021 Elsevier Ltd. (**b**) Schematic illustration of the relationship between the crosslinking degree of coating and the nutrients diffusion of coated MAP fertilizer. (**c**) Photographs of CH@5RC (1) formulation and (2) solid film, (3) uncoated and (4) coated MAP granules. Reprinted with permission from ref. [17]. Copyright 2022 Elsevier Ltd.

Kassem et al. [88] reported a waterborne and biodegradable PVA@CNC nanocomposite hydrogel formulation, which was further applied on the surface of NPK fertilizer granules to form the microlayer in the Wurster chamber of a fluidized bed dryer (Figure 6a). It was found that such a coating layer extended the NPK nutrients' release time from 3 days for uncoated fertilizer to 30 days for PVA@CNC-coated fertilizer in soil medium. Furthermore, the crushing strength of fertilizer was increased after PVA@CNC coating. The improved water-retention capacity of the soil amended with PVA@CNC-coated NPK fertilizer demonstrated the intrinsic water-saving property of PVA@CNC. Apparently, the facile coating method indeed brings physical and chemical advantages to the conventional fertilizer in terms of structural stability, nutrient release and water management.

In 2022, Kassem et al. [17] developed a waterborne and swellable hydrogel formulation as the wrapping material for monoammonium phosphate (MAP) fertilizer. Specifically, sodium CMC/HEC was prepared and composited with regenerated cellulose (RC, 5%) to obtain CH@5RC formulation and solid film (Figure 6b,c). Two coating thicknesses (27 μm, and 71 μm) and two crosslinking conditions (80 °C/8 h, 120 °C/10 min) were utilized to apply the CH@5RC on MAP fertilizer in a spraying process. The relationship between the crosslinking degree of coating and the nutrient diffusion of coated MAP fertilizer were discussed in this study as well. Results showed that the crushing resistance of coated fertilizer reached up to 90.55 N, and the soil containing the coated fertilizer obtained the better water retention. Specifically, the release time of MAP fertilizer can double and better retain the soil moisture when cellulose serves as the main coating material.

To sum up, the coating and wrapping method can greatly reduce the release rate of nutrients by forming a dense coating or wrapping multiple layers with less porosity on the fertilizer surface [65]. Although the process is complex and of high cost, it can integrate various good performances into the product with excellent water retention and a slow/controlled release speed of the fertilizer by wrapping different types of coatings.

4. Slow-Release Fertilizer Hydrogels and Release Mechanism

Slow-release fertilizers (SRFs) or controlled-release fertilizers (CRFs) are the nutrients that release into the environment in a slow and controlled manner. There is no official differentiation between CRFs and SRFs. The association of American plant food control officials (AAPFCO) has stated that SRF is a fertilizer that delays its nutrient delivery and availability to plants after application [65]. According to the instructions of the European standards committee, if the nutrient release at 25 °C satisfies the following three conditions, it can be classified as SRF: (1) no more than 15% within 24 h; (2) no more than 75% within 28 d; and (3) at least 75% within the specified time frame. Compared to conventional fertilizers, they have significant advantages in reducing nutrient loss, improving nutrient utilization, increasing crop yields, and protecting the environment. Therefore, the application of SRF brings many advantages: (1) it facilitates the supply of nutrients to plants throughout the growing season, (2) it reduces the labor and energy costs required for multiple fertilizer applications, (3) it improves safety due to the reduction of specific toxicity caused by high ion concentrations in the soil, (4) it reduces nutrient losses from the soil, and (5) it alleviates environmental problems (e.g., eutrophication of water body) [89].

Slow-release fertilizer hydrogels (SRFHs) release the nutrients slowly, thus reducing the evaporation losses and frequent irrigation. They not only lessen the environmental impact of conventional fertilizers, but also enhance plant nutrition. Such an approach offers a high-quality solution to water and fertilizer management, which effectively boosts food production in arid/semiarid regions, promoting the sustainable development of environmental agriculture.

4.1. *Slow-Release Fertilizer Hydrogels*

Based on material origins, SRFHs can be categorized into three groups including natural hydrogels, purely synthetic hydrogels, and synthetic–natural hybrid hydrogels. With the increasing emphasis on environmental protection and green chemistry, natural polymers gained popularity due to their safety, biocompatibility and biodegradability. Natural SRFHs are often fabricated by natural polysaccharides such as cellulose, starch, chitosan, alginate and so on [85]. Purely synthetic hydrogels are typically synthesized from hydrophilic monomers or polymers, which are derived from petroleum-based sources, including polyacrylamide (PAAm), polyacrylic acid (PAAc), and the copolymers combining PAAm and PAAc [65]. Despite their advantageous water absorption and water retention, purely synthetic hydrogels still result in significant environmental issues due to their poor biodegradability, and the potential for degradation products to disrupt normal soil function [65,71,73].

By combining the characteristics above, the synthetic–natural SRFHs are developed, which are mainly categorized as organic–inorganic and organic–organic composites [65]. Organic–inorganic SRFHs are created by incorporating the synthetic polymers with inorganic materials. The inorganic components in hydrogels not only reduces the overall costs, but also enhances the properties such as degradability, swelling capacity, barrier performance, mechanical strength, and thermal stability. Common inorganic materials include clay minerals such as attapulgite [90], montmorillonite [91], bentonite [92] and kaolin [93]. These materials can be incorporated into the hydrogels in significant proportions, simultaneously reducing the cost in production and enhancing the swelling capability of hydrogels. Organic–organic SRFHs are crafted by combining the synthetic polymers with natural polymers or fibers. Natural polymers are abundant in nature, which offer many advantages such as mitigating environmental pollution, lowering hydrogel production costs, and enhancing biodegradability and mechanical properties [94].

Usually, agricultural residues are either incinerated or discarded, contributing to environmental pollution and resource wastage. Natural fibers can be derived from these agricultural residues. By grafting or blending the hydrogels with such natural fibers or polymers (e.g., cellulose), it is possible to simultaneously achieve waste utilization, cost reduction, biodegradability, eco-friendliness, and nontoxicity. Moreover, the incorporated fibers can serve as the structural support for enhancing the mechanical strength, water-absorption capacity, and plant growth performance [95]. In summary, the organic components offer elasticity, low density, malleability and toughness, while the inorganic components provide hardness, rigidity and thermal stability.

4.2. Release Mechanism of Slow-Release Fertilizers Hydrogels

Further comprehension of the slow-release mechanism is essential to offer theoretical support for the advancement and implementation of SRF. Since there are variations in nutrient-release mechanisms among different types of SRFs, and a complex interaction of factors (composition, soil moisture, temperature, etc.), the release mechanisms of SRFHs cannot be unilaterally defined. Generally, there are mainly four mathematical models used to explore the release mechanism, including the zero-order, first-order, Korsmeyer–Peppas, and Higuchi kinetic models (Table 1). Specifically, the parameter M_t/M_∞ represents the percentage of fertilizer released from the hydrogel at a specific time t. The constants k_0, k_1, k_H and k correspond to the release constants of the zero-order, first-order, Higuchi, and Korsmeyer–Peppas models, respectively. The release mechanisms are categorized based on the diffusion index (n) as follows: (1) Fickian diffusion mechanism: $n < 0.43$; (2) Non-Fickian diffusion mechanism: $0.43 < n < 0.85$; (3) Case-II transport mechanism: $n > 0.85$.

To date, many efforts have made on the reasonable release mechanisms of SRFHs. Liu et al. [28] prepared a novel SRFH with a macroporous flexible core and a microporous interpenetrating polymer networks (IPN) shell by blending and crosslinking SA, PVA and CNFs in the fertilizer formulations containing NPK (Figure 7a–c). To understand its slow-release mechanism, the authors employed the Korsmeyer–Peppas release kinetic model. The results suggest that the releases of the SA/CNF/NPK and SA/CNF/PVA/NPK hydrogels in water and soil were controlled by Fickian diffusion. However, the release of SA/NPK hydrogel is controlled by erosion as well as Super-Case II transportation. Based on the Korsmeyer–Peppas model, this research group hypothesized the release mechanism of SA/CNF-based hydrogels in the medium (water or soil) as follows.

Firstly, the freeze-dried SA/CNF-based gel slowly swelled due to the water absorption, and converted into hydrogel in the medium. Subsequently, the NPK fertilizer encapsulated in the hydrogel was slowly dissolved. Thirdly, the fertilizer was gradually released into the medium through the dynamic exchange (or diffusion control) between the interior and exterior of the hydrogel. As the swelling ratio increased, the size of the macropores within the 3D network expanded, facilitating the diffusion of the fertilizer into the medium. Finally, the release rate of fertilizer into the medium gradually decreased, and approached a constant value over time. This result indicates that the prepared hydrogel has the potential to be applied in regions with drought-prone conditions or with fertilizer-loss issues, offering possibilities for future applications in precision agriculture and horticulture.

Shaghaleh et al. [3] develop a viewpoint that the practical slow-release model of SRFHs varies at different stages. This research group prepared an aminated-CNF (A-CNF) fertilizer hydrogel by encapsulating ammonium nitrate (AN) in A-CNF hydrogel. According to the literature, its entire duration of sustained AN release was divided into three stages. This division was based on the quantitative analysis of the changes in relative slopes of AN release. Subsequently, the data were fitted with a typical model for each stage (Figure 7d). In the first stage, the fertilizer hydrogel exhibited the fastest AN release within 72 h of incubation. This release behavior was governed by the first-order model (k_1 = 0.068–0.0575). In this stage, A-CNF in the fertilizer hydrogel triggered the swelling procedure subsequent to incubation within the buffer or soil mediums. Consequently, a portion of AN in the hydrogel network undergoes the diffusion, which is influenced by the pH level. Notably, such diffusion predominantly happened from the swollen region surface to the inter-fiber space and the surrounding medium. During the incubation, the swelling process persists. The diameter of A-CNF progressively enlarges until the fusion occurs, obtaining a more compact matrix with less channels for AN release, which leads to the next stage of release mechanism. In the second stage, the fertilizer hydrogel exhibited a controlled release pattern characterized by the slowest AN release (144–504 h), following the zeroth-order mechanism. In this stage, the significant reduction of AN release rate (k_0 = 0.0023–0.0027) leads to the sustained presence of AN until the fertilizer hydrogel initiates its degradation. In the third stage, also called final release (720–1540 h), the polymeric network of hydrogel underwent degradation during its prolonged incubation, especially in soils. In this context, the A-CNFs predominantly diminished, and the polymeric matrix fragmented into oligomeric components. This transformation exposed larger pores, facilitating AN release, which was previously confined within the dense hydrogel matrix. Consequently, the Higuchi model governed this release stage, characterized by a new swifter release rate (k_H = 0.0281–0.0268). The inherent properties of soil notably decelerated the rate of AN release across all release stages and pH levels. Compared to the buffer medium, it resulted in a more consistent and stable AN release. Figure 7e directly depicts the greenhouse pot experiment involved in the fertilizer hydrogel in an experimental agricultural farm.

Shang et al. [85] developed a temperature-responsive SRFH. Notably, the pore structure can offer appropriate sites for loading urea. The release mechanism of urea from hydrogels has been investigated through its release kinetics. The diffusion index (n) of the Korsmeyer–Peppas model reveals that n falls within the range of 0.43 to 0.85 when the temperature is below the lower critical solution temperature (LCST, 25 °C). This range signifies that the urea release from hydrogel follows a non-Fickian diffusion pattern, indicating an anomalous diffusion mechanism. In this period, the hydrogel needs to absorb water from the soil, which will permeate the hydrogel chains to facilitate nutrient dissolution. Subsequently, the release process commences once the polymer chain achieves relaxation. Specifically, 80% cumulative release within 12 h can be ascribed to the swelling of hydrogel at lower temperatures. In this state, the molecular chains within the network become entangled, thereby impeding the diffusion and release of fertilizer through the pores. When the temperature is above the LCST, it follows that $n < 0.43$, indicating that the urea release follows Fickian diffusion, which is controlled by the concentration gradient. At this point, urea in the hydrogel rapidly diffuses across diverse concentration gradients within and outside the system. It leads to an impressive cumulative release up to 98% from the hydrogel within the initial 6 h. This phenomenon could stem from the contraction of molecular chains in the hydrogel network, and the disruption of interactions between the fertilizer and the hydrogel. These factors improve the release of urea from the hydrogel. Consequently, this hydrogel can effectively modulate the release of loaded urea through the temperature stimuli.

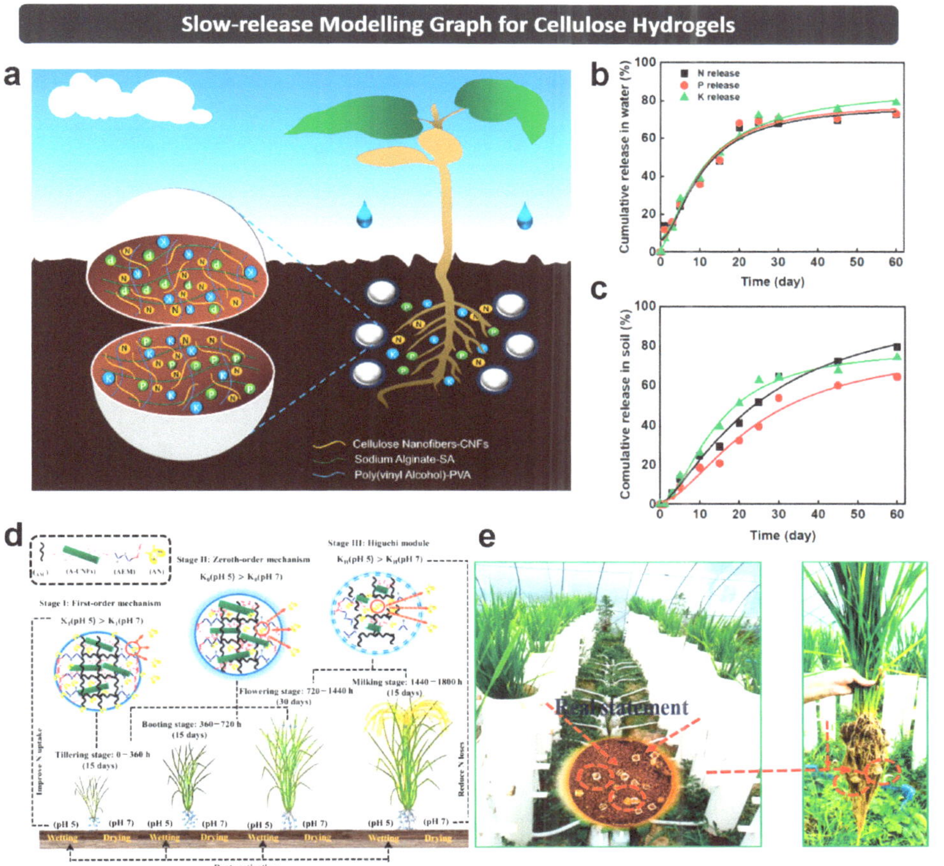

Figure 7. Fertilizer hydrogel: (**a**) schematic diagram, and its NPK release (**b**) in deionized water and (**c**) in soil. The loading rates of SA, CNF, PVA and NPK are 1.5, 0.5, 0.5 and 1.5 wt%, respectively. Reprinted with permission from ref. [28]. Copyright 2021 Elsevier Ltd. (**d**) The relationship between the different rice growth stages and the different AN release stages of fertilizer hydrogel, which are governed by first-order, zeroth-order and Higuchi mechanisms. (**e**) Fertilizer hydrogel in greenhouse pot. Reprinted with permission from ref. [3]. Copyright 2022 Elsevier Ltd.

Guo et al. [30] prepared a novel hydrogel MIL-100(Fe)@CNF-SA, which is involved in the fusion of MOFs with CNF and SA, acting as a carrier for urea. The urea-release mechanism from MIL-100(Fe)@CNF-SA was explored in depth, and the Higuchi model was selected as the most suitable model based on its high correlation coefficient. Specifically, due to its abundant carboxyl groups, the hydrogel after burying in soil starts adsorbing water and undergoes swelling gradually, leading to the expansions of the 3D network structure and volume of hydrogel. At this point, the free water enters the interior of the hydrogel. The urea molecule dissolves, and slowly diffuses out of the network pores through the dynamic exchange of free water. Then, the urea release can be slowed by the internal tortuous paths of hydrogels imparted by MIL-100(Fe), thus decreasing the release rate of fertilizer hydrogel. Finally, the extension of the hydrogel network slows down, which is accompanied by the saturation of hydrogel, leading to a reduction in urea release.

In summary, the slow-releasing capability of SRFHs offers outstanding water-retention performance, extending the efficacy of fertilizers, and decreasing the need for frequent applications. Furthermore, cellulose and its derivatives can contribute to the biodegradability of SRFHs, mitigating environmental concerns due to excessive fertilizer, and the promoting sustainable agricultural development.

5. Stimulus-Responsive Cellulose Hydrogels Serving as Slow-Release Fertilizers

Crops require different nutrients during various growth stages under different temperature conditions. Conventional hydrogels are insensitive to environmental conditions, and their swelling rate cannot change with external conditions [7,94]. Therefore, stimulus-responsive hydrogels emerged. These innovative hydrogels are able to detect subtle physicochemical changes in the external environment. After being triggered by external changes, such as pH, temperature, light, electricity, magnetism, force, etc., the stimulus-responsive hydrogels autonomously undergo stress responses [96]. Specifically, their structural properties, water absorption, solubility and other physicochemical traits will be in response to environmental variations.

Embedding fertilizer within stimulus-responsive cellulose hydrogels (so-called stimulus-responsive cellulose fertilizer hydrogels) can adjust the release behavior of fertilizer based on the fluctuations in surrounding soil. Moreover, it can deliver fertilizer precisely according to the plant's nutrient demands in the right quantity at the optimal time [3]. Stimulus-responsive cellulose hydrogels not only compensate for the limitations of traditional fertilizers (e.g., insensitivity), but also provide a practical strategy for modern agriculture. This review provided a brief summary of stimulus-responsive cellulose fertilizer hydrogels as shown in Table 2. Predictably, they would achieve the slow-release of fertilizer, reduce fertilizer pollution, and minimize water resource wastage.

Table 2. Nutrient loading species, combination ways, water-retention and holding ratio, and cumulative release of stimulus-responsive cellulose fertilizer hydrogels.

Stimuli	Nutrient	Combination Ways	Water-Retention and Holding Ratio	Swelling Capacity	Cumulative Release	Refs.
pH	Urea	In situ encapsulation	Water-holding ratio of 50% and complete loss of water after 30 d in soil.	Superior moisture content of 96.28%.	Only 50% of urea released in the 30th day.	[15]
	NPK		Water-retention ratio was above 80% at 60 d.	High equilibrium swelling capacity (60~70 g/g).	Cumulative release of NPK from hydrogels within 30 d was 67.90%, 70.78% and 71.12% in water; 64.52%, 53.72% and 64.08% in soil.	[28]
	N		-	-	AN release was 3.00 and 2.69 mg·day^{-1} at pH 5.5, while 0.92 and 0.55 mg·day^{-1} at pH 7.4 for 58 and 65 days in buffers and soil.	[3]
Temperature	Urea	In situ encapsulation	Water retention reached 50% in ~10 h.	Swelling ratio of hydrogels was 2056% at ~53 h.	When the temperature is lower than LCST, the cumulative release within 12 h was 80%, while it could reach 98% when the temperature was raised to 37 °C.	[85]
Temperature and pH	Urea	Solubilization and adsorption	Water retention increased to 77.53%, and remained at 22.78% at 30 d in soil.	Swelling capacity reached 37 g/g.	The cumulative release rates of MC-10% was 40.84% on 30 d.	[31]

5.1. pH-Responsive Cellulose Fertilizer Hydrogels

A pH-responsive hydrogel is a type of polymer which will experience a phase transition (e.g., conformation, water absorption, solubility, volume, mechanical property, etc.) according to the pH change in its surroundings. Typically, these hydrogels contain ionizable acidic or alkaline groups. Broadly, pH-responsive hydrogels can be categorized into anionic (alkaline), cationic (acidic), and amphoteric hydrogels [3]. When the pH in the vicinity of hydrogel reaches a certain threshold, its protonation of ionizable groups occurs, which will lead to the hydrogen bonding and the electrostatic interaction among the molecular chains. Meanwhile, the concentration and type of ionic species change inside and outside the cellulose hydrogel simultaneously, which will cause a variation in osmotic pressure. As a consequence, the hydrogel begins to swell or shrink, exhibiting a macroscopic change in volume. In the exploration of SRF, pH-responsive cellulose hydrogels have attracted particular attention, because soil pH variation is very common. Specifically, the soil is slightly acidic, while dry soil is in an alkaline state in the alternating wetting and drying (AWD) cycle. Furthermore, soil in different geographical regions exhibits various pH values. The pH-responsive cellulose hydrogels can undergo volume and release behavior shifts in response to the pH variation, suggesting their potential in the field of controlled fertilizer release [15].

Shaghaleh et al. [3] prepared a pH-responsive slow-release N fertilizer hydrogel (^{pH}RSRNFH) based on A-CNF and poly(acrylamide-co-aminoethyl methacrylate hydrochloride) (PAM-PAEM) through direct AN fertilizer encapsulation. The pH range of neutral soil under AWD irrigation ranged between acidic/neutral and alkaline conditions. After culturation in buffer and soil for 58 and 65 days, the AN release of ^{pH}RSRNFH was 3.00 and 2.69 mg·day^{-1} at a pH of 5.5, and 0.92 and 0.55 mg·day^{-1} at a pH of 7.4 (Figure 8b). The results showed that products had the greatest solubilization, water uptake and AN release at a low pH range of the surroundings. At this point, the plants have the most demand for N fertilizer. This is because the plant root system was activated by the increasing soil moisture during the wetting variation. In contrast, the lowest and slowest AN release was observed in dry soil in neutral and alkaline conditions, which is associated with the low nutrient demand of the plant root system. This mode of fertilizer application enables the necessary management of N fertilizer in accordance with plant demand. Therefore, the adoption of this innovative, pH-responsive and continuous AN delivery system certainly will contribute to sustainable development of agriculture, which offers a high-quality approach to N/water management, ensuring that the optimal N application rates are matched with suitable water-holding capacity and biodegradation rates.

The practical applications of cellulose hydrogels in agricultural scenarios and irrigation are intermittent. Therefore, we anticipate them to exhibit a responsive property to external stimuli, as well as repeatable shrinking and swelling capabilities. Wang et al. [15] designed the pH-responsive CNF/SA/MOF hydrogels (CAMs) with an impressive reusability. Specifically, the CAMs exhibited significant shrinkage in acidic conditions, which means the CAMs possess high absorption and substantial swelling properties in neutral and alkaline conditions. The results showed that the expansion of CAMs after loading urea (U-CAM) coincided with the gradual release of urea. Compared with other products, U-CAM demonstrated excellent swelling performance in an alkaline environment (Figure 8c). To explore its practical applications, the urea release from U-CAM was also investigated at pH = 3, 11. In Figure 8d, the cumulative release of CAM after loading 10% urea (U-CAM-10%) was the lowest, displaying a better slow-release performance. Moreover, its cumulative release was lower at a pH of 11 than at a pH of 3 (Figure 8e). The urea release of U-CAM-10% is satisfactory at a pH of 11, demonstrating that U-CAM is well suited for use in alkaline environments. As expected, U-CAM-10% exhibited the best performance in water adsorption even after three cycles of test at a pH of 11. In terms of reusability, CAM exhibited the ability to shrink in acidic environments, and swell in alkaline conditions, further proving its favorable pH-responsive behavior and repeatable shrinking and swelling capacities.

Overall, the enhanced performance of CAMs in alkaline environments further indicates their potential in semiarid and arid regions.

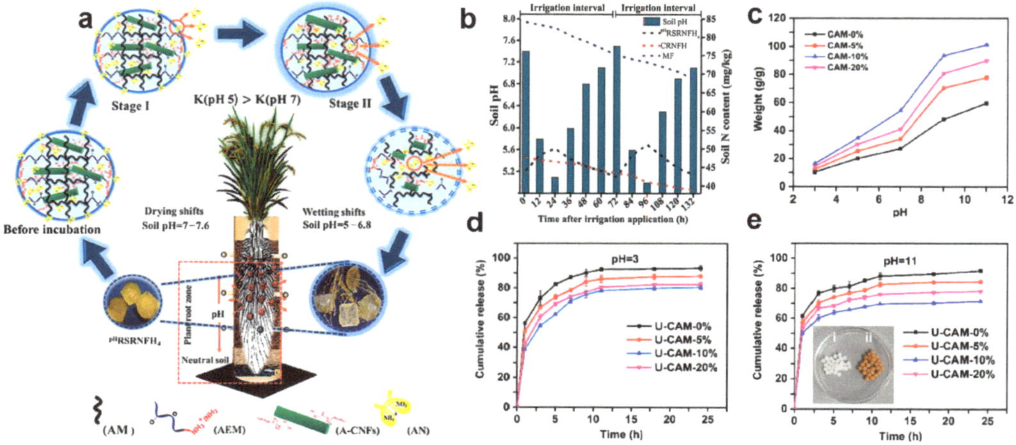

Figure 8. pH-responsive cellulose-based SRFH: (**a**) AN release of ^{PH}RSRNFH varied at different pH, (**b**) N content and pH variations of soil containing ^{PH}RSRNFH and other samples during ADW irrigation application. Reprinted with permission from ref. [3]. Copyright 2022 Elsevier Ltd. (**c**) CAMs display different solubilization capacities depending on pH variations. (**d**,**e**) Cumulative urea release of U-CAMs at a pH of 3 and 11. Reprinted with permission from ref. [15]. Copyright 2021 Elsevier Ltd.

5.2. *Temperature-Responsive Cellulose Fertilizer Hydrogels*

As a crucial and frequently encountered environmental factor, temperature can be readily controlled, and exerts a substantial influence on crop growth. Temperature-responsive cellulose hydrogels exhibit unique sensitivity to temperature variations such as volume, water absorption, swelling, light transmittance and structural properties [97]. They contain specific proportions of hydrophilic and hydrophobic groups, the hydrogen bonding of which can be influenced by the temperature variation, deciding the hydrophilicity or hydrophobicity of the product. The critical phase transition temperature refers to the temperature when a thermosensitive hydrogel shifts from shrinking to swelling or from swelling to shrinking [98]. Temperature-responsive hydrogels are typically classified into two categories: lower critical solution temperature (LCST) and upper critical solution temperature (UCST) types of hydrogels. LCST-type hydrogels generally possess both hydrophilic and hydrophobic groups [99]. For LCST-type hydrogels, when the temperature is higher than LCST, the interaction between hydrophilic and hydrophobic groups weakens, and the hydrophobic groups become dominant. This results in the structural instability of the hydrogel, leading to its absorption loss and shrinkage [100]. Conversely, when the temperature is lower than LCST, the hydration within the hydrogel intensifies, resulting in the swelling of hydrogel. This temperature-responsive behavior is also referred to as a negative temperature response [101]. For UCST-type hydrogels, when the temperature is lower than UCST, the hydrophobic groups dominate, leading to hydrogel shrinkage. When the temperature is higher than UCST, the hydrophilic groups become dominant, triggering the structure and performance variations, which results in the water absorption and expansion of the hydrogel, leading a macroscopic swelling appearance. Apparently, such temperature-responsive behavior is referred to as a positive temperature response [102].

Shang et al. [85] developed the temperature-responsive hydrogels by N-vinylcaprolactam (NVCL) aqueous dispersion through the emulsification of CMC and acrylamide (AM). Additionally, urea was loaded as a fertilizer into the hydrogels (Figure 9a–e). As the hydrophilic groups increase, the solubility of hydrogel at 20 °C can reach 2056% (Figure 9a,b). All hydrogels display a common pattern: elevated ESR at low temperature and reduced ESR at high temperature. The trend can be explained by the presence of temperature-sensitive PNVCL in hydrogel, which exhibits an extended state when the ambient temperature is below the LCST. When the hydrogel was heated above the LCST (25 °C), its PNVCL segment will undergo curling. This transition predominantly exposes the hydrophobic groups, and promotes the increasing hydrogen bonding within the polymer segments, hindering the transport of water molecules and resulting in the reduced ESR. In Figure 9c, the cumulative fertilizer release of the product below the LCST (over 12 h) reaches 80%. At temperatures of 37 °C, the hydrogel exhibited a more pronounced release behavior, reaching a cumulative release up to 98% within the initial 6 h (Figure 9c). A similar behavior was observed in the soil (Figure 9d). This proves that the urea release of hydrogel was effectively controlled through the temperature stimulation (Figure 9e).

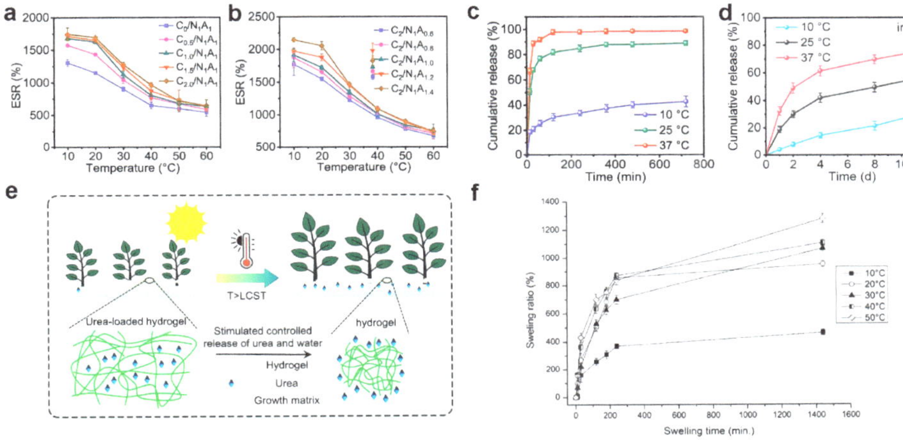

Figure 9. Temperature-responsive SRFH of CMC/poly(NVCL-co-AM): (**a**,**b**) ESR at different temperatures; cumulative urea release in (**c**) water and (**d**) soil; (**e**) temperature-response schematic. Reprinted with permission from ref. [85]. Copyright 2023 Elsevier Ltd. (**f**) Swelling ratio of hydrogel H5CA immersing in distilled water at different temperatures. Reprinted with permission from ref. [29]. Copyright 2021 MDPI Ltd.

Durpekova et al. [29] developed a novel hydrogel by crosslinking cellulose derivatives (HEC and CMC) with citric acid (CA, 15 wt%). The hydrogel H15CA that was integrated into the soil notably exhibited a higher water retention, which surpassed the original soil by nearly 30%. Compared with hydrogel H5CA (5 wt% CA), the urea and KNO_3 released more slowly from hydrogel H15CA. Figure 9f demonstrates the relationship between the temperature variation and the swelling capacity of hydrogel. At a higher temperature (50 °C), a significant increase in the swelling ratio was observed within the initial few hours, followed by a gradual rise. After exposure to the media for 24 h, it reached the equilibrium in swelling. This illustrates that the best swelling capacity appeared at the highest temperatures, which might be attributed to the thermal expansion of the hydrogel network and the disruption of hydrogen bonding between polymer molecules. In contrast, the least water uptake was observed when it was immersed in the distilled water at a low temperature. Under a low-temperature condition (10 °C), its water uptake speed in the swelling process was slow, the swelling ratio of which was stabilized after 3 h.

5.3. Salt-Responsive Cellulose Fertilizer Hydrogels

Saline soils are not suitable for growing crops, and some areas even cannot support any plant growth. In China, saline and alkaline land is widely distributed in the northeastern, northern and north-western regions, with a soil area of about 340,000 km^2, of which 124,000 km^2 can be used for agricultural production after improvement [103]. Songnen Plain is the main production area for grain and livestock in Northeast China [104,105]. However, the farmers in this region avoid irrigating their land because of the high salinity and mineralization of the local water. The main salts found in Songnen Plain are sodium carbonate (Na_2CO_3) and sodium bicarbonate ($NaHCO_3$) [106]. The population of Northeast China continues to grow, but the amount of land available for cultivation decreases dramatically. Factors such as the new crown pneumonia and the Russian–Ukrainian conflict increase the cost of food production, threatening the global food security seriously. Therefore, saline soils have been considered as a potentially valuable land resource [107]. In arid inland areas, over-irrigation not only wastes the water, but also causes fertilizer loss, soil salinization and waterlogging. Currently, the fertilizer and water loss are significant issues in agricultural research [108]. Fertilizer retention agents are of high cost, have a difficult degradation and decay by mildew, and have a complicated operation and poor water absorption. Therefore, it is an important goal to develop novel fertilizer adsorbents that are low cost, salt resistant, water absorbent and available under saline conditions in the future.

Qi et al. [106] synthesized the carboxymethyl cellulose fertilizer microspheres (CFM) with uniform pore structure, high porosity, good biodegradable capability and excellent fertilizer absorbency. Figure 10 illustrates the impact of saline solution on the water absorption of CFM. The absorption capacity of CFM varied based on the cations in saline solution, and the sequence from high to low is Na^+, Ca^{2+} and Fe^{3+} (Figure 10a). Then, the water absorption of CFM was investigated in the saline solution containing various anions. From Figure 10b, the water absorption of CFM in the solution containing the polyatomic monovalent cation (NH^{4+}) was lower than the monovalent cations (K^+, Na^+). This is because NH^{4+} may have reacted with –OH on the CFM chain to form ammonia, resulting in a denser internal structure of CFM and a decrease in water absorption capacity. The water absorbency of CFM in anionic solution following the order from highest to lowest is Cl^-, $SO_4{}^{2-}$ and $PO_4{}^{3-}$ (Figure 10c). Finally, the water absorption of CFM in NaCl, KCl, NH_4Cl, $CaCl_2$, $FeCl_3$, Na_2SO_4 and Na_3PO_4 solution was determined (Figure 10d–f). The water-absorption capacity of CFM decreases with the increasing salt concentration due to the osmotic pressure difference of the 3D network inside and outside the CFM. A smaller difference in osmotic pressure means that less solution enters the CFM. Since the salt concentration increases, the osmotic pressure difference decreases, the network structure of CFM shrinks, and its water-absorption capacity decreases [106]. The water absorptions of CFM in $NaHCO_3$ and Na_2CO_3 solution were 1532 $g{\cdot}g^{-1}$ and 1496 $g{\cdot}g^{-1}$, respectively. In conclusion, CFM has good salt tolerance and responsiveness, which enhances the soil quality and crop yields, offering the potential to improve the saline soils in Songnen Plain.

Li et al. [109] developed a novel hydrogel based on wheat straw cellulose (WSC) with IPNs for the slow release of nitrogen and phosphorus nutrients. They investigated the swelling kinetics and fertilizer-release behavior of product in various saline solutions. Its water absorbency in NaCl, KCl and Na_2SO_4 solutions increased as the soaking time extended, and reached a swelling equilibrium within 45 min. However, the water absorbency of product in $CaCl_2$ solution peaked initially, and then decreased to a specific value. Its swelling capacity in saline solution in order from largest to smallest is NaCl, KCl, Na_2SO_4 and $CaCl_2$ solution (10 mmol/L), and its water absorbency towards different cation influence in order from smallest to largest is Ca^{2+}, K^+ and Na^+. Notably, its total release amount and rate of fertilizer in saline solution were ranked from largest to smallest as NaCl, KCl and $CaCl_2$. Meanwhile, its fertilizer release towards different cation influence followed the order from smallest to largest is Ca^{2+}, K^+ and Na^+. The fertilizer release behavior of product is similar to its water-absorption behavior. These findings indicated that the swelling capacity of product was better under neutral condition. The cations and anions affecting the swelling capacity of product in order from largest to smallest is Na^+, K^+, Ca^{2+} and Cl^-. The sensitivity of product towards the saline solution suggests its potential application in saline soils.

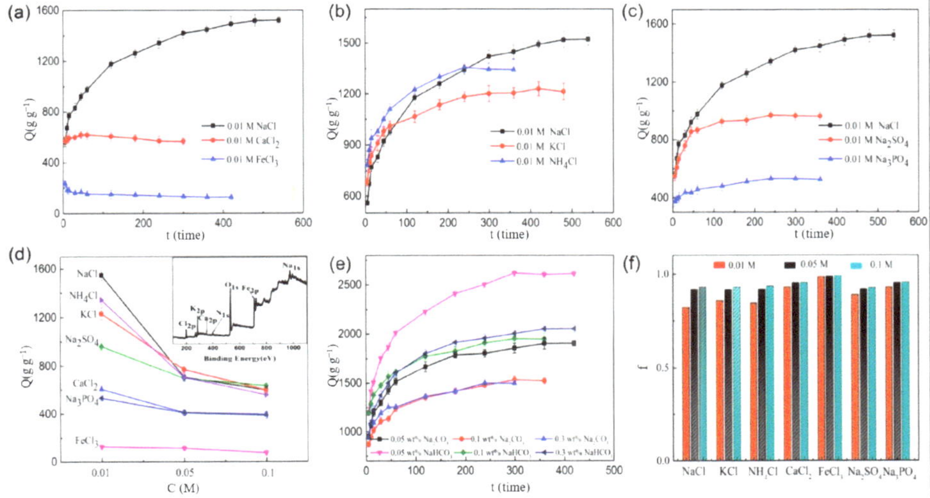

Figure 10. Salt-responsive cellulose SRFH of CFM: its swelling capacity in different saline solutions. (**a**) Cations with same valence; (**b**) cationic solutions; (**c**) anionic solutions; (**d**) equilibrium swelling under different salt concentrations, (Insert in (**d**) was component analysis); (**e**) equilibrium swelling in a Na_2CO_3 (or $NaHCO_3$) solution; (**f**) salt sensitivity factor of CFM for various salts. (*Q* represents adsorption capacity; *C* represents concentration of solution; *f* represents sensitivity factor). Reprinted with permission from ref. [106]. Copyright 2019 Elsevier Ltd.

5.4. Multi-Responsive Cellulose Fertilizer Hydrogels

During actual crop growth, cellulose hydrogels in the soil will encounter a variety of conditions. Multi-responsive cellulose hydrogel is a versatile material designed to react to a spectrum of stimuli such as temperature, pH, salt concentration, light, etc. Typically, such a hydrogel consists of a combination of polymers such as PVA and PAA, which possess multiple functional groups (e.g., photosensitive groups, acid and base groups) [110]. Lin et al. [31] developed an MC hydrogel through the synergistic combination of TEMPO-oxidized CNFs and NVCL via free-radical polymerization. Notably, NVCL and CNF exhibited temperature and pH responsiveness, respectively. Even on the 30th day, MC-10% hydrogel still maintained a notable water retention (22.78%) in the soil, consistently demonstrating the reproducible shrinking and swelling traits in diverse response scenarios.

Figure 11a,b vividly illustrates the ESR decline of MC hydrogel at the increasing temperature. It indicated that MC hydrogels absorbed water more easily at a temperature lower than LCST (25 °C). However, the higher temperature (~30 °C) favored crop growth [111]. Thanks to the remarkable temperature-responsive features of MC hydrogel, its fertilizer release rate was elevated at a warmer temperature, which guaranteed the ample supply of nutrients for sustained crop growth. Additionally, the MC hydrogel exhibited good pH responsiveness. As shown in Figure 11c,d, the swelling ratio of hydrogel escalated as the pH ranged from 3 to 11. Meanwhile, the fertilizer release rate of MC hydrogel was reduced in the alkaline environment, which enhanced the fertilizer utilization efficiency in the environment for crop growth, especially at a pH of approximately 7.5. After the application of MC hydrogel in practice, the tillers and leaves increased, and the photosynthetic rate of wheat improved (Figure 11e). In summary, this temperature- and pH-responsive cellulose hydrogel contributed to crop growth, enhancing its good availability, serving as the SRF.

Since the global population is increasing and environmental pollution worsens, the area of arable land is diminishing. Moreover, factors including water use from rivers and lakes, underground irrigation, and fertilizer application contribute to the excessive accumulation of salt ions in the soil, resulting in soil salinization. This process further reduces the available agricultural land, and directly jeopardizes food security and crop production. Consequently, it is of immense significance to strategically develop a multi-responsive cellulose SRFH that possesses excellent biodegradability, water-retention capability, salt tolerance, and adaptation to the intricate and ever-changing environments in arid and semiarid regions. This endeavor aligns with the goals of the national agricultural strategy, which will certainly mitigate the issues for farming in these challenging regions.

Idrissi et al. [112] developed an innovative nanocomposite hydrogel with pH- and salt-responsive properties, which was achieved by conducting an in situ free-radical copolymerization involving SA, acrylic acid (AA) and AM in aqueous medium by utilizing N,N′-methylene bis-acrylamide (MBA) as a crosslinker and citric acid-functionalized cellulose nanocrystal (C-CNC) as nano-filler. As illustrated in Figure 11f, the hydrogels and their nanocomposites after loading urea displayed a minimal dissolution at a pH of 2, reached their maximum solubility at a pH of 6, and gradually decreased in their solubility with a further increase in pH (pH > 9). In addition, nanocomposite hydrogels were immersed in various saline media, including NaCl, $CaCl_2$ and $FeCl_3$. As depicted in Figure 11g, it is evident that the swelling phenomenon gradually diminishes with the increasing saline solution concentration. Furthermore, the absorption capacity of saline solution decreased with the increasing cation charge, following the order of $Na^+ > Ca^{2+} > Fe^{3+}$. In Figure 11h, it can be observed that the nanocomposite hydrogel with C-CNC exhibited a better reswelling capability than the hydrogel without C-CNC. The water retention experiment result showed that the soil containing Hyd/C-CNC could retain more water for a month (Figure 11i). The product prepared by the formulation containing 25 wt% of urea could release 86.37 ± 1.86% within 15 days (Figure 11j). The as-obtained nitrogen-rich superabsorbent exhibited enhanced water absorbency (412 ± 4 g/g), while also possessing the characteristic of slow-release nitrogen (Figure 11k). This finding suggests that the application of such a superabsorbent could alleviate the drought stress in arid and semiarid regions by improving the moisture retention of soil.

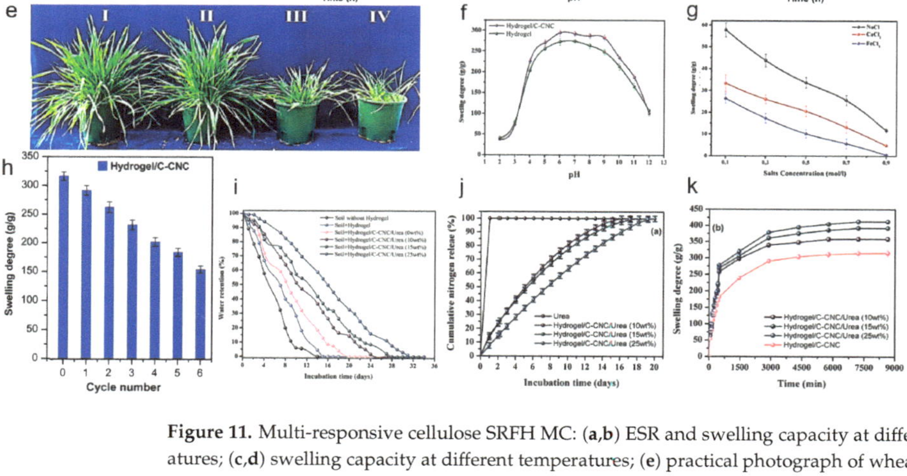

Figure 11. Multi-responsive cellulose SRFH MC: (**a**,**b**) ESR and swelling capacity at different temperatures; (**c**,**d**) swelling capacity at different temperatures; (**e**) practical photograph of wheat containing

(I) MC-10%, (II) MC-0%, (III) pure urea and (IV) nothing. Reprinted with permission from ref. [111]. Copyright 2021 Elsevier Ltd. (**f**) pH effect on the swelling degree of hydrogel samples. (**g**) Water absorbency of hydrogel in saline solution. (**h**) Swelling property of hydrogel during consecutive swelling/drying cycles. (**i–k**) Water retention, cumulative nitrogen release and swelling degree of different hydrogels. Reprinted with permission from ref. [112]. Copyright 2022 Elsevier Ltd.

6. Summary and Prospect

Agriculture is the cornerstone of the national economy. An inappropriate or excessive use of traditional fertilizers leads to low nutrient utilization in crops, reducing crop yields, and increasing fertilization costs. In addition, nutrient leaching, denitrification, surface runoff and volatilization can cause adverse impacts to the environment, such as water and air pollution. It is particularly serious in the arid and semiarid regions with minimal precipitation, unobvious seasonal distribution, and low crop yields. The synergy effect between hydrogel and fertilizer enables the regulation of water and nutrients within a system. Research on environmentally friendly and efficient intelligent fertilizers is an essential prerequisite for the advancement of modern agriculture. This paper provides a comprehensive review of recent research progress concerning stimulus-responsive cellulose hydrogels in the applications of slow-release fertilizer. Although many significant advancements have been made, some challenges still demand further exploration and resolution.

(1) Food security serves as the fundamental pillar of national security. Although plants are the main source of cellulose, comprehensive assessments are still necessary to evaluate the toxicity of as-prepared cellulose hydrogels for SRF application. Additionally, it is imperative to develop eco-friendly organic solvents and crosslinking agents to ensure the nontoxicity of cellulose hydrogels throughout their entire life cycle.

(2) The biocompatibility and biodegradability of cellulose hydrogels are the key merits for SRF development. Generally, the biodegradation of cellulose in soil was once deemed advantageous, while it was also a potential drawback to a certain extent (e.g., susceptibility to microbes and enzymes), which presents a significant challenge and may not necessarily bring the anticipated benefits.

(3) The influences of external conditions on soil property, plant growth, and nutrient-release performance are still unclear. Despite the promising applications in SRF, the as-prepared stimulus-responsive cellulose hydrogels after loading fertilizers cannot control their nutrient release fully. This limitation hinders the synchronization of plant nutrient demands with the fertilizer release cycles. Therefore, the release mechanisms and rules of intelligent cellulose fertilizer hydrogels still need deep investigation.

(4) Particular emphasis should be placed on controlling hydrogel structure and composition, regulating release quantities, developing stimulus-responsive systems, and comprehending the impact of environmental sensitivity on swelling capacity and release kinetics. These efforts are necessary to cater to the diverse requirements for crop growth in varying environments, which will define the focal points and directions of future research.

(5) A combination method of cellulose hydrogel and fertilizer is of paramount importance for the slow-release performance of SRF. Many products prioritize the envelope or coating, but often neglect their combination. The solubilization and adsorption methods for loading nutrients as much as possible still remains limited, resulting in suboptimal fertilizer slow release. It is imperative to establish a comprehensive evaluation standard for SRF to provide robust guidance for industry advancement.

In conclusion, significant progress has been achieved in stimulus-responsive cellulose SRFHs, while it is crucial to transit these advancements into practical production technology for the agricultural field. Future endeavors should involve the utilization of advanced instruments, materials and technologies to enhance the preparation process of intelligent cellulose fertilizer hydrogels. Moreover, emphasis should be placed on recycling and maximizing the value of biomass resources, which indicates the immense potential of stimulus-responsive cellulose SRFHs for sustainable agricultural development and ecological environmental protection.

Author Contributions: Z.L.: conceptualization, visualization, writing—original draft preparation; M.Z.: writing—review and editing, funding acquisition, project administration. All authors have read and agreed to the published version of the manuscript.

Funding: The research was supported by the Natural Science Foundation of Jilin Province (YDZJ202201ZYTS441), National Natural Science Foundation of China (32130073), Jilin Province Development and Reform Commission Industrial Innovation Special Fund Project (2023C038-2), Science and Technology Innovation Development Plan of Jilin City (20200104083), State Key Laboratory of Bio-Fibers and Eco-Textiles (Qingdao University) (KFKT202213) and Fundamental Research Funds for Wood Material Science and Engineering Key Laboratory of Jilin Province.

Institutional Review Board Statement: This manuscript does not contain any studies involving human or animal subjects.

Data Availability Statement: No new data were created or analyzed in this study.

Conflicts of Interest: The authors declare no conflict of interest.

References

1. Chen, Y.; Li, W.; Zhang, S. A multifunctional eco-friendly fertilizer used keratin-based superabsorbent as coatings for slow-release urea and remediation of contaminated soil. *Prog. Org. Coat.* **2021**, *154*, 106158.
2. Yang, S.; Wang, L.S.; Akhtar, K.; Ahmad, I.; Khan, A. Optimizing Nitrogen Fertilization and Variety for Millet Grain Yield and Biomass Accumulation in Dry Regions. *Agronomy* **2022**, *12*, 2116.
3. Shaghaleh, H.; Hamoud, Y.A.; Xu, X.; Wang, S.; Liu, H. A pH-responsive/sustained release nitrogen fertilizer hydrogel based on aminated cellulose nanofiber/cationic copolymer for application in irrigated neutral soils. *J. Clean. Prod.* **2022**, *368*, 133098.
4. Sanchez-Zarco, X.G.; Mora-Jacobo, E.G.; Gonzalez-Bravo, R.; Mahlknecht, J.; Ponce-Ortega, J.M. Water, energy, and food security assessment in regions with semiarid climates. *Clean Technol. Environ. Policy* **2020**, *22*, 2145–2161.
5. Ferraz-Almeida, R.; Albuquerque, C.J.B.; Camargo, R.; Lemes, E.M.; de Faria, R.S.; Lana, R.M.Q. Sorghum-Grass Intercropping Systems under Varying Planting Densities in a Semi-Arid Region: Focusing on Soil Carbon and Grain Yield in the Conservation Systems. *Agriculture* **2022**, *12*, 1762.
6. Li, M.; Fu, Q.; Singh, V.P.; Liu, D.; Li, T.; Li, J. Sustainable management of land, water, and fertilizer for rice production considering footprint family assessment in a random environment. *J. Clean. Prod.* **2020**, *258*, 120785.
7. Pourjavadi, A.; Heydarpour, R.; Tehrani, Z.M. Multi-stimuli-responsive hydrogels and their medical applications. *New J. Chem.* **2021**, *45*, 15705–15717.
8. Li, T.; Lv, S.-Y.; Chen, J.; Gao, C.-M.; Zhang, S.-F.; Liu, M.-Z. Progress in Polymer-based Environment-responsive Fertilizers. *Acta Polym. Sin.* **2018**, *3*, 336–348.
9. Wang, L.; Zhao, W.; Zhao, Y.N.; Li, W.; Wang, G.D.; Zhang, Q. Enzymatically-mineralized double-network hydrogels with ultrahigh mechanical strength, toughness, and stiffness. *Theranostics* **2023**, *13*, 673–684.
10. Kassem, I.; Ablouh, E.-H.; El Bouchtaoui, F.-Z.; Hannache, H.; Ghalfi, H.; Sehaqui, H.; El Achaby, M. Cellulose Nanofibers/Engineered Biochar Hybrid Materials as Biodegradable Coating for Slow-Release Phosphate Fertilizers. *ACS Sustain. Chem. Eng.* **2022**, *10*, 15250–15262.
11. Zhang, H.; Liang, H.; Xing, L.; Ding, W.; Geng, Z.; Xu, C. Cellulose-based slow-release nitrogen fertilizers: Synthesis, properties, and effects on pakchoi growth. *Int. J. Biol. Macromol.* **2023**, *244*, 125413. [CrossRef] [PubMed]
12. Lu, J.; Cheng, M.; Zhao, C.; Li, B.; Peng, H.; Zhang, Y.; Shao, Q.; Hassan, M. Application of lignin in preparation of slow-release fertilizer: Current status and future perspectives. *Ind. Crops Prod.* **2022**, *176*, 114267. [CrossRef]
13. Qiao, D.; Liu, H.; Yu, L.; Bao, X.; Simon, G.P.; Petinakis, E.; Chen, L. Preparation and characterization of slow-release fertilizer encapsulated by starch-based superabsorbent polymer. *Carbohydr. Polym.* **2016**, *147*, 146–154. [CrossRef]
14. Yuan, W.J.; Li, S.H.; Guan, H.H.; Zhang, S.; Zhang, Y.X.; Zhang, M.; Yu, Y.; Chen, X.Y. Preparation and Properties of a Novel Biodegradable Composite Hydrogel Derived from Gelatin/Chitosan and Polylactic Acid as Slow-Release N Fertilizer. *Polymers* **2023**, *15*, 997. [CrossRef]
15. Wang, Y.; Shaghaleh, H.; Hamoud, Y.A.; Zhang, S.; Li, P.; Xu, X.; Liu, H. Synthesis of a pH-responsive nano-cellulose/sodium alginate/MOFs hydrogel and its application in the regulation of water and N-fertilizer. *Int. J. Biol. Macromol.* **2021**, *187*, 262–271. [CrossRef] [PubMed]
16. Bhaladhare, S.; Das, D. Cellulose: A fascinating biopolymer for hydrogel synthesis. *J. Mater. Chem. B* **2022**, *10*, 1923–1945. [CrossRef]
17. Kassem, I.; Ablouh, E.-H.; El Bouchtaoui, F.-Z.; Kassab, Z.; Hannache, H.; Sehaqui, H.; El Achaby, M. Biodegradable all-cellulose composite hydrogel as eco-friendly and efficient coating material for slow-release MAP fertilizer. *Prog. Org. Coat.* **2022**, *162*, 106575. [CrossRef]

18. Du, H.; Liu, W.; Zhang, M.; Si, C.; Zhang, X.; Li, B. Cellulose nanocrystals and cellulose nanofibrils based hydrogels for biomedical applications. *Carbohydr. Polym.* **2019**, *209*, 130–144. [CrossRef]
19. Hu, D.-N.; Sun, Y.-F.; Tao, L.; Yuan, J.-Y.; Sui, X.-F.; Wei, Y. Environmentally Responsive Hydrogels Based on Cellulose. *Acta Polym. Sin.* **2020**, *51*, 880–889.
20. Shao, C.; Yang, J. Dynamics in Cellulose-Based Hydrogels with Reversible Cross-Links. *Self-Heal. Self-Recover. Hydrogels* **2020**, *285*, 319–354.
21. Chen, C.; Xi, Y.; Weng, Y. Recent Advances in Cellulose-Based Hydrogels for Tissue Engineering Applications. *Polymers* **2022**, *14*, 3335. [CrossRef]
22. Tofanica, B.-M.; Belosinschi, D.; Volf, I. Gels, Aerogels and Hydrogels: A Challenge for the Cellulose-Based Product Industries. *Gels* **2022**, *8*, 497. [CrossRef]
23. Wang, B.; Peng, Q.; Yan, Y.; Ding, Y.; Wang, Z. Biomimetic, strong, and tough hydrogels by integrating cellulose nanocrystals into polymer networks. *Ind. Crops Prod.* **2020**, *158*, 112973. [CrossRef]
24. Xie, J.; Yang, Y.; Gao, B.; Wang, Y.; Li, Y.C.; Cheng, D.; Xiao, T.; Li, K.; Fu, Y.; Xu, J.; et al. Magnetic-Sensitive Nanoparticle Self-Assembled Superhydrophobic Biopolymer-Coated Slow-Release Fertilizer: Fabrication, Enhanced Performance, and Mechanism. *ACS Nano* **2019**, *13*, 3320–3333. [CrossRef] [PubMed]
25. Tanpichai, S.; Phoothong, F.; Boonmahitthisud, A. Superabsorbent cellulose-based hydrogels cross-liked with borax. *Sci. Rep.* **2022**, *12*, 8920. [CrossRef]
26. Liu, Y.; Wang, J.; Chen, H.; Cheng, D. Environmentally friendly hydrogel: A review of classification, preparation and application in agriculture. *Sci. Total Environ.* **2022**, *846*, 157303. [CrossRef]
27. Hou, Y.; Deng, B.; Wang, S.; Ma, Y.; Long, X.; Wang, F.; Qin, C.; Liang, C.; Yao, S. High-Strength, High-Water-Retention Hemicellulose-Based Hydrogel and Its Application in Urea Slow Release. *Int. J. Mol. Sci.* **2023**, *24*, 9208. [CrossRef] [PubMed]
28. Liu, S.; Wu, Q.; Sun, X.; Yue, Y.; Tubana, B.; Yang, R.; Cheng, H.N. Novel alginate-cellulose nanofiber-poly(vinyl alcohol) hydrogels for carrying and delivering nitrogen, phosphorus and potassium chemicals. *Int. J. Biol. Macromol.* **2021**, *172*, 330–340. [CrossRef] [PubMed]
29. Durpekova, S.; Di Martino, A.; Dusankova, M.; Drohsler, P.; Sedlarik, V. Biopolymer Hydrogel Based on Acid Whey and Cellulose Derivatives for Enhancement Water Retention Capacity of Soil and Slow Release of Fertilizers. *Polymers* **2021**, *13*, 3274. [CrossRef] [PubMed]
30. Guo, L.; Wang, Y.; Wang, M.; Shaghaleh, H.; Hamoud, Y.A.; Xu, X.; Liu, H. Synthesis of bio-based MIL-100(Fe)@CNF-SA composite hydrogel and its application in slow-release N-fertilizer. *J. Clean. Prod.* **2021**, *324*, 129274. [CrossRef]
31. Lin, X.; Guo, L.; Shaghaleh, H.; Hamoud, Y.A.; Xu, X.; Liu, H. A TEMPO-oxidized cellulose nanofibers/MOFs hydrogel with temperature and pH responsiveness for fertilizers slow-release. *Int. J. Biol. Macromol.* **2021**, *191*, 483–491. [CrossRef]
32. Liu, Y.; Ahmed, S.; Sameen, D.E.; Wang, Y.; Lu, R.; Dai, J.; Li, S.; Qin, W. A review of cellulose and its derivatives in biopolymer-based for food packaging application. *Trends Food Sci. Technol.* **2021**, *112*, 532–546. [CrossRef]
33. Song, S.; Li, H.; Liu, P.; Peng, X. Applications of cellulose-based composites and their derivatives for microwave absorption and electromagnetic shielding. *Carbohydr. Polym.* **2022**, *287*, 119347. [CrossRef] [PubMed]
34. Cheng, W.; Zhu, Y.; Jiang, G.; Cao, K.; Zeng, S.; Chen, W.; Zhao, D.; Yu, H. Sustainable cellulose and its derivatives for promising biomedical applications. *Prog. Mater. Sci.* **2023**, *138*, 101152. [CrossRef]
35. Yang, J.; An, X.; Liu, L.; Tang, S.; Cao, H.; Xu, Q.; Liu, H. Cellulose, hemicellulose, lignin, and their derivatives as multi-components of bio-based feedstocks for 3D printing. *Carbohydr. Polym.* **2020**, *250*, 116881. [CrossRef] [PubMed]
36. Li, Y.; Zhou, Y.; Muhammad, Y.; Zhou, J.; Guo, Z.; Tan, H.; Guo, S. Nanocellulose and Its Derivatives toward Advanced Lithium Sulfur Batteries. *ACS Mater. Lett.* **2021**, *3*, 1130–1142. [CrossRef]
37. Han, X.; Chen, L.; Yanilmaz, M.; Lu, X.; Yang, K.; Hu, K.; Liu, Y.; Zhang, X. From nature, requite to nature: Bio-based cellulose and its derivatives for construction of green zinc batteries. *Chem. Eng. J.* **2023**, *454*, 140311. [CrossRef]
38. Habibi, Y.; Lucia, L.A.; Rojas, O.J. Cellulose Nanocrystals: Chemistry, Self-Assembly, and Applications. *Chem. Rev.* **2010**, *110*, 3479–3500. [CrossRef]
39. Bao, Y.; He, J.; Song, K.; Guo, J.; Zhou, X.; Liu, S. Functionalization and Antibacterial Applications of Cellulose-Based Composite Hydrogels. *Polymers* **2022**, *14*, 769. [CrossRef]
40. Pandey, A. Pharmaceutical and biomedical applications of cellulose nanofibers: A review. *Environ. Chem. Lett.* **2021**, *19*, 2043–2055. [CrossRef]
41. Olad, A.; Doustdar, F.; Gharekhani, H. Fabrication and characterization of a starch-based superabsorbent hydrogel composite reinforced with cellulose nanocrystals from potato peel waste. *Colloids Surf. A Physicochem. Eng. Asp.* **2020**, *601*, 124962. [CrossRef]
42. Zhang, M.; Zhou, N.; Shi, Y.; Ma, Y.; An, C.; Li, J.; Jiang, W.; Wang, C.; Shi, J. Construction of Antibacterial, Superwetting, Catalytic, and Porous Lignocellulosic Nanofibril Composites for Wastewater Purification. *Adv. Mater. Interfaces* **2022**, *9*, 2201388. [CrossRef]
43. Padhi, S.; Singh, A.; Routray, W. Nanocellulose from agro-waste: A comprehensive review of extraction methods and applications. *Rev. Environ. Sci. Bio/Technol.* **2023**, *22*, 1–27. [CrossRef]
44. Zeng, G.; Zhang, L.; Qi, B.; Luo, J.; Wan, Y. Cellulose esterification with carboxylic acid in deep eutectic solvent pretreatment inhibits enzymatic hydrolysis. *Bioresour. Technol.* **2023**, *380*, 129085. [CrossRef]

45. Ye, X.; Peng, H.; Liu, X.; Xiong, H.; Wang, N.; Yang, F.; Kong, Y.; Yang, Z.; Lei, Z. Preparation and fertilizer retention/anti-leakage performances of superabsorbent composite based on hydroxypropyl methyl cellulose. *Carbohydr. Polym.* **2021**, *274*, 118636. [CrossRef] [PubMed]
46. Makhliyo, K.; Abdumutolib, A.; Sirojiddin, S.; Nurbek, A.; Khaydar, Y.; Jiang, G. Preparation of oxidized nanocellulose by using potassium dichromate. *Cellulose* **2023**, *30*, 5657–5668. [CrossRef]
47. Chen, Y.-C.; Chen, Y.-H. Thermo and pH-responsive methylcellulose and hydroxypropyl methylcellulose hydrogels containing K_2SO_4 for water retention and a controlled-release water-soluble fertilizer. *Sci. Total Environ.* **2019**, *655*, 958–967. [CrossRef]
48. Padmanabhan, S.K.; Lamanna, L.; Friuli, M.; Sannino, A.; Demitri, C.; Licciulli, A. Carboxymethylcellulose-Based Hydrogel Obtained from Bacterial Cellulose. *Molecules* **2023**, *28*, 829. [CrossRef]
49. Kudzin, M.H.; Mrozinska, Z.; Festinger, N.; Urbaniak, P.; Drabowicz, J. Stability of cellulose phosphates in alkalic solutions. *Phosphorus Sulfur Silicon Relat. Elem.* **2022**, *197*, 630–633. [CrossRef]
50. Georgouvelas, D.; Abdelhamid, H.N.; Li, J.; Edlund, U.; Mathew, A.P. All-cellulose functional membranes for water treatment: Adsorption of metal ions and catalytic decolorization of dyes. *Carbohydr. Polym.* **2021**, *264*, 118044. [CrossRef]
51. Zhang, M.; Shi, Y.; Wang, R.; Chen, K.; Zhou, N.; Yang, Q.; Shi, J. Triple-functional lignocellulose/chitosan/Ag@TiO_2 nanocomposite membrane for simultaneous sterilization, oil/water emulsion separation, and organic pollutant removal. *J. Environ. Chem. Eng.* **2021**, *9*, 106728. [CrossRef]
52. Mokhtari, A.; Sabzi, M.; Azimi, H. 3D porous bioadsorbents based on chitosan/alginate/cellulose nanofibers as efficient and recyclable adsorbents of anionic dye. *Carbohydr. Polym.* **2021**, *265*, 118075. [CrossRef]
53. El-Sherbiny, I.M.; Yacoub, M.H. Hydrogel scaffolds for tissue engineering: Progress and challenges. *Glob. Cardiol. Sci. Pract.* **2013**, *2013*, 316–342. [CrossRef]
54. Zhang, M.; Xu, S.; Wang, R.; Che, Y.; Han, C.; Feng, W.; Wang, C.; Zhao, W. Electrospun nanofiber/hydrogel composite materials and their tissue engineering applications. *J. Mater. Sci. Technol.* **2023**, *162*, 157–178. [CrossRef]
55. Lavrador, P.; Esteves, M.R.; Gaspar, V.M.; Mano, J.F. Stimuli-Responsive Nanocomposite Hydrogels for Biomedical Applications. *Adv. Funct. Mater.* **2021**, *31*, 2005941. [CrossRef]
56. Kaur, P.; Agrawal, R.; Pfeffer, F.M.; Williams, R.; Bohidar, H.B. Hydrogels in Agriculture: Prospects and Challenges. *J. Polym. Environ.* **2023**, *31*, 3701–3718. [CrossRef]
57. Bachra, Y.; Grouli, A.; Damiri, F.; Zhu, X.X.; Talbi, M.; Berrada, M. Synthesis, Characterization, and Swelling Properties of a New Highly Absorbent Hydrogel Based on Carboxymethyl Guar Gum Reinforced with Bentonite and Silica Particles for Disposable Hygiene Products. *ACS Omega* **2022**, *7*, 39002–39018. [CrossRef] [PubMed]
58. Xiang, X.; He, Q.; Xia, S.; Deng, Z.; Zhang, H.; Li, H. Study of capacitance type flexible electronic devices based on polyacrylamide and reduced graphene oxide composite hydrogel. *Eur. Polym. J.* **2022**, *171*, 111200. [CrossRef]
59. Pita-López, M.L.; Fletes-Vargas, G.; Espinosa-Andrews, H.; Rodríguez-Rodríguez, R. Physically cross-linked chitosan-based hydrogels for tissue engineering applications: A state-of-the-art review. *Eur. Polym. J.* **2021**, *145*, 110176. [CrossRef]
60. Hu, W.K.; Wang, Z.J.; Xiao, Y.; Zhang, S.M.; Wang, J.L. Advances in crosslinking strategies of biomedical hydrogels. *Biomater. Sci.* **2019**, *7*, 843–855. [CrossRef]
61. Sun, W.; Hu, Y.; Cheng, Y.; Yang, S.; Kang, Z. Effect of cross-linking methods on stress relaxation of PVA/PAM-co-PAA-based hydrogels. *Int. J. Polym. Anal. Charact.* **2021**, *26*, 330–341. [CrossRef]
62. Foudazi, R.; Zowada, R.; Manas-Zloczower, I.; Feke, D.L. Porous Hydrogels: Present Challenges and Future Opportunities. *Langmuir* **2023**, *39*, 2092–2111. [CrossRef] [PubMed]
63. Zhu, C.; Bettinger, C.J. Photoreconfigurable Physically Cross-Linked Triblock Copolymer Hydrogels: Photodisintegration Kinetics and Structure-Property Relationships. *Macromolecules* **2015**, *48*, 1563–1572. [CrossRef]
64. Shitrit, Y.; Bianco-Peled, H. Insights into the formation mechanisms and properties of pectin hydrogel physically cross-linked with chitosan nanogels. *Carbohydr. Polym.* **2021**, *269*, 118274. [CrossRef]
65. Ramli, R.A. Slow release fertilizer hydrogels: A review. *Polym. Chem.* **2019**, *10*, 6073–6090. [CrossRef]
66. Wu, C.; Dan, Y.; Tian, D.; Zheng, Y.; Wei, S.; Xiang, D. Facile fabrication of MOF(Fe)@alginate aerogel and its application for a high-performance slow-release N-fertilizer. *Int. J. Biol. Macromol.* **2020**, *145*, 1073–1079. [CrossRef]
67. Szafulera, K.; Wach, R.A.; Olejnik, A.K.; Rosiak, J.M.; Ulanski, P. Radiation synthesis of biocompatible hydrogels of dextran methacrylate. *Radiat. Phys. Chem.* **2018**, *142*, 115–120. [CrossRef]
68. Yang, M.; Guo, W.; Liu, S.; Zhang, B.; Chen, Y.; Wang, Y. Highly stretchable gamma-irradiated poly (vinyl alcohol)/Tannic acid composite hydrogels with superior transparency and antibacterial activity. *J. Polym. Res.* **2021**, *28*, 412. [CrossRef]
69. Chansoria, P.; Asif, S.; Polkoff, K.; Chung, J.; Piedrahita, J.A.; Shirwaiker, R.A. Characterizing the Effects of Synergistic Thermal and Photo-Cross-Linking during Biofabrication on the Structural and Functional Properties of Gelatin Methacryloyl (GelMA) Hydrogels. *ACS Biomater. Sci. Eng.* **2021**, *7*, 5175–5188. [CrossRef]
70. Manaila, E.; Demeter, M.; Calina, I.C.; Craciun, G. NaAlg-g-AA Hydrogels: Candidates in Sustainable Agriculture Applications. *Gels* **2023**, *9*, 316. [CrossRef] [PubMed]
71. Yan, R.; Yan, Y.; Gong, J.; Ma, J. Superb tough hydrogels based on macromolecule micelles as cross-linkers. *Mater. Res. Express* **2019**, *6*, 055322. [CrossRef]
72. Xu, S.J.; Zhang, P.; Ma, W.Z.; Yang, H.C.; Cao, Z.; Gong, F.H.; Zhong, J. High Water Resistance Polyvinyl Alcohol Hydrogel Film Prepared by Melting Process Combining with Citric Acid Cross-Linking. *Polym. Sci. Ser. B* **2022**, *64*, 198–208.

73. Zhang, M.; Ren, X.; Duan, L.; Gao, G. Joint double-network hydrogels with excellent mechanical performance. *Polymer* **2018**, *153*, 607–615. [CrossRef]
74. Sapula, P.; Bialik-Was, K.; Malarz, K. Are Natural Compounds a Promising Alternative to Synthetic Cross-Linking Agents in the Preparation of Hydrogels? *Pharmaceutics* **2023**, *15*, 253. [CrossRef]
75. Demeter, M.; Calina, I.; Vancea, C.; Sen, M.; Kaya, M.G.A.; Manaila, E.; Dumitru, M.; Meltzer, V. E-Beam Processing of Collagen-Poly(N-vinyl-2-pyrrolidone) Double-Network Superabsorbent Hydrogels: Structural and Rheological Investigations. *Macromol. Res.* **2019**, *27*, 255–267. [CrossRef]
76. Shen, Y.; Wang, H.; Liu, Z.; Li, W.; Liu, Y.; Li, J.; Wei, H.; Han, H. Fabrication of a water-retaining, slow-release fertilizer based on nanocomposite double-network hydrogels via ion-crosslinking and free radical polymerization. *J. Ind. Eng. Chem.* **2021**, *93*, 375–382. [CrossRef]
77. Wei, P.; Yu, X.; Fang, Y.; Wang, L.; Zhang, H.; Zhu, C.; Cai, J. Strong and Tough Cellulose Hydrogels via Solution Annealing and Dual Cross-Linking. *Small* **2023**, *19*, e2301204. [CrossRef] [PubMed]
78. Lee, J.S.; Park, H.S.; Kim, Y.J.; Kim, J.-H. Hybrid double-network hydrogel based on poly(aspartic acid) and poly(acryl amide) with improved mechanical properties. *J. Appl. Polym. Sci.* **2018**, *135*, 45925. [CrossRef]
79. Xu, X.; Jerca, V.V.; Hoogenboom, R. Bioinspired double network hydrogels: From covalent double network hydrogels via hybrid double network hydrogels to physical double network hydrogels. *Mater. Horiz.* **2021**, *8*, 1173–1188. [CrossRef]
80. Nath, P.C.; Debnath, S.; Sharma, M.; Sridhar, K.; Nayak, P.K.; Inbaraj, B.S. Recent Advances in Cellulose-Based Hydrogels: Food Applications. *Foods* **2023**, *12*, 350. [CrossRef]
81. Sorze, A.; Valentini, F.; Dorigato, A.; Pegoretti, A. Development of a Xanthan Gum Based Superabsorbent and Water Retaining Composites for Agricultural and Forestry Applications. *Molecules* **2023**, *28*, 1952. [CrossRef] [PubMed]
82. Xu, W.Y.; Molino, B.Z.; Cheng, F.; Molino, P.J.; Yue, Z.L.; Su, D.D.; Wang, X.J.; Willfor, S.; Xu, C.L.; Wallace, G.G. On Low-Concentration Inks Formulated by Nanocellulose Assisted with Gelatin Methacrylate (GelMA) for 3D Printing toward Wound Healing Application. *ACS Appl. Mater. Interfaces* **2019**, *11*, 8838–8848. [CrossRef]
83. Das, D.; Bhattacharjee, S.; Bhaladhare, S. Preparation of Cellulose Hydrogels and Hydrogel Nanocomposites Reinforced by Crystalline Cellulose Nanofibers (CNFs) as a Water Reservoir for Agriculture Use. *ACS Appl. Polym. Mater.* **2023**, *5*, 2895–2904. [CrossRef]
84. Xue, Y.B.; Zhu, S.N.; Schultze-Kraft, R.; Liu, G.D.; Chen, Z.J. Dissection of Crop Metabolome Responses to Nitrogen, Phosphorus, Potassium, and Other Nutrient Deficiencies. *Int. J. Mol. Sci.* **2022**, *23*, 9079. [CrossRef] [PubMed]
85. Shang, H.Y.; Yang, X.X.; Liu, H. Temperature-responsive hydrogel prepared from carboxymethyl cellulose-stabilized N-vinylcaprolactam with potential for fertilizer delivery. *Carbohydr. Polym.* **2023**, *313*, 120875. [CrossRef]
86. Vu, C.T.; Wu, T. MgAl-layered double hydroxides/sodium alginate beads for nitrate adsorption from groundwater and potential use as a slow-release fertilizer. *J. Clean. Prod.* **2022**, *379*, 134508. [CrossRef]
87. Do Nascimento, D.M.; Nunes, Y.L.; Feitosa, J.P.A.; Dufresne, A.; Rosa, M.d.F. Cellulose nanocrystals-reinforced core-shell hydrogels for sustained release of fertilizer and water retention. *Int. J. Biol. Macromol.* **2022**, *216*, 24–31. [CrossRef]
88. Kassem, I.; Ablouh, E.H.; El Bouchtaoui, F.Z.; Kassab, Z.; Khouloud, M.; Sehaqui, H.; Ghalfi, H.; Alami, J.; El Achaby, M. Cellulose nanocrystals-filled poly (vinyl alcohol) nanocomposites as waterborne coating materials of NPK fertilizer with slow release and water retention properties. *Int. J. Biol. Macromol.* **2021**, *189*, 1029–1042. [CrossRef]
89. Marcinczyk, M.; Oleszczuk, P. Biochar and engineered biochar as slow- and controlled-release fertilizers. *J. Clean. Prod.* **2022**, *339*, 130685. [CrossRef]
90. Kenawy, E.-R.; Seggiani, M.; Cinelli, P.; Elnaby, H.M.H.; Azaam, M.M. Swelling capacity of sugarcane bagasse-g-poly(acrylamide)/attapulgite superabsorbent composites and their application as slow release fertilizer. *Eur. Polym. J.* **2020**, *133*, 109769. [CrossRef]
91. Bauli, C.R.; Lima, G.F.; de Souza, A.G.; Ferreira, R.R.; Rosa, D.S. Eco-friendly carboxymethyl cellulose hydrogels filled with nanocellulose or nanoclays for agriculture applications as soil conditioning and nutrient carrier and their impact on cucumber growing. *Colloids Surf. A Physicochem. Eng. Asp.* **2021**, *623*, 126771. [CrossRef]
92. Flores-Cespedes, F.; Villafranca-Sanchez, M.; Fernandez-Perez, M. Alginate-Bentonite-Based Hydrogels Designed to Obtain Controlled-Release Formulations of Dodecyl Acetate. *Gels* **2023**, *9*, 388. [CrossRef] [PubMed]
93. Chen, M.; Chen, X.; Zhang, C.; Cui, B.; Li, Z.; Zhao, D.; Wang, Z. Kaolin-Enhanced Superabsorbent Composites: Synthesis, Characterization and Swelling Behaviors. *Polymers* **2021**, *13*, 1204. [CrossRef] [PubMed]
94. Jiang, Y.; Wang, Y.; Li, Q.; Yu, C.; Chu, W. Natural Polymer-based Stimuli-responsive Hydrogels. *Curr. Med. Chem.* **2020**, *27*, 2631–2657. [CrossRef]
95. Zainal, S.H.; Mohd, N.H.; Suhaili, N.; Anuar, F.H.; Lazim, A.M.; Othaman, R. Preparation of cellulose-based hydrogel: A review. *J. Mater. Res. Technol.* **2021**, *10*, 935–952. [CrossRef]
96. Khan, F.; Atif, M.; Haseen, M.; Kamal, S.; Khan, M.S.; Shahid, S.; Nami, S.A.A. Synthesis, classification and properties of hydrogels: Their applications in drug delivery and agriculture. *J. Mater. Chem. B* **2022**, *10*, 170–203. [CrossRef]
97. Xian, S.; Webber, M.J. Temperature-responsive supramolecular hydrogels. *J. Mater. Chem. B* **2020**, *8*, 9197–9211. [CrossRef]
98. Abdollahi, A.; Roghani-Mamaqani, H.; Razavi, B.; Salami-Kalajahi, M. The light-controlling of temperature-responsivity in stimuli-responsive polymers. *Polym. Chem.* **2019**, *10*, 5686–5720. [CrossRef]

99. Tang, L.; Gong, L.; Zhou, G.; Liu, L.; Zhang, D.; Tang, J.; Zheng, J. Design of low temperature-responsive hydrogels used as a temperature indicator. *Polymer* **2019**, *173*, 182–189. [CrossRef]
100. Martwong, E.; Tran, Y. Lower Critical Solution Temperature Phase Transition of Poly(PEGMA) Hydrogel Thin Films. *Langmuir* **2021**, *37*, 8585–8593. [CrossRef]
101. Nie, L.; Li, J.; Lu, G.; Wei, X.; Deng, Y.; Liu, S.; Zhong, S.; Shi, Q.; Hou, R.; Sun, Y.; et al. Temperature responsive hydrogel for cells encapsulation based on graphene oxide reinforced poly(N- isopropylacrylamide)/hydroxyethyl-chitosan. *Mater. Today Commun.* **2022**, *31*, 103697. [CrossRef]
102. Du, W.; Zhang, J.; Zhao, Z.; Zhang, X. Preparation of novel temperature-responsive double-network hydrogel reinforced with aramid nanofibers. *Compos. Commun.* **2020**, *22*, 100438. [CrossRef]
103. Zhang, Z.; Feng, S.; Luo, J.; Hao, B.; Diao, F.; Li, X.; Jia, B.; Wang, L.; Bao, Z.; Guo, W. Evaluation of Microbial Assemblages in Various Saline-Alkaline Soils Driven by Soluble Salt Ion Components. *J. Agric. Food Chem.* **2021**, *69*, 3390–3400. [CrossRef] [PubMed]
104. Meng, Q.-F.; Li, D.-W.; Zhang, J.; Zhou, L.-R.; Ma, X.-F.; Wang, H.-Y.; Wang, G.-C. Soil properties and corn (*Zea mays* L.) production under manure application combined with deep tillage management in solonetzic soils of Songnen Plain, Northeast China. *J. Integr. Agric.* **2016**, *15*, 879–890. [CrossRef]
105. Zhang, B.; Song, X.; Zhang, Y.; Han, D.; Tang, C.; Yu, Y.; Ma, Y. Hydrochemical characteristics and water quality assessment of surface water and groundwater in Songnen plain, Northeast China. *Water Res.* **2012**, *46*, 2737–2748. [CrossRef]
106. Qi, H.; Ma, R.; Shi, C.; Huang, Z.; Liu, S.; Sun, L.; Hu, T. Novel low-cost carboxymethyl cellulose microspheres with excellent fertilizer absorbency and release behavior for saline-alkali soil. *Int. J. Biol. Macromol.* **2019**, *131*, 412–419. [CrossRef]
107. Chang, C.; Tian, L.; Tian, Z.; McLaughlin, N.; Tian, C. Change of soil microorganism communities under saline-sodic land degradation on the Songnen Plain in northeast China. *J. Plant Nutr. Soil Sci.* **2022**, *185*, 297–307. [CrossRef]
108. Zhang, Y.; Li, F.; Lu, Z.; Pei, Z.; Zhao, H.; Shen, Q.; Hong, M. Organic amendments effects on soil aggregation and carbon sequestration in saline-alkaline croplands in China. *Agron. J.* **2023**, *115*, 2083–2095. [CrossRef]
109. Li, X.; Li, Q.; Xu, X.; Su, Y.; Yue, Q.; Gao, B. Characterization, swelling and slow-release properties of a new controlled release fertilizer based on wheat straw cellulose hydrogel. *J. Taiwan Inst. Chem. Eng.* **2016**, *60*, 564–572. [CrossRef]
110. Hua, B.; Wei, H.; Hu, C.; Zhang, Y.; Yang, S.; Wang, G.; Guo, T.; Li, J. Preparation of pH/temperature-responsive semi-IPN hydrogels based on sodium alginate and humic acid as slow-release and water-retention fertilizers. *Polym. Bull.* **2023**. [CrossRef]
111. Zhang, Q.; Zhang, W.; Yu, Y.; Li, T.; Yu, L. Modeling the Impact of Atmospheric Warming on Staple Crop Growth in China in the 1960s and 2000s. *Atmosphere* **2021**, *12*, 36. [CrossRef]
112. El Idrissi, A.; El Gharrak, A.; Achagri, G.; Essamlali, Y.; Amadine, O.; Akil, A.; Sair, S.; Zahouily, M. Synthesis of urea-containing sodium alginate-g-poly(acrylic acid-co-acrylamide) superabsorbent-fertilizer hydrogel reinforced with carboxylated cellulose nanocrystals for efficient water and nitrogen utilization. *J. Environ. Chem. Eng.* **2022**, *10*, 108282. [CrossRef]

MDPI AG
Grosspeteranlage 5
4052 Basel
Switzerland
Tel.: +41 61 683 77 34

Polymers Editorial Office
E-mail: polymers@mdpi.com
www.mdpi.com/journal/polymers

www.ingramcontent.com/pod-product-compliance
Lightning Source LLC
Chambersburg PA
CBHW041353210926
51850CB00013B/3072